Lecture Notes in Control and Information Sciences 430

Editors

Professor Dr.-Ing. Manfred Thoma
Institut fuer Regelungstechnik, Universität Hannover, Appelstr. 11, 30167 Hannover, Germany
E-mail: thoma@irt.uni-hannover.de

Professor Dr. Frank Allgöwer
Institute for Systems Theory and Automatic Control, University of Stuttgart, Pfaffenwaldring 9, 70550 Stuttgart, Germany
E-mail: allgower@ist.uni-stuttgart.de

Professor Dr. Manfred Morari
ETH/ETL I 29, Physikstr. 3, 8092 Zürich, Switzerland
E-mail: morari@aut.ee.ethz.ch

AF785926

Series Advisory Board

P. Fleming
University of Sheffield, UK

P. Kokotovic
University of California, Santa Barbara, CA, USA

A.B. Kurzhanski
Moscow State University, Russia

H. Kwakernaak
University of Twente, Enschede, The Netherlands

A. Rantzer
Lund Institute of Technology, Sweden

J.N. Tsitsiklis
MIT, Cambridge, MA, USA

For further volumes:
http://www.springer.com/series/642

Hongjiu Yang, Yuanqing Xia, Peng Shi,
and Ling Zhao

Analysis and Synthesis of Delta Operator Systems

Springer

Authors
Dr. Hongjiu Yang
Institute of Electrical Engineering
Yanshan University
Qinhuangdao
China

Prof. Yuanqing Xia
School of Automation
Beijing Institute of Technology
Beijing
China

Prof. Peng Shi
Department of Computing and
Mathematical Sciences
University of Glamorgan
Pontypridd
UK

Dr. Ling Zhao
College of Mechanical Engineering
Yanshan University
Qinhuangdao
China

ISSN 0170-8643
e-ISSN 1610-7411
ISBN 978-3-642-28773-2
e-ISBN 978-3-642-28774-9
DOI 10.1007/978-3-642-28774-9
Springer Heidelberg New York Dordrecht London

Library of Congress Control Number: 2012933239

© Springer-Verlag Berlin Heidelberg 2012
This work is subject to copyright. All rights are reserved by the Publisher, whether the whole or part of the material is concerned, specifically the rights of translation, reprinting, reuse of illustrations, recitation, broadcasting, reproduction on microfilms or in any other physical way, and transmission or information storage and retrieval, electronic adaptation, computer software, or by similar or dissimilar methodology now known or hereafter developed. Exempted from this legal reservation are brief excerpts in connection with reviews or scholarly analysis or material supplied specifically for the purpose of being entered and executed on a computer system, for exclusive use by the purchaser of the work. Duplication of this publication or parts thereof is permitted only under the provisions of the Copyright Law of the Publisher's location, in its current version, and permission for use must always be obtained from Springer. Permissions for use may be obtained through RightsLink at the Copyright Clearance Center. Violations are liable to prosecution under the respective Copyright Law.
The use of general descriptive names, registered names, trademarks, service marks, etc. in this publication does not imply, even in the absence of a specific statement, that such names are exempt from the relevant protective laws and regulations and therefore free for general use.
While the advice and information in this book are believed to be true and accurate at the date of publication, neither the authors nor the editors nor the publisher can accept any legal responsibility for any errors or omissions that may be made. The publisher makes no warranty, express or implied, with respect to the material contained herein.

Printed on acid-free paper

Springer is part of Springer Science+Business Media (www.springer.com)

Fast sampling is valuable in automation and control systems

Delta operator deals with fast sampling very effectively

For researchers devoted to Delta Operator Systems

Preface

In today's digital control, signal processing and communications, image processing and wide band communications, it is necessary to design discrete-time procedures that operate on data obtained by sampling continuous time processes at rates that are rapid relative to the process dynamics. Conventional design of discrete-time filters and systems utilizes the shift operator. High-speed signal transmission in digital communications and control systems has become common in recent years due to advances in technology that allow rapid signal processing and storage of the large quantities of associated data. Communication receivers and control systems require processing of these fast-sampled signals in estimating the communication channel or identifying system dynamics. However, in cases where high sampling rates are needed, shift-operator-based designs may exhibit unacceptable numerical problems. Goodwin and Middleton illustrated the application and benefit of the delta operator for the design and implementation of controllers and filters.

Two major advantages are claimed for the use of delta operator parameterizations: a theoretically interesting unified formulation of continuous-time and discrete-time control theory which entails a better understanding of discrete-time control under fast sampling, and a range of practically interesting numerical advantages connected with finite word length effects. Furthermore, sampling-periods are explicit parameters in delta operator systems(DOSs), it is easy to observe and analyze the control effect with different sampling periods. As the structure of DOSs is different from that of tradition control systems, there exist various specific problems in DOSs, for examples, sliding mode control(SMC), time delay, filter, finite frequency and networked control. Therefore, analysis and synthesis of DOSs are of great importance.

In Chapter 1, a motivation on the research and its history and an overview on recent development of DOSs are provided.

Then, this book will present five parts:

Part I: Some development of SMC for DOSs is introduced. In Chapter 2, a discrete robust sliding mode observer and an observer-based SMC problem are investigated for DOSs with parameter uncertainties and nonlinear exogenous disturbances. In Chapter 3, a discrete robust adaptive sliding mode controller is presented for a DOS with parameter uncertainties and exogenous disturbances.

Part II: Some new results on DOSs with time delays are presented. In Chapter 4, the stabilization problem of DOSs with time-varying delays and linear fractional uncertainties is investigated. In Chapter 5, a bridge between robust fuzzy control and DOS framework is made.

Part III: Two kinds of filters are designed for different DOSs. In Chapter 6, a robust H_∞ filter is designed for DOSs with time-varying delays and Markovian jump parameters. In Chapter 7, a general delta operator Kalman filter(DOKF) is designed to estimate the full-state vectors of DOSs and convergence analysis of the designed DOKF is also given.

Part IV: We generalize the Kalman-Yakubovič-Popov(KYP) lemma with finite frequency conditions in delta domain and some new results by using the KYP are presented. In Chapter 8, some results on low frequency positive real control for DOSs are investigated by using the generalized KYP lemma. In Chapter 9, the generalized KYP lemma with three kinds of finite frequency conditions in delta domain are described and a finite frequency fault detection filter with H_∞ and H_- performance indexes is designed.

Part V: With the development of networked control systems(NCSs), the research on DOSs over networks is of great interest by using some results of the preceding chapters. In Chapter 10, NCSs are describes by a class of Markovian jump DOSs with mode-dependent time delays and a stochastic asymptotically stable criterion is given for the NCSs. In Chapter 11, a new linear switched DOS with delay is proposed to describe NCSs with both packets dropout and network-induced delays, control method is also given in this chapter. In Chapter 12, a Markov jump linear DOS is proposed to describe NCSs by using a DOKF and considering Quality-of-Service(QoS) of networks, some sufficient conditions for the existence of a guaranteed cost controller is also given for the NCSs. In Chapter 13, the results of obtained generalized KYP lemma in delta domain are used for the purpose of stability analysis for high frequency NCSs.

The supports from the National Basic Research Program of China (973 Program) (2012CB720000), the National Natural Science Foundation of China under Grant (60974011), Program for New Century Excellent Talents in University of People's Republic of China (NCET-08-0047), the Ph.D. Programs Foundation of Ministry of Education of China (20091101110023), and Program for Changjiang Scholars and Innovative Research Team in University, and Beijing Municipal Natural Science Foundation (4102053), respectively, are gratefully acknowledged. This work was also partially supported by the National Nature Science Foundation of China under Grant (60804011, 60904001), and the Engineering and Physical Sciences Research Council, UK (EP/F029195). This work was supported in part by the National Basic Research Program of China (973 Program) under Grant 2012CB821200 and the

National Natural Science Foundation of China under Grant 61004021. This work was also supported in part by the Science Fund for Distinguished Young Scholars of Hebei Province, the Program for New Century Excellent Talents in the University of China, and the National Natural Science Foundation of China under Grant 60974018 and 61104033.

Haigang District, Qinhuangdao, China — Hongjiu Yang
Haidian District, Beijing, China — Yuanqing Xia
Trefforest, Glamorgan, United Kingdom — Peng Shi
Haigang District, Qinhuangdao, China — Ling Zhao
July 2011

Symbols and Acronyms

T	sampling period
j	scalar $\sqrt{-1}$
R	field of real numbers
R^n	n-dimensional real Euclidean space
$R^{n\times m}$	space of $n \times m$ real matrices
C	field of complex numbers
I	identity matrix
O	zero matrix
A	system matrix
A^{-1}	inverse of matrix A
A^T	transpose of matrix A
A^*	complex conjugate transpose of matrix A
$A^\dagger$	Moore-Penrose inverse of matrix A: $A^T(AA^T)^{-1}$
$A \geq 0$	symmetric positive semi-definite
$A > 0$	symmetric positive definite
$A \leq 0$	symmetric negative semi-definite
$A < 0$	symmetric negative definite
lim	limit
min	minimum
max	maximum
LMI	linear matrix inequality
DOS	delta operator system
DOKF	delta operator Kalman filter
NCS	networked control system
SMC	sliding mode control
ZOH	zero-order hold
QoS	quality-of-service
SISO	single input single output
MIMO	multiple input multiple output
tr(A)	trace of matrix A
det(A)	determinant of matrix A

rank(A)	rank of matrix A
$0_{n\times m}$	zero matrix of dimension $n \times m$
$\lambda(A)$	eigenvalue of matrix A
$\lambda_{\min}(A)$	minimum eigenvalue of matrix A
$\lambda_{\max}(A)$	maximum eigenvalue of matrix A
$\sigma(A)$	singular value of matrix A
$\sigma_{\min}(A)$	minimum singular value of matrix A
$\sigma_{\max}(A)$	maximum singular value of matrix A
sign(x)	sign of x
$\|x\|$	absolute value (or modulus) of x
$\\|x\\|$	Euclidean norm
$\\|P\\|$	induced norm $\sup_{\\|x\\|=1} \\|Px\\|$
$\forall$	for all
$\in$	belong to
$\to$	tend to, or mapping to (case sensitive)
$\otimes$	matrix Kronecker product
$\sum$	sum
$\mathbf{E}\{\cdot\}$	mathematical expectation operator
sup	supremum
inf	infimum
Re$(\cdot)$	real parts of the arguments
Im$(\cdot)$	imaginary parts of the arguments
Ω	sample space
$\mathbf{F}$	algebra of events
$\mathbf{P}$	probability measure
$\mathbf{H}_n$	set of $n \times n$ Hermitian matrices
$*$	symmetric terms in a symmetric matrix
diag$\{X_1, X_2, \cdots, X_m\}$	diagonal matrix with X_i as its ith diagonal element
σ: $\mathbf{C} \times \mathbf{H}_2 \to R$	form function

$$\sigma(\lambda, \Pi) = \begin{bmatrix} \lambda \\ 1 \end{bmatrix}^* \Pi \begin{bmatrix} \lambda \\ 1 \end{bmatrix}.$$

Contents

Chapter 1
Introduction

1.1 Background

It is well known that discrete systems are suitable for computer realization and continuous systems are convenient for theoretical analysis. Usually, the discrete-time model is expressed in the form of the shift operator. However, the first disadvantage of using shift operator is an inconvenience not like the continuous-time operator. The second weakness of the normal shift operating systems are the problem of crowding poles within the boundary of stability circle at small sampling interval and the difficulties of truncation and round-off errors. Goodwin constructed delta operator instead of traditional shift operator for sampling continuous systems in [82] and [170]. The proposed method can unify some previous related results of the continuous and discrete systems into the framework of the DOSs framework. In contrast to the discrete shift operator, the discrete delta operator approximates the Euler derivative. The shorter the sampling period is, the better the system performances are for discrete time control systems. Therefore, the delta operator method is significantly less sensitive than the standard shift operator method at high sampling rate. This leads to a quasi-continuous time s-domain model for high frequencies and gives a better representation of the underlying physical model. That is the poles approach the system poles and the zeros approaches the system zeros for fast sampling. The zeros introduced by the sample and hold circuit tend to minus infinity.

Delta operator based implementations have gained considerable interest due to its excellent performance in finite-word-length computers under fast sampling. Since its introduction, the delta-operator-based designs have been studied extensively. A direct model reference adaptive control algorithm for discrete-time systems arising from rapid sampling of continuous-time, stable invertible plants has been described in [123]. For control system analysis and design, and high-speed realtime computation, the transformations between the delta and forward shift operator transfer function models of dynamic

H. Yang et al.: Analysis and Synthesis of Delta Operator Systems, LNCIS 430, pp. 1–18.
springerlink.com © Springer-Verlag Berlin Heidelberg 2012

physical systems have been highlighted in [177]. The computational formulation, properties and applications of delta operator model of dynamic physical systems have been illustrated in [176], which also presented that the delta operator model tends to the analog system model in the fast sampling limit. The relationships between optimal realization sets in shift and delta operator have been established in [142]. In [193], a tabular method for the investigation of real polynomial root distribution with respect to a circle in a complex plane has been presented. In [194], given a continuous-time system, a technique to directly obtain an approximate delta-operator formulated discrete-time system (δ-system) has been presented. The main design features of the loop transfer recovery controller both at the input and at the output of the plant has been briefly over viewed in [222] using the delta operator formulation. A systematic approach to a class of approximations to the pulse transfer function of a system consisting of a ZOH and a linear continuous-time plant has been presented in [15]. In [33], the problem of digital controller structures realized has been addressed by using delta operator and the relevant issues of closed-loop stability subject to finite word length implementation. In [141], a new structure which is a generalization of the transposed direct-form II structures in the conventional shift operator and the prevailing delta operator has been derived based on a polynomial-operator approach. In [307], an approach inspired from the δ-operator has been proposed to implement distributed delay in linear control laws using rational transfer functions. In [134], the authors presented a least square error criterion based procedure to design 2-D digital filters in δ-domain that utilized the symmetry properties of the magnitude specification. To improve the state matching performance for that case, a refined LMI-based digital redesign has been proposed in [137] by using delta operator. The robustness problem for some DOSs with parametric uncertainties has been first presented in [215]. Furthermore, the H_∞ control in delta domain has been firstly presented in [171]. A delta-domain version of the generic H_∞ discrete-time algorithm has been derived in [83] and examined in [42]. Based on a J-lossless coprime factorisations, an approach to H_∞ controller design in delta domain has been presented in [218]. In control systems design, delta operator or named Euler operator is defined as

$$\delta = (q-1)/\mathtt{T}$$

where q is shift operator which means $qx(k) = x(k+1)$, $\mathtt{T}$ is a sampling period. In summary, delta operator offers the same flexibility as the shift operator, but also has three principal advantages:

i The sampling-period is an explicit parameter in DOSs, thus it is easy to observe and analyze the control effect with different sampling periods.
ii The model parameters in DOSs will approach to the ones in continuous system as $\mathtt{T} \to 0$, so as the control results.

iii The delta operator model has the advantage of better numerical properties at high sampling rates in finite word length computer especially.

1.2 Methods and Issues

1.2.1 Description of DOSs

Firstly, the definition of δ transformation is given as follows.

Definition 1.1. Let $f(t)$ be a piecewise continuous function signal. Then

$$\mathrm{T}\sum_{n=0}^{\infty} f(t)(1+\mathrm{T}\delta)^{-n}\mid_{t=n\mathrm{T}},\ (|1+\mathrm{T}\delta|>R)$$

is the δ transformation of function $f(t)$, denoted as

$$F(\delta)=\Delta[f(t)]=\mathrm{T}\sum_{n=0}^{\infty} f(t)(1+\mathrm{T}\delta)^{-n}\mid_{t=n\mathrm{T}},\ (|1+\mathrm{T}\delta|>R)$$

Here, R/T is the absolute convergence radius of the complex series formula, where $\Delta[\cdot]$ denotes the transformation, δ denotes the variable and $f^*(t)$ is the inverse transformation of $F(t)$ denoted as $f^*(t)=\Delta[F(\delta)]$.

Throughout this book, we let $t_k = k\mathrm{T}$ for the sake of convenience in the following analysis and the delta operator for the state variable $x(t_k)$ is defined by:

$$\delta x(t_k)=\begin{cases}\frac{dx(t)}{dt}, & \mathrm{T}=0\\ \frac{x(t_k+\mathrm{T})-x(t_k)}{\mathrm{T}}, & \mathrm{T}\neq 0\end{cases} \tag{1.1}$$

Form the definition of delta operator, it can be obtained that $\delta x(t_k)\to\dot{x}(t_k)$ as $\mathrm{T}\to 0$ and $\delta x(t_k)=\dot{x}(t_k)$ when $\mathrm{T}=0$. When $\mathrm{T}=1$, there exists $\delta x(t_k)=x(t_{k+1})-x(t_k)$ which is equivalent to traditional shift operator. Also, the convergence process of $\delta x(t_k)\to\dot{x}(t_k)$ is a monotonous process. Hence, if the sampling period is more than 1, the numerical advantage of delta operator will be less than shift operator. That is why the sampling period should be less than 1 to obtain the advantage of the delta operator approach.

By using (1.1), a delta operator system(DOS) is described as follows

$$\delta x(t_k)=f(x(t_k)) \tag{1.2}$$

where $f(\cdot)$ is a linear or nonlinear function.

Definition 1.2. Let $V(x(t_k))$ is a Lyapunov functional in delta domain. DOS (1.2) is asymptotically stable, if the following conditions hold:
(i) $V(x(t_k)) \geq 0$, with equality if and only if $x(t_k) = 0$;
(ii) $\delta V(x(t_k)) = [V(x(t_k + \mathtt{T})) - V(x(t_k))]/\mathtt{T} < 0$;

For Lyapunov functional $V(x(t_k))$ both in z-domain and s-domain, the condition (i) $V(x(t_k)) \geq 0$ in Definition 1.2 is given. In condition (ii), when $\mathtt{T} = 1$, there exists

$$\delta V(x(t_k))|_{\mathtt{T}=1} = \left[V(x(k+1)) - V(x(k))\right]/1 = \Delta V(x(k)) < 0.$$

On the other hand, when $\mathtt{T} \to 0$, there exists

$$\lim_{\mathtt{T}\to 0} \delta V(x(t_k)) = \lim_{\mathtt{T}\to 0} \frac{V(x(t_k + \mathtt{T})) - V(x(t_k))}{\mathtt{T}} = \frac{dV(x(t))}{dt} < 0.$$

It can be seen that the Lyapunov functional in δ-domain can be reduced to the traditional Lyapunov functional in z-domain or s-domain when sampling period is 1 or tends to 0.

In general, a frequency range is visualized as a curve or curves on the complex plane. In the following, we give the mapping relationships in s-domain, z-domains and δ-domain, respectively. Following the relations

$$z = e^{s\mathtt{T}},\ s = \frac{1}{\mathtt{T}} \ln z,\ \rho = \frac{z-1}{\mathtt{T}} = \frac{e^{s\mathtt{T}} - 1}{\mathtt{T}},\ s = \frac{1}{\mathtt{T}} \ln(1 + \mathtt{T}\rho)$$

Let $s = \sigma + j\omega$, then there exist $\rho = \frac{z-1}{\mathtt{T}} = \frac{\varrho e^{j\theta}-1}{\mathtt{T}}$ and $z = \varrho e^{j\theta}$, in which $\varrho = e^{\sigma \mathtt{T}}$ and $\theta = \omega \mathtt{T}$ hold. When $\sigma = 0$, the imaginary axis in s-plane is mapped to the unit circle with the center at origin in z-plane, and also mapped to the circle with center $-\frac{1}{\mathtt{T}}$ and radius $\frac{1}{\mathtt{T}}$ in δ-plane, that is $|z| = |1 + \mathtt{T}\rho| = 1$, $|\rho + \frac{1}{\mathtt{T}}| = \frac{1}{\mathtt{T}}$. We are familiar with the stability regions on the complex plane of s-domain and z-domain. Fig. 1.1 shows the stability regions in s-plane, δ-plane and z-plane. The stability region for the delta implementation

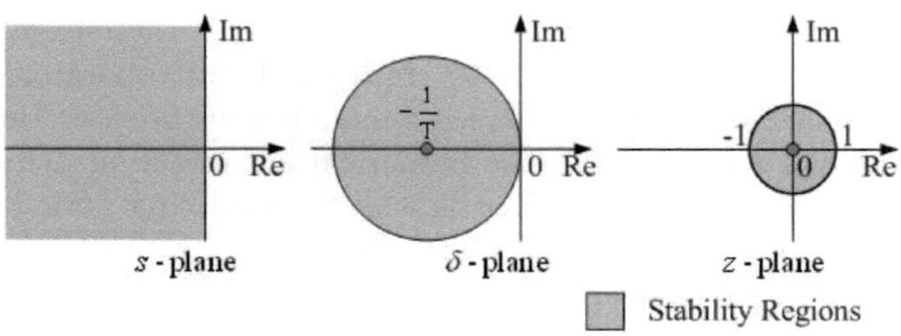

Fig. 1.1 Stability regions for the continuous Laplace plane, the discrete delta and shift planes.

will once again grow to approach that of the Laplace domain (i.e., the whole left hand plane). Whereas, the stability region of the shift implementation is fixed, causing the clustering at 1 as the sample rate increases.

In order to show the advantage of delta operator approach, we consider the following continuous time linear system:

$$\dot{x}(t) = Ax(t) + Bu(t), \quad y(t) = Cx(t)$$

where $x(t) \in R^n$ is the state variable, $u(t) \in R^m$ is the control input; $y(t) \in R^p$ is the model output. A, B and C are known constant real matrices with appropriate dimensions. Utilizing the shift operator approach, the discrete system of sampling result of the above continuous system in z-domain is obtained as follows:

$$qx(k\mathrm{T}) = A_z x(k\mathrm{T}) + B_z u(k\mathrm{T}), \quad y(k\mathrm{T}) = C_z x(k\mathrm{T})$$

where

$$A_z = e^{A\mathrm{T}}, \ B_z = \int_0^{\mathrm{T}} e^{A(\mathrm{T}-s)} B ds, \ C_z = C$$

Using delta operator approach, the relevant discrete system in δ-domain is

$$\delta x(k\mathrm{T}) = A_\delta x(k\mathrm{T}) + B_\delta u(k\mathrm{T}), \quad y(k\mathrm{T}) = C_\delta x(k\mathrm{T}),$$

where

$$A_\delta = (A_z - I)/\mathrm{T}, \ B_\delta = B_z/\mathrm{T}, \ C_\delta = C$$

As $\mathrm{T} \to 0$, there exist

$$\lim_{\mathrm{T}\to 0} A_z = I, \ \lim_{\mathrm{T}\to 0} B_z = O, \ \lim_{\mathrm{T}\to 0} A_\delta = A, \ \lim_{\mathrm{T}\to 0} B_\delta = B$$

which have shown that the delta operator model has the advantage of better numerical properties at high sampling rates.

1.2.2 Numerical Example

The numerical example shows the characteristic of discrete-time systems and DOSs in sampling continuous-time systems.

Example 1.3. Consider the following continuous-time system in s-domain as:

$$\dot{x}(t) = \begin{bmatrix} -1 & 0 \\ 1 & -2 \end{bmatrix} x(t) + \begin{bmatrix} 0 \\ 1 \end{bmatrix} u(t), \tag{1.3}$$

$$y(t) = \begin{bmatrix} 1 & 1 \end{bmatrix} x(t). \tag{1.4}$$

By using shift operator and delta operator in sampling the continuous-time system, respectively, we obtain the relevant different discrete-time system in z-domain and δ-domain. When $\mathtt{T} = 1$, there exist

$$\begin{aligned} x((k+1)\mathtt{T}) &= \begin{bmatrix} 0.3679 & 0 \\ 0.2325 & 0.1353 \end{bmatrix} x(k\mathtt{T}) + \begin{bmatrix} 0 \\ 0.4323 \end{bmatrix} u(k\mathtt{T}), \\ \delta x(k\mathtt{T}) &= \begin{bmatrix} -0.6321 & 0 \\ 0.2325 & -0.8647 \end{bmatrix} x(k\mathtt{T}) + \begin{bmatrix} 0 \\ 0.4323 \end{bmatrix} u(k\mathtt{T}). \end{aligned}$$

When $\mathtt{T} = 0.55$, there exist

$$\begin{aligned} x((k+1)\mathtt{T}) &= \begin{bmatrix} 0.5769 & 0 \\ 0.2441 & 0.3329 \end{bmatrix} x(k\mathtt{T}) + \begin{bmatrix} 0 \\ 0.3336 \end{bmatrix} u(k\mathtt{T}), \\ \delta x(k\mathtt{T}) &= \begin{bmatrix} -0.7692 & 0 \\ 0.4438 & -1.2130 \end{bmatrix} x(k\mathtt{T}) + \begin{bmatrix} 0 \\ 0.6065 \end{bmatrix} u(k\mathtt{T}). \end{aligned}$$

When $\mathtt{T} = 0.1$, there exist

$$\begin{aligned} x((k+1)\mathtt{T}) &= \begin{bmatrix} 0.9048 & 0 \\ 0.0861 & 0.8187 \end{bmatrix} x(k\mathtt{T}) + \begin{bmatrix} 0 \\ 0.0906 \end{bmatrix} u(k\mathtt{T}), \\ \delta x(k\mathtt{T}) &= \begin{bmatrix} -0.9516 & 0 \\ 0.8611 & -1.8127 \end{bmatrix} x(k\mathtt{T}) + \begin{bmatrix} 0 \\ 0.9063 \end{bmatrix} u(k\mathtt{T}). \end{aligned}$$

From the above results, it is easy to see that why the virtue of DOSs in sampling continuous-time systems come out when the sampling is fast, and two output curve graphs of the above systems with different sampling periods are given in Figs. 1.2 and 1.3.

1.3 Problems Studied in This Book

The following five major issues for DOSs will be studied in this book: SMC, time delay, filter, finite frequency and NCSs. The discrete sliding mode approach requires high speed discontinuous action to steer the states of a system into a sliding surface and to maintain subsequent motion on this surface. Time delay is commonly encountered in various engineering systems and is frequently a source of instability and poor performance. Filter design methods are the key of state estimation which is applicable to virtually all areas of engineering and science. Each design specification is often given not for the entire frequency range but rather for a finite frequency range of relevance. The Internet is changing the way of life and the Internet will change the way of control. Therefore, NCSs are devoted by using some preceding research results on DOSs.

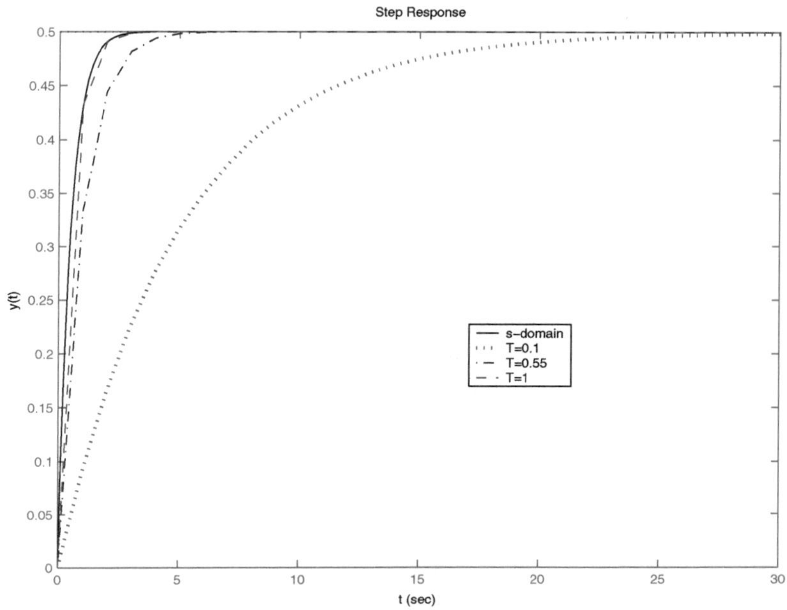

Fig. 1.2 Trajectories of the output in z-domain

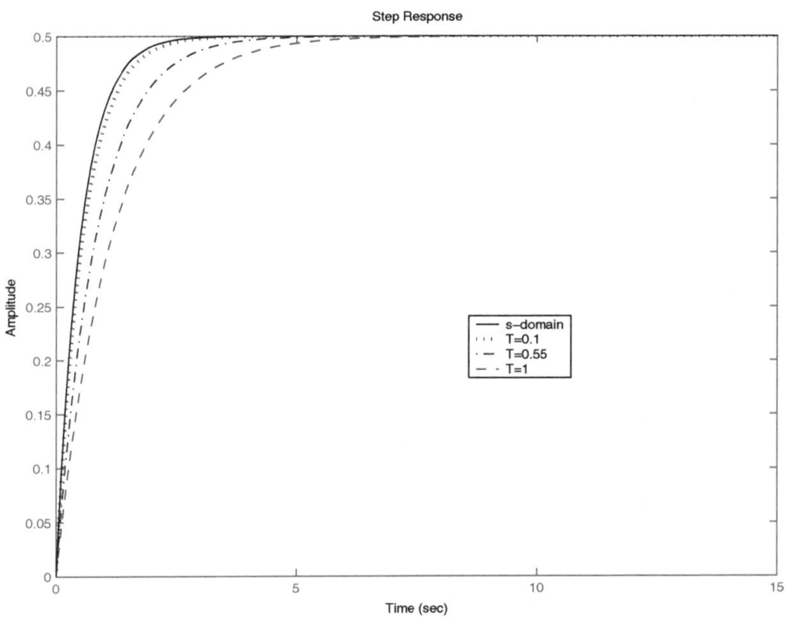

Fig. 1.3 Trajectories of the output in δ-domain

1.3.1 SMC of DOSs

Since its early appearance in the 1950s, SMC has proven to be an effective robust control strategy for incompletely modeled or uncertain systems [182]. Traditionally, SMC as a general design tool for robust control systems has been gained much attention. The SMC problem of uncertain linear discrete systems with input delay has been considered in [260]. State estimation and SMC of Markovian jump singular systems have been considered in [246]. In addition, the problem of SMC of uncertain state continuous systems with state delay has been investigated in [211, 259]. The SMC problem has been considered for a class of DOS with time delays [254]. Recently, sliding mode observers have been used for estimating states of discrete or continuous systems many times. An effective method has been presented to design sliding mode observers for uncertain control systems in [226]. A sliding-mode observer has been obtained for estimation of vehicle dynamic parameters in [173]. A robust sliding mode observer has been investigated for nonlinear systems in [273] and a second-order sliding-mode observer has been used for mechanical systems in [43]. An observer-based SMC problem has been studied for state-delayed systems with unmeasurable states and nonlinear uncertainties [182]. Both a sliding mode observer design problem and a observer-based SMC problem have been investigated in [250]. The application of a particular sliding mode observer to the problem of fault detection and isolation has been considered in [51]. In recently years, many results on adaptive control have been obtained, for example [99, 100, 102, 106, 107], and the references cited therein. Adaptive control law could lead to a stable closed-loop system and the deviation from the sliding surface is bounded [213]. Adaptive output control of linear time-varying systems and adaptive control for linear nonstationary objects output have been considered in [16] and [17], respectively. A robust adaptive control scheme which has been applied to a practical autopilot design for feedback-linearized Skid-to-Turn missiles with aerodynamic uncertainties [38]. The adaptive SMC has been discussed for the piezo-actuated stage [35]. Robust adaptive SMC for fuzzy systems has been considered in [295]. A robust adaptive variable structure output feedback control for uncertain dynamics systems with explicitly considering nonlinearity and dead-zone control has been proposed in [96]. A systematic and simple SMC with adaptive tuning has been proposed for the robust control of nonlinear systems with uncertain parameters [110]. In [27], the delta operator approach has been used to develop an adaptive SMC for a class of discrete systems. However, there are also a lot of space to be improved on adaptive SMC for DOSs.

An SMC system can be considered as a variable-structure one with the control structure intentionally switched using some high speed-switching feedback control. The basic idea of SMC is to drive the system state onto some predefined switching surface, and maintain the trajectory on it for all subsequent time. The motion when confined to the switching surface acquires certain desired properties such as prescribed decay speed, good transient performance, and, especially, complete insensitivity to matched uncertainties or external disturbances. The salient advantages of SMC are: (i) its robustness against a large class of perturbations or model uncertainties; and (ii) the possibility of stabilizing some complex non-linear systems which are difficult to stabilize by continuous state feedback laws. It can be seen that sliding mode theory has been developed mostly in the continuous time domain. However, when the continuous-time sliding mode approach is implemented on a digital computer, this observer-based control has to be considered in discrete-time sliding mode structure. It is well known that the discrete sliding mode approach requires high speed discontinuous action to steer the states of a system into a sliding surface and to maintain subsequent motion on this surface. However, this advantage can only be achieved if infinitely fast switching of control signals is achieved, or locally infinitely high gain control action is available. On the other hand, due to the widespread use of computers for control purpose, the design of discrete time SMC has received considerable amount of research in recent years. In the discrete-time system, the control input is calculated once in every sampling interval and is held constant during the sampling period. This means that the structure of an discrete-time SMC may be changed only at discrete instants, which is in contrast to its continuous counterpart whose structure switching may be made at any instant, once the state trajectories cross the switching surface. Thus, the finite sampling rate results in that the system state in discrete-time SMC may approach the switching surface but is generally unable to stay on it. As a result, it would move about the surface, which yields a sliding-like mode, termed as quasi-sliding mode. Moreover, this discrete-time sliding mode is inherently different from which may appear in continuous systems due to non-ideal behavior of the analog components and it may make the system unstable [203]. In contrast to the discrete shift operator, the delta operator approach means that the Euler derivative can lead to a quasi-continuous time s-domain model for high sampling frequencies [54]. Hence, based on the delta operator approach, the reaching law proposed is able to reduce the chattering due to the implementation of variable structure controller, and possesses the desired characteristics of robustness and good performance.

1.3.2 Delays in DOSs

Time-delay occurs in many dynamical systems such as biological systems, chemical systems, metallurgical processing systems, nuclear reactor, long transmission lines in pneumatic, hydraulic systems and electrical networks. Especially, in recent years, time-delay which exists in NCSs has brought more complex problem into a new research area [143, 144, 145, 146, 147]. Frequently, it is a source of the generation of oscillation, instability and poor performance. Considerable effort has been applied to different aspects of linear time-delay systems during recent years [98, 101, 105, 103, 249]. The introduction of the delay factor renders the system analysis more complicated, in addition to the difficulties caused by the perturbation or uncertainties in the control of time-delay systems. Time delay is commonly encountered in various engineering systems and is frequently a source of instability and poor performance. Therefore, the problem of stability analysis and control of time-delay systems has attracted much attention, and considerable effort has been applied to different aspects of linear time-delay systems during recent years [253]. For a time-delay system, the delay-independent initial stability condition has been provided in [233]. It is well known that delay-independent stability conditions are simpler to apply, while delay-dependent stability conditions are less conservative especially in the case when the time delay is small. Therefore, recent years have witnessed a resurgence of research interests in developing delay-dependent stability conditions of time-delay systems. Based on the Newton-Leibniz formula a lot of delay dependent stability results have been obtained, see [25, 26, 149, 159] for example, and the references cited therein. by using Park's inequality, an improved delay-dependent stability condition has been reported in [186]. It should be noted that Park's inequality has been further improved by using the Moon's inequality in [172]. Besides of Park's inequality and Moon's inequality, another one named Jensen's inequality, which has been used to deal with different kinds of time delay systems in order to obtain delay-dependent results, see [84, 87, 248] for example, and the reference therein.

It is well known the transformed model is not equivalent to the original time-delay system, and may lead to conservatism. In order to reduce such potential conservatism, a method based on descriptor systems has been introduced in [57] to perform the stability analysis and controller design for system with time-varying delays. This method uses a descriptor system model to derive delay-dependent stability conditions, which is equivalent to the original time-delay system. The descriptor system approach has been widely used to deal with various problems of time delay systems in order to provide delay-dependent results, see [57, 58, 127, 139, 279] for example. Since the free-weighting matrix method has been proposed from the Newton-Leibniz formula in [91], it plays an important role in deriving delay-dependent stability conditions [64, 90, 91, 290]. It is obvious that free-weighting matrix approach is effective in reducing conservatism in existing delay dependent results, however, since many matrix variables will be introduced, which makes stability criteria very complicated. Furthermore, some useful terms are ignored when estimating the upper bound of the derivative of Lyapunov functional [59, 89, 91, 251]. Therefore, there is room for further investigation. A delay-dependent stability condition, which is less conservative than the one in [89], has been given in [125]. Based on Jensen's inequality, the author of [206] provided a delay-dependent condition to reduce the complication of the free-weighting matrix method. By choosing an appropriate Lyapunov functional and using Finsler's Lemma, some new delay-dependent stability criteria have been obtained in [253]. It is worthy mentioning that the developed results in [253], compared with recently published results, such as [89, 126, 125, 206, 207] involve fewer matrix variables but have less conservatism.

The Takagi-Sugeno(T-S) fuzzy model described by a family of fuzzy IF-THEN rules was firstly introduced in [223]. In other words, it formulates the complex nonlinear systems into a framework that interpolates some affine local models by a set of fuzzy membership functions. Based on this framework, a systematic analysis and design procedure for complex nonlinear systems can be possibly developed in view of the powerful control theories and techniques in linear systems. Using the T-S fuzzy model, a complex dynamic model can be composed of a set of local linear subsystems via fuzzy rule. Therefore, many important results on T-S fuzzy systems have been reported such as in [296, 297, 303], and references therein. An adaptive fuzzy neural network control problem has been considered via a T-S fuzzy model for a robot manipulator including actuator dynamics in [234]. An adaptive fuzzy sliding control method has been used for a double-pendulum-and-cart system in [228]. A modeling and control problem of fuzzy discrete event systems has been investigated in [154]. Stability analysis and controller design of discrete-time fuzzy large-scale systems based on piecewise Lyapunov functions have been considered in [293]. Relaxed stabilization criteria for discrete-time T-S fuzzy control systems based on a switching fuzzy model and piecewise Lyapunov function has been investigated in [240]. Control synthesis of continuous-time T-S fuzzy systems with local nonlinear models have been considered in [46]. State-based control of fuzzy discrete-event systems has been given in [24]. However, to the best of our knowledge, there have been few papers on the state feedback controller design of DOSs with time-varying delays or H_∞ control for T-S fuzzy DOSs with delays.

1.3.3 Filter for DOSs

The states of a system are those variables that provide a complete representation of the internal condition or status of the system at a given instant of time. State estimation is applicable to virtually all areas of engineering and science. Any discipline that is concerned with the mathematical modeling of its systems is a likely candidate for state estimation. State estimation is interesting to engineers for at least two reasons: 1) An engineer needs to estimate the system states in order to implement a state-feedback controller. 2) An engineer needs to estimate the system states because those states are interesting in their own right. Since the Kalman filtering theory has been introduced in [132], much effort has been devoted to the problem of estimating an output error signal through a filter structure. Recently, the problems of the filtering has been studied by a number of authors, see for example [9, 71, 72, 178] and the references therein. The Kalman filtering theory has been well studied for more than three decades, and has been successfully applied in various branches of science and engineering such as the areas of control design and signal processing. Using an extended Kalman filter, the load disturbance rejection response has been remarkably improved by compensation of estimated load torque [131]. In [92], a machine drive technique using a Kalman filter has been proposed to estimate both instantaneous speed and disturbance load torque. In [8], some extended-Kalman-filter-based estimation algorithms have been developed in induction motors. In [7], the Kalman filter has been used for the estimation of rotor currents in an induction motor. It evaluated the applicability of a Kalman filter to the identification of an autonomous underwater vehicle [229]. By using an extended Kalman filter to estimate the fault-related parameters, a sensor failure detection and identification scheme for a closed-loop inverted pendulum system has been described in [80].

Kalman filtering assumes that the message generating process has a known dynamics and that the exogenous inputs have known statistical properties. Unfortunately, these assumptions limit the utility of minimum variance estimators in situations where the message model and/or the noise descriptions are unknown. Furthermore, H_∞ filter provides a bound for the worst-case estimation error without the need for knowledge of noise statistics, it has been demonstrated that H_∞ filtering has the advantage of being less sensitive to uncertainties of the underlying system. Therefore, robust H_∞ filtering problem has to be studied when parameter uncertainty arises in a system model [75]. The aim of H_∞ filtering problem is basically to find a full or reduced-order filter such that the associated filtering error system satisfies a prescribed H_∞ norm-bound constraint. Much work has been done for H_∞ filtering problem, for example [5, 79, 242, 243] and so on. It has also been well recognized that time-delay exists commonly in dynamic systems and is frequently a source of instability and poor performance. Therefore, the last 10 years have witnessed significant advances in dealing with analysis and design problems for time-delay systems. In particular, the H_∞ filtering problem has been thoroughly studied for various time-delay systems, see e.g., [74, 185, 276, 277] for some recent papers. The class of Markovian jump systems is an important family of systems subject to abrupt variations. The Markovian jump systems have finite modes, which may jump from one to another at different times and between different modes. A system with such a "jumping" character may be modeled as a hybrid system, and the parameter jumps among different modes can be seen as discrete events. Note that the concept of Markovian jump filter had already been used in some papers, see e.g. [6, 20, 209, 275]. However, to the best of our knowledge, there are few papers considering delta operator filter for Markovian jump DOSs with time delays up to now.

A delta operator Riccati difference equation has been gained for discrete-time Kalman filter in [202], however, a detailed delta operator Kalman filter algorithm has been not given in the results. Hence, it is a valuable idea to apply discrete-time Kalman filter to estimate full-state for DOSs. A robust Kalman filter for systems under norm bounded uncertainties has been given in [256]. Kalman filter has been also used for networked data fusion [262]. Kalman filter-based identification of unknown exogenous input of stochastic linear systems has been investigated via pseudo-measurement approach in [183]. An optimization method of linear observations for Kalman filter has been given based on a generalized water filling theorem in [225]. Kalman filter has been also used in GPS/SINS integrated navigation fusion algorithm in [36]. A two-stage Kalman filter method has been given for chaos systems in [130]. Both a Kalman-filter-based iterative algorithm and a recursive frequency estimator using linear prediction have been considered in [305]. In [122], the synthesis of two observer-based nonlinear control algorithms has been performed within the globally linearizing control framework. Robust Kalman filtering for continuous systems with Markovian jumping parameters has been given in [160, 208]. The convergence analysis of an extended Kalman filter for nonlinear deterministic systems has been presented in [21]. A problem of optimal linear Kalman filtering with packet losses has been considered in [109]. A generalization structure of the transposed direct-form II filter structures in shift operator and delta operator has been derived in [141]. A least square error criterion based procedure to design 2-D digital filters in delta-domain has been presented in [134]. Therefore, it is an interesting idea and stimulating to apply a DOKF in delta domain to estimate the full-state of DOSs.

1.3.4 Finite Frequency DOSs

It is well known that each design specification is often given not for the entire frequency range but rather for a finite frequency range of relevance. The weighting method has proven useful to deal with the finite frequency requirements [150]. The current state of the art for fixing the incompatibility is to introduce the so-called weighting functions. A low/band/high-pass filter would be added to the system in series as a weight that emphasizes a particular frequency range and then the design parameters are chosen such that the weighted system norm is small. The weighting method has proven useful in practice, but suffers from nontrivial deficiencies. Firstly, the additional weights tend to increase the system complexity, and the amount of increased complexity is positively correlated with the complexity of the weights. Secondly, the process of selecting appropriate weights is tedious and can be time-consuming, especially when the designer has to shoot for a good trade-off between the complexity of the weights and the accuracy in capturing desired specifications. Finally, the weighting method is effective only for gain specifications and it is not clear how to treat constraints on phase properties. Another alternative approach has been also used to grid the frequency axis [286]. In this case, infinitely many frequency domain inequalities are approximated by a finite number of frequency domain inequalities at selected frequency points. This approach has a practical significance especially when the system is well damped and the frequency response is expected to be "smooth", i.e., no sharp peaks. The resulting computational burden often seems moderate. However, it is difficult to determine a priori how fine the grid should be to achieve a certain performance. The method also suffers from the lack of a rigorous performance guarantee in the design process as the violation of the specifications may occur at a frequency between grid points. Different from weighting function and frequency gridding methods, a generalized KYP lemma which gives a direct treatment of the finite frequency performance and obtain a linear matrix inequality result in the continuous-time setting has been proposed in [119]. Thus it can avoid computational burden and guarantee gain property performances simultaneously [115]. There are several results in the literature along this line, for examples [114, 116, 117], and so on.

Originated from network theory, the positive real lemma has many important applications in control-system study for either continuous and discrete state-space systems. Since the notion of positive realness was introduced in [3], many researchers have considered the positive real control problem for linear systems such as in [161, 162, 270], and references therein. The notion of a positive real system is one of the oldest in system, circuit, and control theory. Within control theory, positive real systems have played a major role in stability theory. The books by Popov [192], Narendra and Taylor [175], and Anderson and Vongpanitlerd [3] provide a good overview of positive real systems in control and systems theory. The recent book by Joshi [129] and the references cited there should be consulted for applications of positive real systems ideas to flexible space structures. The main motivation for the study of the positive real control problem stems from robust and nonlinear control. More precisely, in feedback interconnections, the positive realness of a certain loop transfer function will ensure the overall stability of feedback systems if uncertainty or nonlinearity can be characterized by a positive real system. Some frequency domain conditions for strictly positive real functions have been given in [112]. The finite frequency positive realness approach for dynamical system design from a control perspective has been given in [118].

In recent years, fault detection in dynamic systems has attracted considerable attention from many researchers due to the increasing demand for reliability and safety in industrial processes. The basic idea of fault detection is to construct a residual signal and compare it with a predefined threshold. If the residual exceeds the threshold, an alarm is generated. Because noises and disturbances may result in significant changes in the residual and lead to false alarms, fault detection observers have to be robust, namely, it is insensitive or even invariant to noise and disturbances. Some results on fault detection and isolation by using frequency domain approach have been reported in literatures such as [56, 302]. Problems related to the integrated design of robust fault detection systems in time-frequency domain have been studied in [288]. A nonlinear output frequency response function has been used to fault detection in [188]. A high-resolution frequency estimation method for three-phase induction machine fault detection has been given in [135]. Usually, H_∞ norm optimization technique is used to design the robust fault detection observers, the way of this technique is to introduce a performance index and formulate the fault detection as an optimization problem. But the technique may result in significant changes in the residual and lead to false alarms, and it is contrary to main objective of fault detection because the H_- norm measures the maximum effect of an input on an output. Based on H_-, which is defined as the smallest nonzero singular value of the transfer function matrix from fault to residual at the particular frequency, the study on minimum singular value aiming to maximize the minimum effect of faults on the residual output of a fault detection observer has gained much attention. On the one hand, the LMI methodology has been under intensive research and widely used for various kinds of robust control and filtering problems. One advantage of the LMI approach is the relative ease in incorporating additional design objectives into the formulation. Hence, LMI formulations for the H_- and mixed H_-/H_∞ problems are of interest. Based on these performance evaluation criteria, multiple objective optimal filter problems such as H_-/H_∞ have attracted a great deal of research interests in the recent years, for example [93, 121, 151]. There are also some results on fault finite frequency ranges detection filters with H_∞ and H_- performance indices for discrete systems [235] and continuous systems [237], respectively.

1.3.5 Networked Control via DOSs

Control systems are called NCSs on condition that control loops are closed via a real-time shared media network which is generally viewed as an unreliable communication channel [252]. The advantages of NCSs are a flexible architecture and a reduction of installation and maintenance cost. Several practical factors, such as limited bandwidth, packets dropout and packet delay, make the analysis and design of NCSs difficult. The network-induced delays occur when sensors, actuators and controllers exchange data across networks. Network packets dropout occasionally happens in NCSs when there are node failures or message collisions. Most network protocols are equipped with transmission-retry mechanisms, however, they can only retransmit for a limited time. After this time is expired, the packets are dropped. Recently, network systems are widely used in several areas, and some considerable attentions have been directed to NCSs. NCSs is defined as a feedback control system that the control loops are closed through a real-time network [40, 60]. Although the networks make it convenient to control large distributed systems, there also exist many control issues, which are not considered by conventional control theory, such as packets dropout and network-induced delays, sampling and transmitting methods, out of order delivery and congestion. The general theory for NCSs has received much attention in recent years. The general theory for NCSs have attracted much research interest in recent years, see for example, [73, 108, 167] for the network-induced delay issue, [97, 244, 304] for packets dropout in digital communication networks, [153, 200, 205] for the H_∞ control performance of NCSs, [157, 166] for Markovian characterization of NCSs, [155, 156, 306] for networked predictive control systems. A control performance using the H_∞ gain has been measured for NCSs [205]. The stabilization problems of NCSs with random delays described by discrete-time systems and continuous-time systems have been investigated in [301] and [108], respectively. The stability analysis of discrete-time NCSs subject to packet loss has been concerned in [97]. Teleoperation over the Internet with/without velocity signal has been considered in [104]. Mean-square asymptotic stability for a closed-loop NCS with a new delay characterization is derived in [73]. Under a multiple-packet transmission policy, stability analysis for discrete-time NCSs is concerned in [97]. Using a switched linear systems approach, a stability analysis of NCSs has been given in [48]. It has been pointed out in [136] that signals transmitted from one point to another may experience a few segments of networks which can possibly induce successive delays with different properties due to the variable network transmission conditions in practical situations. It can be summed up from above works that network-induced delays and packets dropout are two key problems to be dealt with in NCSs. It is worthy pointing out that the values of packets dropout and network-induced delays in NCSs are derived from the values of the input delay by the viewpoint of a logic ZOH [272]. As is known that the shorter the sampling period is, the better the system performance is for NCSs. However, a short sampling period will increase the possibility of network congestion. Felicitously, the sampling-period in DOSs is an explicit parameter which can be chosen as an appropriate value based on networks load conditions. Therefore, it may be obtained better control effect by using delta operator approach than shift operator approach in sampling NCSs.

Furthermore, Markovian jump systems have attracted a lot of researchers from control and operations research communities [95, 208, 211, 245, 247] and so on. The stability and stabilization problems for both continuous-time and discrete-time Markovian jump systems have been investigated in [298]. Some results on Markovian jump systems with time delays have also been given in [299, 300]. The kind of Markovian jump systems has been widely used in NCSs. A stabilization problem for NCSs with Markovian characterization has been investigated in [157]. The kind of mode-dependent time delays, which means the combination of switching mechanism and time delays, is more in line with the actual situation than others for networked control models [271]. To the best of our knowledge, there have been few papers on stability problems of Markovian jump systems with mode-dependent time delays via delta operator approach, and less papers using the kind of Markovian jump DOSs describe a NCS.

In a NCS, all signals are sampled by a sampler [94], and the sampling rate for each signal may be varying from sample to sample according to actual situations. The kind of sampling periods has been called as time-varying sampling periods or nonuniform sampling periods. The nonuniform sampled-data control has been extensively studied in the past years. A stabilization problem has been investigated for a T-S fuzzy system with nonuniform uncertain sampling [65]. The problem of reconstructing a periodic bandlimited signal from nonuniform samples has been considered in [168]. For limited bandwidth networks, a nonuniform sampling strategy may be achieved better performance than a uniform sampling strategy [220]. In [287], the sampling period is the same as the node transmission period, which is unknown and time-varying in advance. Time-varying observers and state-feedback controllers have been designed for computer control under time-varying sampling period [201]. The problem of designing H_∞ controllers for NCSs has been studied by using an active-varying sampling period method in [241], however, the sampling period is chosen in a finite set and can not switch randomly. In [77], the problem of robust H_∞ control has been investigated for sampled-data systems with probabilistic sampling, but the control plant has been not discretized so that the result is hard to be applied in NCS frameworks.

In the aforementioned results, the effects of the delay and packet dropout on the NCSs have been focused more on the data transmission rate but less on the dynamics of the networked systems. According to actual situations, all signals are sampled by a sampler in NCSs [94, 187]. For limited bandwidth networks, a nonuniform sampling strategy may be achieved better performance than a uniform sampling strategy [220]. The nonuniform sampled-data NCSs have been extensively studied such as in [201, 241, 287]. However, most mentioned results let the sampling period be the same as the node transmission period and the QoS of communication network has not been investigated effectively. Since the network throughput between different client terminals and digital archives are different, a random search by key-word matching only may cause the user to waste time waiting for media retrieval. A better idea is to shorten network-induced delay and reduce packet dropout by taking into account the system QoS criterion, i.e., the network throughput. As a method of remote control, a QoS-based remote control scheme for a NCS has been presented in [39, 138]. In this paper, we will separate sample period and transmission period in network sensor, and choose an appropriate transmission time according to QoS of communication networks instantaneously. On the other hand, guaranteed cost control is a useful method in the systems which is not only stable but also guarantees an adequate level of performance, since it was firstly presented in [28]. There are also some valuable results on guaranteed cost control in NCSs, for example [55, 267, 294].

Motivated by the widespread use of NCSs, the sampled-data control theory has been well-developed in the last two decades, see [34, 287] and [220]. A robust sampled-data H_∞ control problem has been considered with stochastic sampling in [77]. It is possible to use the tools of linear sampled-data systems for analysis and design linear NCSs [61]. Note that most of published literatures on stability analysis of NCSs are defined in the whole frequency domain. Based on the famous Shannon sampling theory, sampling frequency is much larger than system frequency. In the view for frequency domain, the poles are located in the stable boundary and the discrete systems will lose stability in finite word length computer using the traditional shift operator for high frequency sampling. For the reason of that high frequency sampling will increase the possibility of network congestion, high frequency constricts of NCSs is more affected by time delays and packets dropout than low frequency constricts. Therefore, designing a high frequency system to describe the NCS satisfying stability performance by using delta operator approach is also a key task in this topic.

1.4 Summary

Delta operator is a new discretization method that plays the important role in high-speed signal processing and digital sampled control. Delta operator approach can not only avoid the ill-conditioned problems caused by the conventional shift operator based theory when fast sampling, but also the delta operator formulated model approaches the corresponding continuous time model as the sampling interval goes to zero, thus the design method for continuous time systems can be directly applied to discrete time systems. In this book, these five issues are discussed about DOSs: SMC, delays, filter, finite frequency, NCSs. The results in this book will no doubt advance the study of DOSs and potential applications in the areas of control.

Part I
SMC of DOSs

Chapter 2
Observer Based on SMC for DOSs

2.1 Introduction

Over the past decades, much attention has been focused on the problems of robust control for uncertain standard state-space systems, refer to [10, 11, 68, 73, 278] and so on. Furthermore, the state observer design problem has been also paid considerable attention at the same time, for example [163, 164, 165, 280]. A robust non-fragile H_∞ observer-based control has been investigated for continuous-time delayed systems [31]. Based on input-output linearization method, a kind of state observers has been presented for nonlinear systems in [128]. Some T-S fuzzy observers have been proposed based on some relaxed stability conditions [158]. Variable structure control strategy by using sliding-mode concept has been widely studied and developed for control and state estimation problems since the works of Utkin in [232]. A robust predictor based SMC for discrete systems has been designed in [260]. The problem of SMC for continuous systems has been investigated in [259] and [211]. The robust adaptive SMC for discrete systems has been given in [265]. Sensorless SMC of induction motors has been considered in [195]. Moreover, some problems of SMC and identification for uncertain stochastic systems have been investigated in [13].

Recently, sliding mode observers are often used for estimating states of control systems. For example, an effective method to design sliding mode observers was presented for uncertain control systems [43, 173, 226, 273]. A sliding mode observer has been used for nonlinear robust fault reconstruction and estimation in [283]. A class of fuzzy sliding mode observers have been investigated in [14]. An observer-based SMC problem has been studied for state-delayed systems in [182]. Both a sliding mode observer and a observer-based SMC have been investigated in [250]. The application of a particular sliding mode observer to fault detection and isolation has been considered in [51]. The dynamic discrete-time response does not converge smoothly to its continuous counterpart by using these algorithms [170]. Delta operator

H. Yang et al.: Analysis and Synthesis of Delta Operator Systems, LNCIS 430, pp. 21–34.
springerlink.com © Springer-Verlag Berlin Heidelberg 2012

method has been introduced to avoid the ill-conditioned problems under fast sampling [82]. Delta operator require smaller word length than shift operator in fixed-point digital control processors [113]. The relationships of optimal realization sets between shift operator and delta operator have been established in [142]. A little results have been reported on SMC by using delta operator in literatures for example [27, 254]. It is an useful idea by using delta operator method in sliding mode observers for practical application and theoretical analysis. This has motivated our study.

2.2 Problem Statement

In this chapter, consider the following uncertain DOS:

$$\delta x(t_k) = (A + \Delta A(t_k))x(t_k) + B(u(t_k) + f(x(t_k), t_k)) \tag{2.1}$$

$$y(t_k) = Cx(t_k) \tag{2.2}$$

where $x(t_k) \in R^n$ is the state variable, $u(t_k) \in R^m$ is control input, $y(t_k) \in R^p$ is the system output, $f(x(t_k), t_k) \in R^m$ is an unknown nonlinear function. Furthermore, there exists a known scalar function $\rho(t_k, y)$ such that $f(x(t_k), t_k)$ satisfies $\|f(x(t_k), t_k)\| \leq \rho(t_k, y)$. A, B and C are real constant matrices with appropriate dimensions. We assume that there exists that the eigenvalues of matrix A are negative and $rank(B) = m$. The parametric uncertainty is described by $\Delta A = EF(t_k)H$, where E and H are known constant real matrices with appropriate dimensions. $F(t_k)$ is a unknown time-varying matrix satisfying $F^T(t_k)F(t_k) \leq I$.

Before ending this section, the following lemmas will be used to prove our main results.

Lemma 2.1. [196] *The property of delta operator, for any time function $x(t_k)$ and $y(t_k)$, there exists*

$$\delta(x(t_k)y(t_k)) = \delta x(t_k)y(t_k) + x(t_k)\delta y(t_k) + T\delta x(t_k)\delta y(t_k).$$

Lemma 2.2. [269] *For some given matrices Υ, G and H of appropriate dimensions and with Υ symmetric, then*

$$\Upsilon + GF(t_k)H + H^T F^T(t_k)G^T \leq 0,$$

with $F^T(t_k)F(t_k) \leq I$, if and only if there exists a scalar $\varepsilon > 0$ such that

$$\Upsilon + \varepsilon^{-1}H^T H + \varepsilon GG^T \leq 0.$$

Lemma 2.3. [120] *Given matrices $G \in R^{n\times n}$ and $U \in R^{n\times m}$. Assume that U has full rank $m < n$ and $G = G^T$. Then, $\sigma U^T U - G > 0$ for a scalar σ if* *and only if $\tilde{U}^T G \tilde{U} < 0$ where $\tilde{U}$ is any matrix whose columns form the basis of the null space of U^T.*

Lemma 2.4. [189] *Let E, $F(t_k)$ and H be real matrices of appropriate dimensions, with $F^T(t_k)F(t_k) < I$, then we have that for any scalar $\varepsilon > 0$*

$$EF(t_k)H + H^T F^T(t_k)E^T \leq \varepsilon^{-1}EE^T + \varepsilon H^T H.$$

Lemma 2.5. *For any real vectors a and b of compatible dimensions: $2a^T b < a^T a + b^T b$.*

2.3 Main Results

2.3.1 Stability Analysis of Sliding Mode

Firstly, a sliding variable is defined as follows:

$$s(x(t_k), t) = B^T P^{-1} x(t_k) \tag{2.3}$$

with $P \in R^{n\times n}$ is a positive definite matrix to be designed. Similar to the methods in [182], we have

$$z(t_k) = \begin{bmatrix} z_1(t_k) \\ z_2(t_k) \end{bmatrix} = Wx(t_k),\ W = \begin{bmatrix} (\tilde{B}^T P\tilde{B})^{-1}\tilde{B}^T \\ (\tilde{B}^T P\tilde{B})^{-1}\tilde{B}^T P^{-1} \end{bmatrix} \tag{2.4}$$

with $z_1(t_k) \in R^{n-m}$ and $z_2(t_k) \in R^m$, $\tilde{B}$ is any basis of the null space of B^T, i.e. $\tilde{B} \in \{M : B^T M = 0\}$. Note that the transformation (2.4) corresponds to the standard normal form presented in [50]. It is easily shown that

$$s(t_k) = B^T P^{-1} B z_2(t_k),\ W^{-1} = \begin{bmatrix} P\tilde{B}\ B \end{bmatrix} \tag{2.5}$$

Then, a reduced-order sliding mode dynamics on the sliding surface $s(z(t_k), t_k) = 0$ with dimension of $n - m$ can be obtained by the state transformation of $z(t_k) = Wx(t_k)$. It can be described as:

$$\delta z_1(t_k) = (\tilde{B}^T P\tilde{B})^{-1}\tilde{B}^T(A + \Delta A(t_k))P\tilde{B}z_1(t_k) \tag{2.6}$$

Finally, we are ready for the following result that ensures the asymptotic stability of the sliding mode dynamics of (2.6) in delta domain.

Theorem 2.6. *The reduced-order sliding mode dynamics (2.6) is asymptotical stable in delta domain, if there exist a matrix $P > 0$, and positive scalars α_1 and α_2 such that the following LMI holds*

$$\begin{bmatrix} AP + PA^T + \alpha_1 EE^T - \alpha_2 BB^T & A^T P & PH^T \\ PA & (T - 2)P - \alpha_2 BB^T & 0 \\ HP & 0 & -\alpha_1 I \end{bmatrix} < 0 \quad (2.7)$$

Proof. Construct a Lyapunov functional in delta domain as follows:

$$V_1(t_k) = z_1^T(t_k)(\tilde{B}^T P \tilde{B}) z_1(t_k) \quad (2.8)$$

By using Lemma 2.1 and taking the delta operator manipulations of $V(t_k)$ along the trajectory of DOS (2.6), we can obtain that:

$$\begin{aligned} \delta V_1(t_k) &= \delta^T(z_1(t_k))(\tilde{B}^T P \tilde{B}) z_1(t_k) + z_1^T(t_k)(\tilde{B}^T P \tilde{B})\delta(z_1(t_k)) \\ &\quad + \mathrm{T}\delta^T(z_1(t_k))(\tilde{B}^T P \tilde{B})\delta(z_1(t_k)) \\ &= z_1^T(t_k)(A + \Delta A(t_k))^T(\tilde{B}^T P \tilde{B}) z_1(t_k) \\ &\quad + z_1^T(t_k)(\tilde{B}^T P \tilde{B})(A + \Delta A(t_k)) z_1(t_k) \\ &\quad + \mathrm{T}\delta^T(z_1(t_k))(\tilde{B}^T P \tilde{B})\delta(z_1(t_k)) \end{aligned} \quad (2.9)$$

For the positive definite real matrix $\tilde{B}^T P \tilde{B}$, there exists

$$0 = -2\delta^T(z_1(t_k))\tilde{B}^T P \tilde{B}(\delta z_1(t_k) - (A + \Delta A(t_k)) z_1(t_k)) \quad (2.10)$$

Then, if the terms on the right sides of (2.9)-(2.10) are added into $\delta V_1(t_k)$, we have

$$\delta V_1(t_k) \leq \left[z_1^T(t_k) \; \delta^T(z_1(t_k)) \right] \Pi \left[z_1^T(t_k) \; \delta^T(z_1(t_k)) \right]^T < 0 \quad (2.11)$$

where

$$\Pi = \Upsilon + \bar{G} F(t_k) \bar{H} + \bar{H}^T F^T(t_k) \bar{G}^T < 0 \quad (2.12)$$

with

$$\Upsilon = \begin{bmatrix} \tilde{B}^T(A^T P + PA)\tilde{B} & \tilde{B}^T A^T P \tilde{B} \\ \tilde{B}^T P A \tilde{B} & (\mathrm{T} - 2)\tilde{B}^T P \tilde{B} \end{bmatrix},$$

$$\bar{G} = \begin{bmatrix} \tilde{B}^T E \\ 0 \end{bmatrix}, \; \bar{H} = \begin{bmatrix} (HP\tilde{B})^T \\ 0 \end{bmatrix}^T$$

By using Lemma 2.2 and Schur's complement, $\Pi < 0$ can be written as

$$\begin{bmatrix} \tilde{B}^T & 0 & 0 \\ 0 & \tilde{B}^T & 0 \\ 0 & 0 & I \end{bmatrix} \begin{bmatrix} \Pi_{11} & A^T P & PH^T \\ PA & (\mathrm{T} - 2)P & 0 \\ HP & 0 & -\alpha_1 I \end{bmatrix} \begin{bmatrix} \tilde{B} & 0 & 0 \\ 0 & \tilde{B} & 0 \\ 0 & 0 & I \end{bmatrix} < 0 \quad (2.13)$$

with

$$\Pi_{11} = AP + PA^T + \alpha_1 EE^T$$

Lastly, it is easily shown from Lemma 2.3 that LMI (2.13) is solvable for $P > 0$ and $\alpha_1 > 0$ if and only if LMI (2.7) is solvable for $P > 0$, $\alpha_1 > 0$ and $\alpha_2 > 0$. Based on Definition 1.2, this completes the proof.

2.3.2 Observer Based on SMC

In this section, we shall design a full-order sliding mode observer for system (2.1) with the form of (2.14), which can be described as follows:

$$\delta\hat{x}(t_k) = A\hat{x}(t_k) + Bu(t_k) + Bu_e(t_k) + L(y(t_k) - C\hat{x}(t_k)) \tag{2.14}$$

where $\hat{x}(t_k)$ is the estimate of the state $x(t_k)$, $u_e(t_k)$ is an external discontinuous feedback compensation control, which can eliminate the effect of the nonlinear term $f(x(t_k), t_k))$, and combined with $u(t_k)$, it guarantees the asymptotic stability of the error estimation dynamics, the constant matrix $L \in R^{n\times p}$ is the observer gain matrix to be specified. It guarantees that a sliding mode is attained on the sliding surface in the state-estimation error space given by:

$$s_e(e(t_k), t) = B^T Xe(t_k) \tag{2.15}$$

where the state estimation error is defined as $e(t_k) = x(t_k) - \hat{x}(t_k)$. The positive definite matrix $X \in R^{n\times n}$ is to be designed satisfying the constraint: $B^T X = NC$. For a matrix $N \in R^{m\times p}$ to be chosen later such that

$$s_e(e(t_k), t) = NCe(t_k) = N(y(t_k) - C\hat{x}(t_k)) \tag{2.16}$$

Therefore, by subtracting (2.14) from (2.1), a state estimation error dynamics can be given as follows:

$$\delta e(t_k) = (A - LC + \Delta A(t_k))e(t_k) + \Delta A(t_k)\hat{x}(t_k) \tag{2.17}$$
$$-B(u_e(t_k) - f(x(t_k), t_k))) \tag{2.18}$$

The following robust control term $u_e(t_k)$ is given

$$u_e(t_k) = (\rho(t, y) + \gamma_1)\mathrm{sign}(s_e(t_k)) \tag{2.19}$$

where $\gamma_1 > 0$. A linear sliding surface in the state estimate space is defined as follows:

$$\hat{s}(\hat{x}(t_k), t) = B^T X\hat{x}(t_k) = 0 \tag{2.20}$$

The following input controller is designed as

$$u(t_k) = u_{eq}(t_k) + u_c(t_k) \tag{2.21}$$

with

$$
\begin{aligned}
u_{eq}(t_k) &= -B^T X A \hat{x}(t_k) \\
u_c(t_k) &= -(\rho(t, y) + \gamma_1 + \gamma_2)\mathrm{sign}(\hat{s}(t_k))
\end{aligned}
$$

where $\gamma_2 > 0$ is a real positive scalar to be specified.

The following theorem gives a sufficient condition for asymptotic stability of the overall closed-loop system which composes of state estimate dynamics (2.14) and state estimation error dynamics (2.17).

Theorem 2.7. *The combined closed-loop system which composes of state estimate dynamics (2.14) and state estimation error dynamics (2.17) is asymptotically stable, if there exist matrices Y, N, $X > 0$, and scalars $\varepsilon_1 > 0$, $\varepsilon_2 > 0$, $\varepsilon_3 > 0$ and $\varepsilon_4 > 0$ satisfying*

$$B^T X = NC \tag{2.22}$$

$$\Sigma < 0 \tag{2.23}$$

where

$$
\Sigma = \begin{bmatrix}
\Sigma_{11} & \Sigma_{12} & C^T Y^T & C^T Y^T & XE & XE & 0 & 0 & 0 & 0 & 0 \\
\Sigma_{21} & \Sigma_{22} & 0 & 0 & 0 & 0 & XE & XE & 0 & 0 & 0 \\
YC & 0 & \Sigma_{33} & A^T X & 0 & 0 & 0 & 0 & \Sigma_{39} & \Sigma_{310} & 0 \\
YC & 0 & XA & \Sigma_{44} & 0 & 0 & 0 & 0 & 0 & 0 & \Sigma_{411} \\
E^T X & 0 & 0 & 0 & -\varepsilon_1 I & 0 & 0 & 0 & 0 & 0 & 0 \\
E^T X & 0 & 0 & 0 & 0 & -\varepsilon_2 I & 0 & 0 & 0 & 0 & 0 \\
0 & E^T X & 0 & 0 & 0 & 0 & -\varepsilon_3 I & 0 & 0 & 0 & 0 \\
0 & E^T X & 0 & 0 & 0 & 0 & 0 & -\varepsilon_4 I & 0 & 0 & 0 \\
0 & 0 & \Sigma_{39}^T & 0 & 0 & 0 & 0 & 0 & -I & 0 & 0 \\
0 & 0 & \Sigma_{310}^T & 0 & 0 & 0 & 0 & 0 & 0 & -I & 0 \\
0 & 0 & 0 & \Sigma_{410}^T & 0 & 0 & 0 & 0 & 0 & 0 & -I
\end{bmatrix}
$$

with

$$
\begin{gathered}
\Sigma_{11} = XA - YC + A^T X - C^T Y^T + \varepsilon_1 H^T H + \varepsilon_3 H^T H, \\
\Sigma_{12} = A^T X - C^T Y^T \;\; \Sigma_{21} = XA - YC, \;\; \Sigma_{22} = (T - 2)X, \\
\Sigma_{33} = XA + A^T X + \varepsilon_2 H^T H + \varepsilon_4 H^T H, \;\; \Sigma_{39} = XBB^T, \\
\Sigma_{310} = \sqrt{2} A^T X, \;\; \Sigma_{44} = (T - 2)X, \;\; \Sigma_{410} = XBB^T.
\end{gathered}
$$

Moreover, an observer feedback matrix is given by $L = X^{-1}Y$.

Proof. Choose the following Lyapunov functional in delta domain:

$$V_2(t_k) = e^T(t_k) X e(t_k) + \hat{x}^T(t_k) X \hat{x}(t_k) \tag{2.24}$$

By using Lemma 2.1 and taking the delta operator manipulations of $V_2(t_k)$, results in

$$\begin{aligned}\delta V_2(t_k) &= 2e^T(t_k)X\delta e(t_k) + h\delta^T e(t_k)X\delta e(t_k)\\&\quad +2\hat{x}^T(t_k)X\delta\hat{x}(t_k) + \mathtt{T}\delta^T\hat{x}(t_k)X\delta\hat{x}(t_k)\\&= 2e^T(t_k)X\left[(A - LC + \Delta A(t_k))e(t_k) + \Delta A(t_k)\hat{x}(t_k)\right.\\&\quad \left.-B(u_e(t_k) - f(x(t_k), t_k)))\right] + \mathtt{T}\delta^T e(t_k)X\delta e(t_k)\\&\quad +2\hat{x}^T(t_k)X\left[A\hat{x}(t_k) + Bu(t_k) + Bu_e(t_k)\right.\\&\quad \left.+L(y(t_k) - C\hat{x}(t_k))\right] + \mathtt{T}\delta^T\hat{x}(t_k)X\delta\hat{x}(t_k)\end{aligned} \tag{2.25}$$

Utilizing Lemma 2.4 and Lemma 2.5, it is gotten

$$\begin{aligned}2e^T(t_k)EF(t_k)He(t_k) &\le \varepsilon_1^{-1}e^T(t_k)XEE^TXe(t_k)\\&\quad +\varepsilon_1 e^T(t_k)H^THe(t_k)\end{aligned} \tag{2.26}$$

$$\begin{aligned}2e^T(t_k)EF(t_k)H\hat{x}(t_k) &\le \varepsilon_2^{-1}e^T(t_k)XEE^TXe(t_k)\\&\quad +\varepsilon_2\hat{x}^T(t_k)H^TH\hat{x}(t_k)\end{aligned} \tag{2.27}$$

$$\begin{aligned}-2\hat{x}^T(t_k)XBB^TXA\hat{x}(t_k) &\le \hat{x}^T(t_k)(XBB^T)(BB^TX)\hat{x}(t_k)\\&\quad +\hat{x}^T(t_k)A^TXXA\hat{x}(t_k)\end{aligned} \tag{2.28}$$

Because of the inequality $-\hat{s}(t_k)\text{sign}(s_e(t_k)) \le |\hat{s}(t_k)|$, there exists:

$$\begin{aligned}-2\gamma_1|s_e| - 2\gamma_2|\hat{s}| &\ge -2s_e^T(\rho(t,y) + \gamma_1)\text{sign}(s_e) + 2\|s_e\|\rho(t,y)\\&\quad -2\hat{s}^T(\rho(t,y) + \gamma_1 + \gamma_2)\text{sign}(\hat{s}) + 2\|\hat{s}\|(\rho(t,y) + \gamma_1)\end{aligned} \tag{2.29}$$

Hence, it is obtained that

$$\begin{aligned}\delta V_2(t_k) &\le e^T(t_k)\left[X(A - LC) + (A - LC)^TX + \varepsilon_1 H^TH\right]e(t_k)\\&\quad +e^T(t_k)\left[\varepsilon_1^{-1}XEE^TX + \varepsilon_2^{-1}XEE^TX\right]e(t_k)\\&\quad +\hat{x}^T(t_k)\left[XA + A^TX + (XBB^T)(BB^TX)\right]\hat{x}(t_k)\\&\quad +\hat{x}^T(t_k)\left[A^TXXA + \varepsilon_2 H^TH\right]\hat{x}(t_k)\\&\quad +2\hat{x}^T(t_k)XLCe(t_k) + \mathtt{T}\delta^T e(t_k)X\delta e(t_k)\\&\quad +\mathtt{T}\delta^T\hat{x}(t_k)X\delta\hat{x}(t_k) - 2\gamma_1|s_e| - 2\gamma_2|\hat{s}|\end{aligned} \tag{2.30}$$

For the positive definite real matrix $X > 0$, there exist

$$\begin{aligned}0 &= 2\delta^T e(t_k)X\left[-\delta e(t_k) + (A - LC + \Delta A(t_k))e(t_k)\right.\\&\quad \left.+\Delta A(t_k)\hat{x}(t_k) - B(u_e - f(x(t_k), t_k)))\right]\end{aligned} \tag{2.31}$$

$$\begin{aligned}0 &= 2\delta^T\hat{x}(t_k)X\left[-\delta\hat{x}(t_k) + A\hat{x}(t_k) + Bu(t_k) + Bu_e\right.\\&\quad \left.+L(y(t_k) - C\hat{x}(t_k))\right]\end{aligned} \tag{2.32}$$

Similar with the inequalities (2.26)-(2.29), we obtain

$$
\begin{aligned}
2\delta^T e(t_k)EF(t_k)He(t_k) &\leq \varepsilon_3^{-1}\delta^T e(t_k)XEE^TX\delta e(t_k) \\
&\quad +\varepsilon_3 e^T(t_k)H^THe(t_k) && (2.33)\\
2\delta^T e(t_k)EF(t_k)H\hat{x}(t_k) &\leq \varepsilon_4^{-1}\delta^T e(t_k)XEE^TX\delta e(t_k) \\
&\quad +\varepsilon_4 \hat{x}^T(t_k)H^TH\hat{x}(t_k) && (2.34)\\
-2\delta^T\hat{x}(t_k)XBB^TXA\hat{x}(t_k) &\leq \delta^T\hat{x}(t_k)(XBB^T)(BB^TX)\delta\hat{x}(t_k) \\
&\quad +\hat{x}^T(t_k)A^TXXA\hat{x}(t_k) && (2.35)
\end{aligned}
$$

Then, considering the inequality $-\delta\hat{s}(t_k)\text{sign}(s_e(t_k)) \leq |\delta\hat{s}(t_k)|$, we have

$$
\begin{aligned}
0 \leq\ & 2\delta^T e(t_k)X(A-LC)e(t_k) + \delta^T e(t_k)\left[(\varepsilon_3^{-1}+\varepsilon_4^{-1})XEE^TX + 2X\right]\delta e(t_k) \\
&+\varepsilon_3 e^T(t_k)H^THe(t_k) + \hat{x}^T(t_k)\left[\varepsilon_4 H^TH + A^TXXA\right]\hat{x}(t_k) \\
&+2\delta^T\hat{x}(t_k)XA\hat{x}(t_k) + \delta^T\hat{x}(t_k)\left[2X + (XBB^T)(BB^TX)\right]\delta\hat{x}(t_k) \\
&+2\hat{x}^T(t_k)XLCe(t_k) - 2\gamma_1|\delta s_e| - 2\gamma_2|\delta\hat{s}| \qquad (2.36)
\end{aligned}
$$

Finally, it follows from (2.30) and (2.36) that

$$
\delta V_2(t_k) \leq \xi^T(t_k)\Sigma_1\xi(t_k) < 0, \tag{2.37}
$$

where

$$
\xi^T(t_k) = \left[\, e^T(t_k)\ \delta^T e(t_k)\ \hat{x}^T(t_k)\ \delta^T\hat{x}(t_k)\,\right],
$$

and

$$
\Sigma_1 = \begin{bmatrix} \Sigma_1(1,1) & (A-LC)^TX & C^TL^TX & C^TL^TX \\ X(A-LC) & \Sigma_1(2,2) & 0 & 0 \\ XLC & 0 & \Sigma_1(3,3) & A^TX \\ XLC & 0 & XA & \Sigma_1(4,4) \end{bmatrix} < 0,
$$

with

$$
\begin{aligned}
\Sigma_1(1,1) &= X(A-LC) + (A-LC)X + \varepsilon_1^{-1}XEE^TX + \varepsilon_2^{-1}XEE^TX \\
&\quad +\varepsilon_1 H^TH + \varepsilon_3 H^TH, \\
\Sigma_1(3,3) &= XA + A^TX + (XBB^T)(BB^TX) + 2A^TXXA \\
&\quad +\varepsilon_2 H^TH + \varepsilon_4 H^TH, \\
\Sigma_1(2,2) &= (\mathtt{T}-2)X + \varepsilon_3^{-1}XEE^TX + \varepsilon_4^{-1}XEE^TX, \\
\Sigma_1(4,4) &= (\mathtt{T}-2)X + (XBB^T)(BB^TX).
\end{aligned}
$$

By choosing $L = X^{-1}Y$, it is shown for Schur's complement that $\Sigma_1 < 0$ is equivalent to $\Sigma < 0$. From Definition 1.2, the closed-loop system (2.14) and (2.17) is asymptotically stable in delta domain. This completes the proof.

Remark 2.8. The linear equality condition (2.22) in Theorem 2.7 can not be solved by applying LMI-Toolbox in Matlab environment. To deal with problem, we replace (2.22) with the following form:

$$\operatorname{tr}\left((B^TX - NC)^T(B^TX - NC)\right) = 0 \tag{2.38}$$

Introducing the condition $(B^TX - NC)^T(B^TX - NC) \leq \beta I$, and Schur's complement gives

$$\begin{bmatrix} -\beta I & XB - C^TN^T \\ B^TX - NC & -I \end{bmatrix} \leq 0 \tag{2.39}$$

Hence, the observer-based SMC problem is changed to a problem of finding a global solution of the following minimization problem:

$$\min\ \beta \quad \text{subject to (2.23) and (2.39).} \tag{2.40}$$

Then, the observed-based SMC problem is solvable.

2.3.3 Reachability Analysis of Sliding Surfaces

In this section, we investigate the reachability of sliding surfaces $\hat{s}(\hat{x}(t_k), t) = 0$ in the state-estimate space and $s_e(e(t_k), t) = 0$ in the estimation error space.

Theorem 2.9. *If there exist matrices Y, N, $X > 0$, and scalars $\varepsilon_1 > 0$, $\varepsilon_2 > 0$, $\varepsilon_3 > 0$ and $\varepsilon_4 > 0$ satisfying (2.22) and (2.23), and observer gain matrix $L = X^{-1}Y$, then the SMC law (2.21) guarantees that the sliding motion is attained on the sliding surfaces $s_e(e(t_k), t) = 0$ and $\hat{s}(\hat{x}(t_k), t) = 0$, respectively.*

Proof. Construct a Lyapunov-Krasovskii candidate functional in delta domain as follows:

$$V_3(t_k) = \frac{1}{2}s_e^T(t_k)(B^TXB)^{-1}s_e(t_k) + \frac{1}{2}\hat{s}^T(t_k)(B^TXB)^{-1}\hat{s}(t_k) \tag{2.41}$$

Then, it can be obtained that

$$\begin{aligned} \delta V_3(t_k) = {} & s_e^T(t_k)(B^TXB)^{-1}B^TX\left[(A - LC + \Delta A(t_k))e(t_k)\right. \\ & \left.+\Delta A(t_k)\hat{x}(t_k) - B(u_e - f(x(t_k), t_k))\right] \\ & +\hat{s}^T(t_k)(B^TXB)^{-1}B^TX\left[A\hat{x}(t_k) + Bu(t_k) + Bu_e\right. \\ & \left.+L(y(t_k) - C\hat{x}(t_k))\right] + \frac{\mathrm{T}}{2}\delta^T s_e(t_k)(B^TXB)^{-1}\delta s_e(t_k) \\ & +\frac{\mathrm{T}}{2}\delta^T\hat{s}(t_k)(B^TXB)^{-1}\delta\hat{s}(t_k) \end{aligned} \tag{2.42}$$

Next, let $\Phi(\hat{s}, s_e) = \left[\delta^T s_e(t_k)(B^TXB)^{-1}\delta s_e(t_k) + \delta^T \hat{s}(t_k)(B^TXB)^{-1}\delta \hat{s}(t_k)\right]/2$, there exists

$$\begin{aligned}
\delta V_3(t_k) \leq\; & \|s_e(t_k)\|\|(B^TXB)^{-1}\|\left[\|B^TX(A-LC)\|\|e(t_k)\|\right.\\
& \left.+\|B^TXE\|\|H\|\left(\|e(t_k)\|+\|\hat{x}(t_k)\|\right)\right]\\
& +\|\hat{s}(t_k)\|\left[\|(B^TXB)^{-1}-I\|\|B^TXA\|\|\hat{x}(t_k)\|\right.\\
& \left.+\|(B^TXB)^{-1}\|\|B^TXLC\|\|e(t_k)\|\right]\\
& +\mathtt{T}\Phi(\hat{s}, s_e) - \gamma_1|s_e(t_k)| - \gamma_1|s_e(t_k)|
\end{aligned} \tag{2.43}$$

Note that the term $\mathtt{T}\Phi(\hat{s}, s_e)$ can be suppressed by selecting the sampling interval $\mathtt{T}$ small enough. Let

$$\eta_1 = \max\left\{\|(B^TXB)^{-1}\|\|B^TX(A-LC)\|,\ \|(B^TXB)^{-1}\|\|B^TXE\|\|H\|\right\}$$
$$\eta_2 = \max\left\{\|(B^TXB)^{-1}-I\|\|B^TXA\|,\ \|(B^TXB)^{-1}\|\|B^TXLC\|\right\}$$

In the state space composed of state estimate vector and estimation error vector, define the following domain:

$$\Omega_\varepsilon = \{\eta_1\|e(t_k)\| + \eta_2\|\hat{x}(t_k)\| < \min\{\gamma_1,\ \gamma_2\} - \varepsilon\}$$

with a constant ε satisfying $0 < \varepsilon < \min\{\gamma_1,\ \gamma_2\}$. Therefore, in the domain Ω_ε, from (2.43) we can get

$$\delta V_3(t_k) \leq -\varepsilon|s_e(t_k)| - \varepsilon|\hat{s}(t_k)| < 0$$

Therefore, it is concluded from Definition 1.2 that $s_e(t_k) \to 0$ and $\hat{s}(t_k) \to 0$. Furthermore, it is concluded that both the state estimate trajectories and the estimation error trajectories will have the sliding surface $s_e(t_k) = 0$ and $\hat{s}(t_k) = 0$, respectively, within the region Ω_ε. This completes the proof.

Remark 2.10. In this chapter, we want to make a bridge between sliding mode observer and DOS framework. It is well known that sliding mode observers have a special effect of dealing with nonlinearity and uncertainties. Furthermore, there exists better numerical advantage by using delta operator approach than using shift operator approach for short sampling period. Moreover, the discrete sliding mode approach requires high speed discontinuous action to steer the states of a system into a sliding surface and to maintain subsequent motion on this surface. Hence, the combination of sliding mode observer and delta operator approach is an extensive idea to consider.

2.4 Numerical Example

In this section, an illustrative example is given for testing the design method developed in this chapter.

Example 2.11. The following nonlinear uncertain DOS is used to illustrate the SMC scheme:

$$\delta x(t_k) = (A + EF(t_k)H)\, x(t_k) + B(u(t_k) + f(t_k, x(t_k))). \qquad (2.44)$$

with $\mathtt{T} = 0.001$, $F(t_k) = \mathrm{diag}\{\sin(t_k),\ \cos(t_k),\ \sin(t_k)\}$ and

$$A = \begin{bmatrix} -0.21 & 0.1 & 0 \\ -0.06 & -0.15 & 0 \\ 0 & -0.1 & -0.1 \end{bmatrix}, \ B = \begin{bmatrix} 0.5 & -0.1 \\ 0 & 0.2 \\ -0.3 & 0.1 \end{bmatrix}, \ C = \begin{bmatrix} -0.1 & 0.4 & -0.2 \\ 0.3 & -0.1 & 0.2 \end{bmatrix},$$

$$E = \begin{bmatrix} 0.1 & 0.2 & 0.3 \\ 0.3 & 0.1 & 0.4 \\ 0.2 & 0.2 & 0.2 \end{bmatrix}, \ H = \begin{bmatrix} 0.03 & 0.01 & 0.02 \\ 0.01 & 0.02 & 0.01 \\ 0.04 & 0.03 & 0.01 \end{bmatrix}, \ f(t_k, x(t_k)) = \begin{bmatrix} 0.8\sin(t_k) \\ 0.9\cos(t_k) \end{bmatrix}.$$

Then solving the minimization problem (2.40), there exist the following feasible solutions:

$$X = \begin{bmatrix} 166.4514 & -105.7374 & 174.2921 \\ -105.7374 & 304.5649 & -183.4426 \\ 174.2921 & -183.4426 & 236.8590 \end{bmatrix}, \ Y = \begin{bmatrix} 8.3922 & 59.0877 \\ 88.8585 & 16.3696 \\ -7.6983 & 46.6478 \end{bmatrix},$$

$$N = \begin{bmatrix} 34.0276 & 114.4694 \\ 126.4218 & -25.7374 \end{bmatrix}, \ \begin{array}{l} \varepsilon_1 = 492.2771, \varepsilon_2 = 494.7489, \\ \varepsilon_3 = 284.3793, \varepsilon_4 = 287.5463, \end{array}$$

and $\beta \approx 1.5 \times 10^{-11}$ (hence the linear constraint $B^T X = NC$ is satisfied). Furthermore, the sliding mode observer parameter can be obtained as follows:

$$L = X^{-1}Y = \begin{bmatrix} -0.0262 & 0.4649 \\ 0.5149 & 0.2394 \\ 0.3856 & 0.0402 \end{bmatrix} \qquad (2.45)$$

The observer-based sliding mode controller is given by:

$$\begin{aligned} u(t_k) = &- \begin{bmatrix} -0.105 & 0.08 & 0.03 \\ 0.009 & -0.05 & -0.01 \end{bmatrix} \hat{x}(t_k) \\ &-(\rho(t, y) + \gamma_1 + \gamma_2)\mathrm{sign}(\hat{s}(t_k)), \end{aligned} \qquad (2.46)$$

where $\gamma_1 = 3$ and $\gamma_2 = 4$. Let the initial condition $x(t_k) = [0.85\ 0.85\ 0.85]^T$ and $\hat{x}(t_k) = [0.85\ 0.85\ 0.85]^T$. Figs. 2.1-2.4 illustrate the simulation results.

The state response $x(t_k)$ is given in Fig. 2.1, where $x(t_k)$ is denoted by $x_1(t),\ x_2(t),\ x_3(t)$.

Fig. 2.2 shows the simulation results of state estimate trajectories $\hat{x}(t_k)$ which is denoted by $x1_r(t),\ x2_r(t),\ x3_r(t)$.

Fig. 2.3 shows the sliding surface $\hat{s}(\hat{x}(t_k), t_k)$ in the state estimate space, where $\hat{s}(\hat{x}(t_k), t_k)$ is denoted by $\hat{S}(t)$.

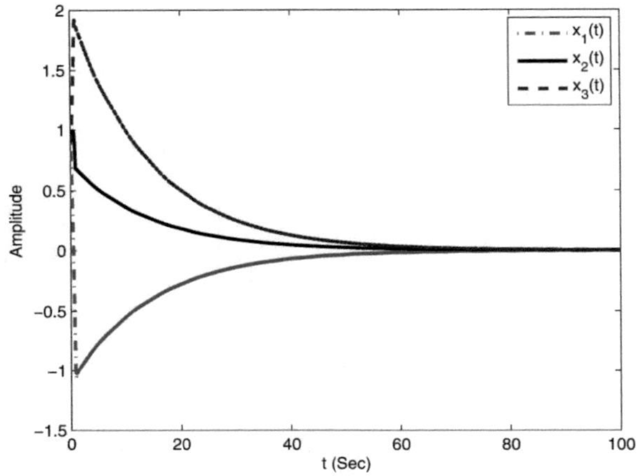

Fig. 2.1 State trajectories $x(t_k)$

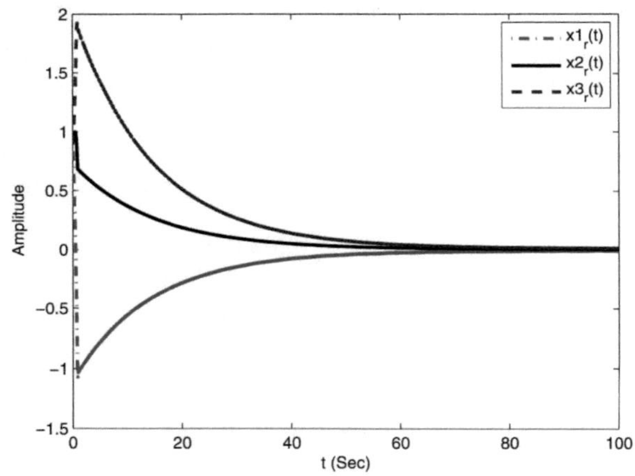

Fig. 2.2 State estimate trajectories $\hat{x}(t_k)$

Fig. 2.4 shows the sliding mode variables $s_e(e(t_k), t_k)$ in the state-estimation error space, where $s_e(e(t_k), t_k)$ is denoted by $S_e(t)$.

It can be seen that the present observer-based SMC scheme are robust against parameter uncertainties and nonlinearities, and the asymptotic stability of closed-loop DOS is guaranteed (2.44).

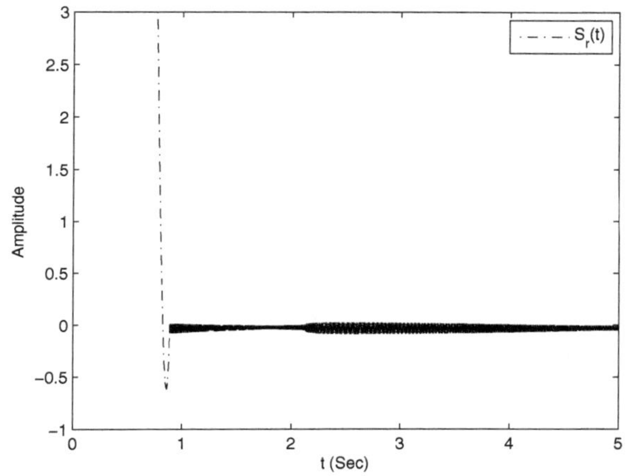

Fig. 2.3 Sliding surface $S_r(t_k)$

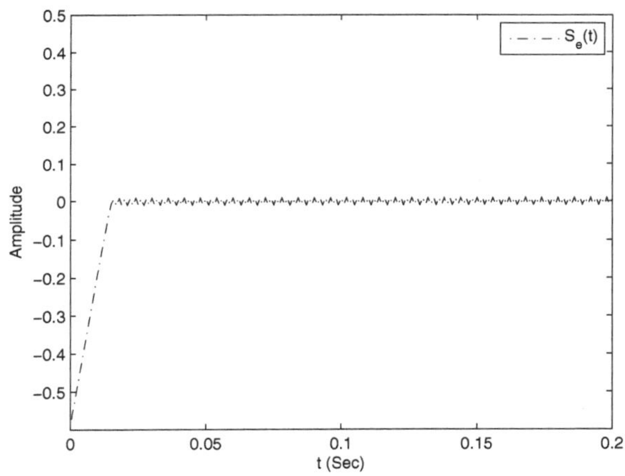

Fig. 2.4 Sliding mode variables $S_e(t_k)$

2.5 Summary

In this chapter, a discrete robust sliding mode observer and an observer-based SMC problem for a class of DOSs with parameter uncertainty and nonlinear exogenous disturbance have been investigated. A novel robust stability condition has been obtained for a sliding mode dynamics by using a class of Lyapunov functional in delta domain. A sliding mode controller has been synthesized by employing the SMC theory combined with the reaching law technique based on the designed sliding mode observer. The robust asymptotical stability has also been discussed for the overall closed-loop system composed of the observer dynamics and the state estimation error dynamics. Lastly, a numerical example has been given to illustrate the effectiveness of the developed techniques.

Chapter 3
Robust Adaptive SMC for Uncertain DOSs

3.1 Introduction

Robustness of a control system is very important since there exist many various uncertainties in practical application. In the theory of robust control, it has been frequently encountered in several class of systems which include NCSs [66, 68], quantized control systems [67], discrete singular hybrid systems [264], input delay systems [289, 291], state delay systems [12, 74, 75, 88] and so on. Goodwin constructed delta operator instead of traditional shift operator for sampling continuous systems [82]. The model parameters in DOS will approach to the ones in continuous system as sampling period approach to zero, so as control results [54]. It has been shown better numerical properties by using delta operator than using shift operator at high sampling rates in [219]. The robustness problem for DOSs with parametric uncertainties was first presented in [215]. Traditionally, SMC as a general design tool for robust control systems has been gained much attention. The problems of SMC for linear systems with time delays have been investigated in [211, 254, 259, 260] and so on.

A robust adaptive control scheme which has been applied to a practical autopilot design for feedback-linearized Skid-to-Turn missiles with aerodynamic uncertainties [38]. Both adaptive output control for time-varying systems and adaptive control for nonstationary objects output have been considered in [16] and [17], respectively. Adaptive output-feedback fuzzy decentralized controllers for large-scale nonlinear systems have been developed in [231]. Adaptive control law could lead to a stable closed-loop system and the deviation from the sliding surface is bounded [213]. A model reference adaptive-SMC has been presented and applied for a variable stiffness actuated system [111]. The adaptive SMC has been discussed for a piezo-actuated stage [35]. By an adaptive law, variable structure control has been applied for fuzzy T-S systems [37]. A robust adaptive variable structure output feedback control for uncertain dynamics systems has been proposed in [96]. A

H. Yang et al.: Analysis and Synthesis of Delta Operator Systems, LNCIS 430, pp. 35–47.
springerlink.com © Springer-Verlag Berlin Heidelberg 2012

systematic and simple SMC with robust adaptive tuning has been proposed for nonlinear systems [110]. Delta operator approach has been used to develop an adaptive SMC for discrete systems in [27]. However, there are also a lot of space to be improved on adaptive SMC for DOSs, which motivates us to make an effort in this chapter.

3.2 Problem Formulation

In this chapter, a class of uncertain linear DOSs under consideration is described by:

$$\delta x(t_k) = (A + \Delta A)x(t_k) + B(u(t_k) + w(t_k, x(t_k))) \tag{3.1}$$

where $x(t_k) \in R^n$ is plant states; $u(t_k) \in R^m$ is control inputs; $w(t_k) \in R^m$ is matched uncertainty disturbances. A and B are two constant matrices with appropriate dimensions and $\text{rank}(B) = m$. The norm-bounded parameter uncertainty is described by $\Delta A = GF(t_k)H$, where G and H are two known constant matrices with appropriate dimensions. $F(t_k)$ is unknown time-varying matrix satisfying $F^T(t_k)F(t_k) \leq I$.

To get a regular form of DOS (3.1), we choose a nonsingular matrix Ψ such that

$$\Psi B = \begin{bmatrix} 0_{(n-m)\times m} \\ B_2 \end{bmatrix}$$

where $B_2 \in R^{m\times m}$ is nonsingular. For convenience, let us choose $\Psi = [U_2\ U_1]^T$ where $U_1 \in R^{n\times m}$ and $U_2 \in R^{n\times(n-m)}$ are two sub-blocks of a unitary matrix resulting from the singular value decomposition of B, i.e.,

$$B = \begin{bmatrix} U_1\ U_2 \end{bmatrix} \begin{bmatrix} \Sigma \\ 0_{(n-m)\times m} \end{bmatrix} V^T$$

where $\Sigma \in R^{m\times m}$ is a diagonal positive-definite matrix and $V \in R^{m\times m}$ which is a unitary matrix. By the state transformation $z(t_k) = \Psi x(t_k)$, system (3.1) has the following regular form

$$\delta z(t_k) = (\overline{A} + \Delta\overline{A})z(t_k) + \begin{bmatrix} 0_{(n-m)\times m} \\ B_2 \end{bmatrix} (u(t_k) + w(t, x(t_k))) \tag{3.2}$$

where $\overline{A} = \Psi A \Psi^{-1}$ and $\Delta\overline{A} = \Psi \Delta A \Psi^{-1}$. System (3.2) can be written as:

$$\delta z_1(t_k) = (\overline{A}_{11} + \Delta\overline{A}_{11})z_1(t_k) + (\overline{A}_{12} + \Delta\overline{A}_{12})z_2(t_k) \tag{3.3}$$

$$\begin{aligned} \delta z_2(t_k) = {} & (\overline{A}_{21} + \Delta\overline{A}_{21})z_1(t_k) + (\overline{A}_{22} + \Delta\overline{A}_{22})z_2(t_k) \\ & + B_2(u(t_k) + w(t_k, x(t_k))) \end{aligned} \tag{3.4}$$

where $z_1(t_k) \in R^{n-m}$, $z_2(t_k) \in R^m$, $B_2 = \Sigma V^T$, $\overline{A}_{11} = U_2^T A U_2$, $\overline{A}_{12} = U_2^T A U_1$, $\Delta\overline{A}_{11} = {U_2}^T \Delta A U_2$, $\Delta\overline{A}_{12} = {U_2}^T \Delta A U_1$, $\overline{A}_{21} = U_1^T A U_2$, $\overline{A}_{22} = U_1^T A U_1$, $\Delta\overline{A}_{21} = {U_1}^T \Delta A U_2$, $\Delta\overline{A}_{22} = {U_1}^T \Delta A U_1$.

It is obvious that system (3.3) represents the sliding motion dynamics of system (3.2). Hence, the corresponding sliding surface can be chosen as follows:

$$S(t_k) = [\ C\ I\ ]\, z(t_k) = C z_1(t_k) + z_2(t_k) = 0 \tag{3.5}$$

where $C \in R^{m\times(n-m)}$ is a sliding surface parameter matrix. Substituting $z_2(t_k) = -Cz_1(t_k)$ to system (3.3), the following sliding motion can be obtained:

$$\delta z_1(t_k) = \tilde{A}(t_k) z_1(t_k) \tag{3.6}$$

where $\tilde{A}(t_k) = \overline{A}_{11} + \Delta\overline{A}_{11} - \overline{A}_{12}C - \Delta\overline{A}_{12}C$.

Assumption 3.1. *The matched uncertainties $\omega(t_k, x(t_k))$ are assumed to satisfy the following condition:*

$$\|\omega(t_k, x(t_k))\| \le c + k_0 \|x(t_k)\|$$

where c and k_0 are two constants, but it may not be easily obtained due to the complexity of the structure of the matched uncertainties.

Definition 3.1. The uncertain sliding motion (3.6) is said to be quadratically stable in delta domain if there exists a positive-definite matrix $P \in R^{(n-m)\times(n-m)} > 0$ and a positive constant $\xi > 0$ such that the Lyapunov functional $V(z_1, t) = z_1^T(t_k) P z_1(t_k) \ge 0$ satisfies

$$\delta V(z_1, t_k) = \left[V(z_1, t_k + \mathrm{T}) - V(z_1, t_k)\right]/\mathrm{T} \le -\xi \lambda_{min}(P) \|z_1(t_k)\|^2. \tag{3.7}$$

Definition 3.2. The solution of DOS (3.1) is uniformly ultimately bounded (UUB) if for all $x(t_0) = x_0$, there exists $\varepsilon > 0$ and $T(\varepsilon, x_0)$, such that $\|x(t_k)\| < \varepsilon$, for all $t \ge t_0 + T$.

The objective in this chapter is to design sliding surface parameter $C \in R^{m\times(n-m)}$ and a reaching motion control law $u(t_k)$ such that

1) Sliding motion (3.6) is quadratically stable;

2) The trajectory of closed-loop DOS (3.2) is convergent into a residual set of the origin with the reaching control law $u(t_k)$.

Before ending this section, the following lemmas will be used to prove our main results in this chapter.

Lemma 3.3. [196] *The property of delta operator: for any time-varying function $x(t_k)$ and $y(t_k)$*

$$\delta(x(t_k)y(t_k)) = \delta x(t_k) y(t_k) + x(t_k)\delta y(t_k) + T\delta x(t_k)\delta y(t_k).$$

where T is a sampling period.

Lemma 3.4. [269] *Given constant matrices G and H and a symmetric constant matrix Υ of appropriate dimensions, the following inequality holds:*

$$\Upsilon + GF(t_k)H + H^TF^T(t_k)G^T \leq 0,$$

where $F(t_k)$ satisfies $F^T(t_k)F(t_k) \leq I$, if and only if for $\varepsilon > 0$

$$\Upsilon + \varepsilon^{-1}H^TH + \varepsilon GG^T \leq 0.$$

3.3 Main Results

The following theorem presents the first result on designing sliding surface.

Theorem 3.5. *The reduced order DOS (3.6) is quadratically stable if there exist a symmetric positive-definite matrix $Z \in R^{(n-m)\times(n-m)}$ and a general matrix $Y \in R^{m\times(n-m)}$ as well as a positive scalar $\alpha > 0$ such that*

$$\begin{bmatrix} (T-2)Z & \overline{A}_{11}Z - \overline{A}_{12}Y & 0 & \alpha U_2^TG \\ * & \Upsilon_{22} & ZU_2^TH^T - Y^TU_1^TH^T & \alpha U_2^TG \\ * & * & -\alpha I & 0 \\ * & * & * & -\alpha I \end{bmatrix} < 0 \qquad (3.8)$$

with

$$\Upsilon_{22} = Z\overline{A}_{11}^T - Y^T\overline{A}_{12}^T + \overline{A}_{11}Z - \overline{A}_{12}Y.$$

Moreover, the sliding surface of system (3.3) is

$$S(t_k) = YZ^{-1}z_1(t_k) + z_2(t_k) = 0. \qquad (3.9)$$

Proof: Take a symmetric positive-definite matrix variable $P \in R^{(n-m)\times(n-m)}$ and choose a Lyapunov functional in delta domain as follows:

$$V(z_1, t) = z_1^T(t_k)Pz_1(t_k),$$

which is positive-definite for all $z_1(t_k) \neq 0$. By using Lemma 3.3 and taking the delta operator manipulations of $V(z_1, t)$ along the trajectory of system (3.6), it can be obtained

$$\begin{aligned} \delta V(z_1, t_k) &= \delta^T(z_1(t_k))Pz_1(t_k) + z_1^T(t_k)P\delta(z_1(t_k)) + \mathrm{T}\delta^T(z_1(t_k))P\delta(z_1(t_k)) \\ &= z_1^T(t_k)\tilde{A}^T(t_k)Pz_1(t_k) + z_1^T(t_k)P\tilde{A}(t_k)z_1(t_k) \\ &\quad + \mathrm{T}\delta^T(z_1(t_k))P\delta(z_1(t_k)). \end{aligned} \qquad (3.10)$$

For the positive definite real matrix P, one has that

$$\begin{aligned} 0 &= -2\delta^T(z_1(t_k))P[\delta(z_1(t_k)) - \tilde{A}(t_k)z_1(t_k)] \\ &= -2\delta^T(z_1(t_k))P\delta(z_1(t_k)) + 2\delta^T(z_1(t_k))P\tilde{A}(t_k)z_1(t_k). \end{aligned}$$

Then, we have

$$\delta V(z_1,t) = \begin{bmatrix} P\delta(z_1(t_k)) \\ Pz_1(t_k) \end{bmatrix}^T \Theta \begin{bmatrix} P\delta(z_1(t_k)) \\ Pz_1(t_k) \end{bmatrix} \tag{3.11}$$

where

$$\Theta = \Upsilon + \bar{G}F(t_k)\bar{H} + \bar{H}^T F^T(t_k)\bar{G}^T \tag{3.12}$$

$$\Upsilon = \begin{bmatrix} (\mathrm{T}-2)Z & \tilde{A}_{11}Z \\ * & \Upsilon(2,2) \end{bmatrix},\ \bar{G} = \begin{bmatrix} U_2^T G \\ U_2^T G \end{bmatrix},\ \bar{H} = \begin{bmatrix} 0 \\ ZU_2^T H^T - Y^T U_1^T H^T \end{bmatrix}^T,$$

with

$$Z = P^{-1},\ \Upsilon(2,2) = Z\tilde{A}_{11}^T + \tilde{A}_{11}Z,\ \tilde{A}_{11} = \overline{A}_{11} - \overline{A}_{12}C.$$

Applying Lemma 3.4, $\Theta < 0$ holds if and only if there exists a scalar $\varepsilon > 0$ such that

$$\Sigma_1 = \Upsilon + \begin{bmatrix} \varepsilon\bar{H}^T & \varepsilon^{-1}\bar{G} \end{bmatrix} \begin{bmatrix} I & 0 \\ 0 & I \end{bmatrix}^{-1} \begin{bmatrix} \varepsilon\,\bar{H} \\ \varepsilon^{-1}\bar{G}^T \end{bmatrix} < 0. \tag{3.13}$$

Hence, by defining $Y = CZ$ and using Schur's complement, $\Sigma_1 < 0$ can be written as $\Sigma_2 < 0$, where

$$\Sigma_2 = \begin{bmatrix} (\mathrm{T}-2)Z & \overline{A}_{11}Z - \overline{A}_{12}Y & 0 & \varepsilon^{-1}U_2^T G \\ * & \Sigma_2(2,2) & \varepsilon(ZU_2^T - Y^T U_1^T)H^T & \varepsilon^{-1}U_2^T G \\ * & * & -I & 0 \\ * & * & * & -I \end{bmatrix},$$

with

$$\Sigma_2(2,2) = Z\overline{A}_{11}^T - Y^T\overline{A}_{12}^T + \overline{A}_{11}Z - \overline{A}_{12}Y$$

Pre-multiplying and post-multiplying Σ_2 by diag$\{I,\ I,\ \varepsilon^{-1},\ \varepsilon^{-1}\}$, and setting $\varepsilon^{-2} = \alpha$, the inequality $\Sigma_2 < 0$ is equivalent to (3.8).

Therefore, the inequality $\Theta < 0$ implies that there exists a sufficiently small scalar $\xi > 0$ such that

$$\Theta + \begin{bmatrix} 0_{(n-m)\times(n-m)} & 0_{(n-m)\times(n-m)} \\ 0_{(n-m)\times(n-m)} & \xi I_{(n-m)\times(n-m)} \end{bmatrix} < 0.$$

It follows from (3.11) and above inequality that $\delta V(z_1,t) \leq -\xi\|Pz_1(t_k)\|^2 \leq -\xi\lambda_{min}(P)\|z_1(t_k)\|^2$. Therefore, from Definition 3.1 the reduced system (3.6) is quadratical stable with $C = YZ^{-1}$. Moreover, the sliding surface of system (3.3) is

$$S(t_k) = [\ C\ I\ ]\, z = YZ^{-1} z_1(t_k) + z_2(t_k) = 0. \tag{3.14}$$

The proof is completed.

Remark 3.6. Considering $\mathtt{T} \to 0$, the simplified results of Lyapunov function in delta domain can be changed to the ones in s-domain as $V(z_1, t) = z_1^T(t_k) P z_1(t_k)$, which is positive-define for all $z_1(t_k) \neq 0$. Taking the derivative of $V(z_1, t)$ and following the proof process of Theorem 3.5, we can get a simplified result in s-domain such as in [254].

Assumption 3.2. *The parameter uncertainty $\Delta\overline{A} z(t_k)$ satisfies the following condition:*

$$\|B_2^{-1} \bar{C} \Delta\overline{A} z(t_k)\| \leq (k - k_0)\|z(t_k)\|.$$

From the equality $\|x(t_k)\| = \|z(t_k)\|$, it can be given that

$$\|B_2^{-1} \bar{C} \Delta\overline{A} z(t_k)\| + \|\omega(t_k, x(t_k))\| \leq c + k\|x(t_k)\| = \rho,$$

where $\bar{C} = [\ C\ I\ ]$, c and k are two constants.

In the next step, we will design an adaptive sliding mode controller such that any trajectory of the closed-loop DOS (3.3) will be convergent to the original.

Theorem 3.7. *Suppose inequality (3.8) in Theorem 3.5 has solutions Z and Y, the linear sliding surface is given by (3.9). Then the trajectory of closed-loop DOS (3.3) can be driven onto the sliding surface in finite time with the following control law (3.15) and evolves in a neighborhood around the sliding surface, and finally converges into a residual set at the origin.*

$$u(t_k) = -B_2^{-1}[KS(t_k) + \bar{C}\overline{A} z(t_k)] + u_{adv}(t_k) \tag{3.15}$$

with

$$u_{adv}(t_k) = \begin{cases} -\frac{B_2^T S(t_k)}{\|B_2^T S(t_k)\|}\hat{\rho}, & \hat{\rho}\|B_2^T S(t_k)\| > \varepsilon, \\ -\frac{B_2^T S(t_k)}{\varepsilon}\hat{\rho}^2, & \hat{\rho}\|B_2^T S(t_k)\| \leq \varepsilon, \end{cases}$$

where $K = diag\{k_i\}$, k_i is a positive constant, and $\hat{\rho} = \hat{c}(t, z) + \hat{k}(t, z)\|x(t_k)\|$. The adaptation laws are given as follows

$$\delta\hat{c}(t, x(t_k)) = q_1(-\varepsilon_0 \hat{c} + \|B_2^T S(t_k)\|)$$
$$\delta\hat{k}(t, x(t_k)) = q_2(-\varepsilon_1 \hat{k} + \|B_2^T S(t_k)\|\|x(t_k)\|)$$

with q_1, q_2, ε_0 and ε_1 are design parameters.

$$B_2^T = \begin{bmatrix} b_1 \ b_2 \cdots b_m \end{bmatrix}^T, \ b_i w = \sum_{j=1}^{m} b_{ij} w_j \leq \sum_{j=1}^{m} |b_{ij}| \overline{w}_j = N_{2i},$$
$$N_1 = \begin{bmatrix} N_{11} \cdots N_{1m} \end{bmatrix}, \ N_2 = \begin{bmatrix} N_{21} \cdots N_{2m} \end{bmatrix}, \ N_{1i} \geq N_{2i}$$

Proof. Note that

$$\begin{aligned} \delta S(t_k) &= \bar{C}\delta z(t_k) \\ &= \bar{C}\left[(\overline{A} + \Delta\overline{A})z(t_k) + \begin{bmatrix} 0_{(n-m)\times m} \\ B_2 \end{bmatrix} (u(t_k) + w(t_k, x(t_k)))\right] \end{aligned} \tag{3.16}$$

Consider the following Lyapunov functional in delta domain:

$$V_s(t_k) = \frac{1}{2}\left[S^T(t_k)S(t_k) + \frac{1}{q_1}\tilde{c}^2 + \frac{1}{q_2}\tilde{k}^2\right] \tag{3.17}$$

where $\tilde{c} = c - \hat{c}(t_k, x(t_k))$ and $\tilde{k} = k - \hat{k}(t_k, x(t_k))$. Taking the delta operator manipulations of $V_s(t_k)$, we have

$$\begin{aligned} \delta V_s(t_k) &= S^T(t_k)\delta S(t_k) - \frac{1}{q_1}\tilde{c}\delta\hat{c} - \frac{1}{q_2}\tilde{k}\delta\hat{k} + \frac{\mathrm{T}}{2}\left[\delta S^T(t_k)\delta S(t_k) + \delta^2\hat{c} + \delta^2\hat{k}\right] \\ &= S^T(t_k)\bar{C}\left[(\overline{A} + \Delta\overline{A})z(t_k) + \begin{bmatrix} 0_{(n-m)\times m} \\ B_2 \end{bmatrix} (u(t_k) + w(t_k, x(t_k)))\right] \\ &\quad - \frac{1}{q_1}\tilde{c}\delta\hat{c} - \frac{1}{q_2}\tilde{k}\delta\hat{k} + \frac{\mathrm{T}}{2}\left[\delta S^T(t_k)\delta S(t_k) + \delta^2\hat{c} + \delta^2\hat{k}\right] \end{aligned} \tag{3.18}$$

It follows from Lemma 3.3 that there exist a scalar β_1 such that

$$\begin{aligned} &S^T(t_k)\bar{C}\Delta\overline{A}z(t_k) \\ &\leq \frac{1}{2}\left[\beta_1^{-1}(H\Psi^{-1}z(t_k))^T \ \beta_1 S^T(t_k)\bar{C}\Psi G\right]\begin{bmatrix} \beta_1^{-1}H\Psi^{-1}z(t_k) \\ \beta_1(S^T(t_k)\bar{C}\Psi G)^T \end{bmatrix} \\ &= \frac{1}{2}\beta_1^2 S^T(t_k)\bar{C}\Psi G G^T \Psi^T \bar{C}^T S(t_k) + \frac{1}{2}\beta_1^{-2} z^T(t_k)\Psi^{-T}H^T H\Psi^{-1}z(t_k) \\ &\leq \beta_2 S^T(t_k)S(t_k) + \beta_3 z^T(t_k)z(t_k) \end{aligned} \tag{3.19}$$

where

$$\beta_2 = \lambda_{max}(\tfrac{1}{2}\beta_1^2\bar{C}\Psi G G^T \Psi^T \bar{C}^T)$$
$$\beta_3 = \lambda_{max}(\tfrac{1}{2}\beta_1^{-2}\Psi^{-T}H^T H\Psi^{-1})$$

Then, letting

$$\Phi(S(t_k), \hat{c}, \hat{k}) = \delta S^T(t_k)\delta S(t_k) + \delta^2\hat{c} + \delta^2\hat{k}$$

there exists

$$\delta V_s(t_k) \leq S^T(t_k)\left[\bar{C}\overline{A}z(t_k) + \bar{C}\Delta\overline{A}z(t_k) + B_2(u(t_k) + \omega(t_k, x(t_k)))\right] \\ -\frac{1}{q_1}\tilde{c}\delta\hat{c} - \frac{1}{q_2}\tilde{k}\delta\hat{k} + \frac{\mathtt{T}}{2}\Phi(S(t_k), \hat{c}, \hat{k}) \tag{3.20}$$

Considering the case of $\|B_2^T S(t_k)\|\hat{\rho} > \varepsilon$, based on the control input (3.15) and Assumption 3.1, it can be obtained

$$\begin{aligned}\delta V_s(t_k) \leq{}& S^T(t_k)\left[-KS(t_k) + \bar{C}\Delta\overline{A}z(t_k) - \frac{B_2 B_2^T S(t_k)}{\|B_2^T S(t_k)\|}\hat{\rho} + B_2\omega(t, x(t_k))\right] \\ &-\frac{1}{q_1}\tilde{c}\delta\hat{c} - \frac{1}{q_2}\tilde{k}\delta\hat{k} + \frac{\mathtt{T}}{2}\Phi(S(t_k), \hat{c}, \hat{k}) \\ \leq{}& -S^T(t_k)KS(t_k) - \|B_2^T S(t_k)\|(\hat{c} + \hat{k}\|x(t_k)\|) \\ &+S^T(t_k)B_2(c + k\|x(t_k)\|) - \tilde{c}(-\varepsilon_0\hat{c} + \|B_2^T S(t_k)\|) \\ &-\tilde{k}(-\varepsilon_1\hat{k} + \|B_2^T S(t_k)\|\|x(t_k)\|) + \frac{\mathtt{T}}{2}\Phi(S(t_k), \hat{c}, \hat{k}) \\ \leq{}& -S^T(t_k)KS(t_k) + \frac{\mathtt{T}}{2}\Phi(S(t_k), \hat{c}, \hat{k}) + \varepsilon_0\tilde{c}\hat{c} + \varepsilon_1\tilde{k}\hat{k} \\ \leq{}& -\gamma S^T(t_k)S(t_k) + z^T(t_k)\bar{C}^T(\gamma I - K)\bar{C}z(t_k) \\ &+\frac{1}{4}(\varepsilon_0 c^2 + \varepsilon_1 k^2) + \frac{\mathtt{T}}{2}\Phi(S(t_k), \hat{c}, \hat{k})\end{aligned} \tag{3.21}$$

On the other hand, if $\|B_2^T S(t_k)\|\hat{\rho} \leq \varepsilon$ with the control law defined in (3.15), there exists

$$\delta V_s(t_k) \leq -\gamma S^T(t_k)S(t_k) + z^T(t_k)\bar{C}^T(\gamma I - K)\bar{C}z(t_k) + \frac{\varepsilon}{4} \\ +\frac{1}{4}(\varepsilon_0 c^2 + \varepsilon_1 k^2) + \frac{\mathtt{T}}{2}\Phi(S(t_k), \hat{c}, \hat{k}) \tag{3.22}$$

Note that, the term $\frac{\mathtt{T}}{2}\Phi(S(t_k), \hat{c}, \hat{k})$ in the above inequality contains the parameter uncertainties, which can be suppressed by a properly selected K. When the parameter uncertainties are large, the sampling interval $\mathtt{T}$ should be selected small such that the term $\frac{\mathtt{T}}{2}\Phi(S(t_k), \hat{c}, \hat{k})$ is small enough. Therefore, it can be suppressed by appropriately selected parameter K such that $\lambda_{max}(\bar{C}^T(\gamma I - 2K)\bar{C}) < -\sigma$, where γ and σ are proper positive scalars. Finally, we can obtain

$$\delta V_s(t_k) < -\gamma S^T(t_k)S(t_k) + \Psi(\varepsilon, \varepsilon_0, \varepsilon_1) \tag{3.23}$$

where

$$\Psi(\varepsilon, \varepsilon_0, \varepsilon_1) = \begin{cases} \frac{1}{4}(\varepsilon_0 c^2 + \varepsilon_1 k^2), & \hat{\rho}\|B_2^T S(t_k)\| > \varepsilon \\ \frac{\varepsilon}{4} + \frac{1}{4}(\varepsilon_0 c^2 + \varepsilon_1 k^2), & \hat{\rho}\|B_2^T S(t_k)\| \leq \varepsilon \end{cases}$$

Based on Definition 3.2 and (3.23), it can be concluded that all signals are uniformly ultimately bounded. That is, sliding motion enters a neighborhood of equilibrium in finite time and remains within it. Similar with the process in [255], this proof can be finished.

Remark 3.8. Since there exist adaptation laws $\delta\hat{c}(t, x(t_k))$ and $\delta\hat{c}(t, x(t_k))$ in control law (3.15), based on Definition 1.1 the converse operation of delta operator $\delta^{-1}(t_k)$ has to be considered for numerical simulation. In this chapter, we utilize a new algorithm which is similar to the forth order Runge-Kutta algorithm to deal with the problem.

3.4 Numerical Example

In this section, an illustrative example is given for testing the design method developed in this chapter.

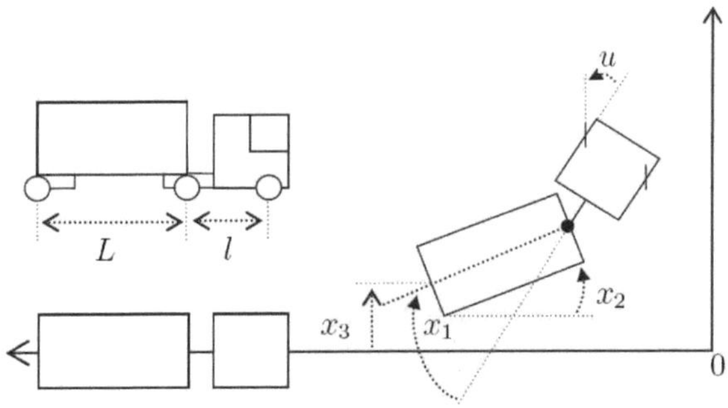

Fig. 3.1 The truck-trailer model

Example 3.9. The proposed method will be applied to design a robust sidling mode controller for truck-trailer system. Consider the truck-trailer model shown in Fig. 3.1, which can be modeled as

$$\begin{aligned}
\dot{x}_1 &= -\frac{v\bar{t}}{(L+\Delta L(t))t_0}x_1 + \frac{v\bar{t}}{lt_0}u \\
\dot{x}_2 &= \frac{v\bar{t}}{(L+\Delta L(t))t_0}x_1 \\
\dot{x}_3 &= \frac{v\bar{t}}{t_0}\sin\left(x_2 + \frac{v\bar{t}}{2(L+\Delta L(t))}x_1\right)
\end{aligned}$$

where $v = -1.0$, $\bar{t} = 2.0$, $t_0 = 0.5$, $L = 5.5$, $l = 2.8$ and $-0.2519 \leq \Delta L \leq 0.2891$. When $x_2 + \frac{v\bar{t}}{2(L+\Delta L(t_k))}x_1$ is about zero, it is easy to see that the truck-trailer system can be sampled to the class of DOS when $\mathtt{T} = 0.01$ as follows:

$$\delta x(t_k) = (A + GF(t_k)H)\, x(t_k) + B(u(t_k) + \omega(t_k, x(t_k))). \qquad (3.24)$$

with

$$A = \begin{bmatrix} 0.7299 & 0 & 0 \\ -0.7299 & 0 & 0 \\ 0.7446 & -4 & 0 \end{bmatrix}, \quad B = \begin{bmatrix} -1.4338 \\ 0.0052 \\ -0.0053 \end{bmatrix},$$
$$F(t_k) = \mathrm{diag}\{\sin(t_k),\ \cos(t_k),\ \sin(t_k)\},\ G = 0.05 \cdot I_{3\times 3},$$
$$H = [h_{3\times 1}\ 0_{3\times 2}]\,,\ h_{3\times 1} = [0.2091\ \ -0.2091\ \ -0.2091]^T.$$

Taking $\Psi = \begin{bmatrix} 0.0036 & 1 & 0 \\ -0.0037 & 0 & 1 \\ -1 & 0.0036 & -0.0037 \end{bmatrix}$, such that $\Psi B = \begin{bmatrix} 0 \\ 0 \\ 1.4338 \end{bmatrix}$, then LMI (9) has feasible solutions:

$$Z = \begin{bmatrix} 0.1554 & 0.0490 \\ 0.0490 & 0.6135 \end{bmatrix}, \ Y = \begin{bmatrix} 0.5556 & -0.1601 \end{bmatrix}, \ \alpha = 1.0142.$$

It follows from Theorem 3.5 that $C = \begin{bmatrix} 3.7524 & -0.5610 \end{bmatrix}$, thus the linear sliding surface is

$$S(t_k) = \begin{bmatrix} 3.7524 & -0.5610 & 1.0000 \end{bmatrix} z(t_k) = 0.$$

From Theorem 3.7, the reaching control law can be taken as follows:

$$u(t_k) = -0.6974 \times \left[KS(t_k) + \begin{bmatrix} 2.2446 & 0.0144 & 3.8885 \end{bmatrix} x(t_k)\right] + u_{adv}(t_k)$$
$$u_{adv}(t_k) = \begin{cases} -\frac{B_2^T S(t_k)}{\|B_2^T S(t_k)\|}\hat{\rho}, & \hat{\rho}\|B_2^T S(t_k)\| > \varepsilon, \\ \\ -\frac{B_2^T S(t_k)}{\varepsilon}\hat{\rho}^2, & \hat{\rho}\|B_2^T S(t_k)\| \leq \varepsilon, \end{cases}$$

where parameter K can be tuned to reduce the chattering on the sliding surface. Figs. 3.2-3.5 are simulation results when choosing $K = 0.4$, $x_1(0) = x_2(0) = x_3(0) = 1$, $S(0) = -2.26$, $u(0) = -9$, $\hat{c} = 0.25$ and $\hat{k} = 0.75$. Obviously, the system is asymptotically stable and the sliding motion trends to the origin in finite time in spite of uncertainties.

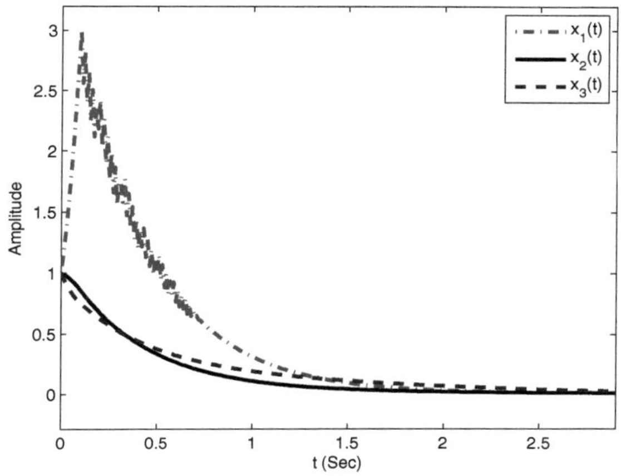

Fig. 3.2 The controlled state trajectories

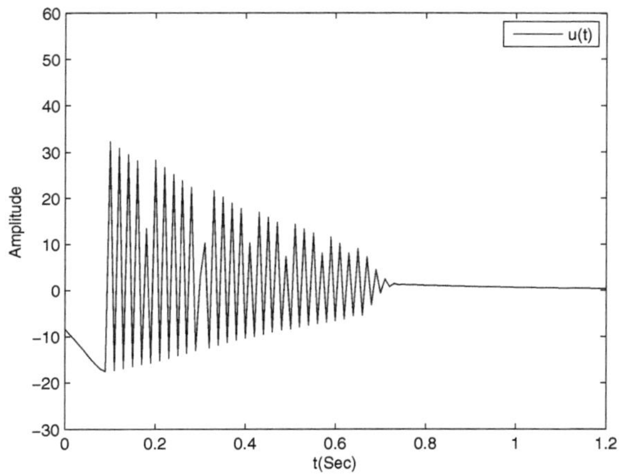

Fig. 3.3 The control signal

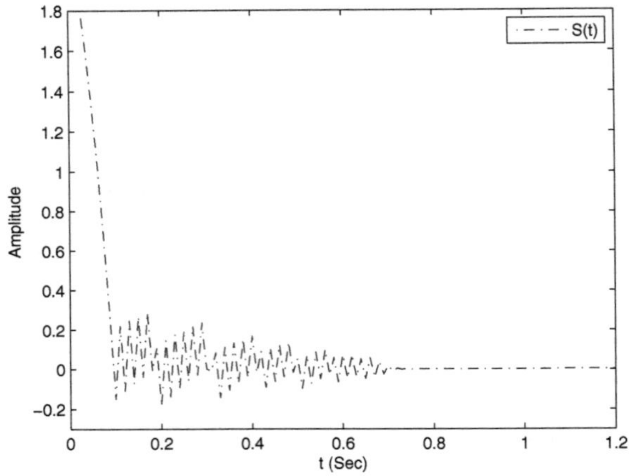

Fig. 3.4 The sliding mode surface.

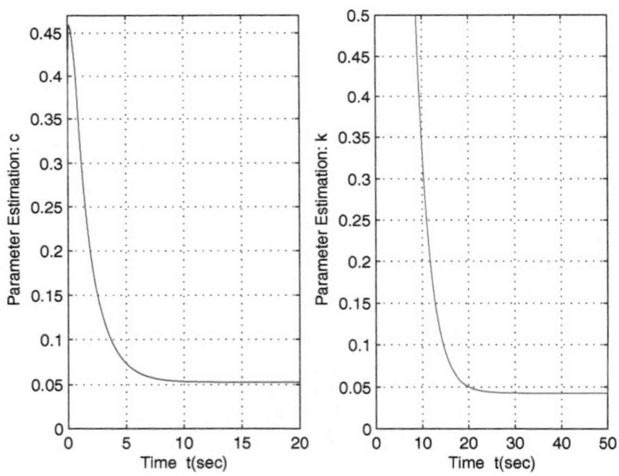

Fig. 3.5 The estimated parameters.

3.5 Summary

In this chapter, a discrete robust adaptive sliding mode controller has been presented for a DOS with parameter uncertainties and exogenous disturbances. The proposed method is adaptive SMC in conjunction with DOSs. The adaptive law has been proposed to reduce the chattering due to the implementation of variable structure controller and possesses the desired characteristics of robustness and good performance. Sufficient conditions for the existence of line sliding surfaces have been derived using LMI method. By solving the conditions, a linear sliding surface has been characterized for the DOS. Furthermore, an adaptive sliding mode controller has been designed to make an bridge between robust adaptive SMC and DOS framework. A numerical example has been given to illustrate the effectiveness of the developed techniques.

Part II
Stabilization of DOSs with Time Delays

Chapter 4
Robust Stabilization for DOSs with Time-Varying Delays

4.1 Introduction

Goodwin constructed delta operator instead of traditional shift operator for sampling continuous systems in [82]. The proposed method can unify some previous related results of the continuous and discrete systems into the framework of the DOSs framework. A class of systems in delta domain has been studied in [32, 140, 266]. The problem of system instability when the sampling time is fast can be solved by using delta operator model [49]. The delta operator model also has the advantage of better numerical properties at high sampling rates [219]. In contrast to the discrete shift operator approximates, the delta operator approach means that the Euler derivative can lead to a quasi-continuous time s-domain model for high sampling frequencies, see for example, [54] and the reference therein. The robustness problem for some DOSs with parametric uncertainties also has been investigated. There are a few papers consider the stability of DOSs such as [214] and [215], and H_∞ control problem has been reported in [52], which proved that the discrete-time controller approach to the continuous-time systems using delta operator.

In time-domain, the direct Lyapunov method is a powerful tool for studying the problems of stability and feedback control for continuous systems and discrete systems. Because time delays may lead to instability of systems, the issues on the state feedback controller design of continuous systems with time-varying delay have received considerable attention [19, 257, 258, 259, 263]. Only a few results reported in the literature are for the class of linear discrete systems with constant time delay, for the reason of that this class of systems can be transformed into equivalent systems without time delays. Furthermore, there are also some papers considering the discrete systems with time-varying delay [18, 65, 75, 221, 274]. However, to the best of our knowledge, there have been few papers on the state feedback controller design of discrete-time systems with time-varying delay via delta operator approach.

H. Yang et al.: Analysis and Synthesis of Delta Operator Systems, LNCIS 430, pp. 51–60.
springerlink.com © Springer-Verlag Berlin Heidelberg 2012

This problem is important and challenging in both theory and practice, which motivated us for this study.

4.2 Problem Statement

The following quasi-continuous uncertain DOS with time-varying delays is considered:

$$\delta x(t_k) = (A + \Delta A(t_k))x(t_k) + (A_d + \Delta A_d(t_k))x(t_k - d(k)) + Bu(t_k) \quad (4.1)$$

where $x(t_k) \in R^q$ is the state variable; $u(t_k) \in R^m$ is control input; the time delay $d(k)$ is a time-varying function that satisfies $0 \leq d_m \leq d(k) \leq d_M$, with $d_m = n_m\mathrm{T}$ and $d_M = n_M\mathrm{T}$, n_m and n_M are two known positive and finite integers. The linear fractional parametric uncertainties $\Delta A(t_k)$ and $\Delta A_d(t_k)$ are time-varying matrices with appropriate dimensions, which are defined as follows:

$$[\Delta A(t_k)\ \Delta A_d(t_k)] = D\hat{F}(t_k)\left[E_1\ E_2\right], \ \hat{F}(t_k) = F(t_k)\left[I - GF(t_k)\right]^{-1} \quad (4.2)$$

where D, G, E_1 and E_2 are known constant real matrices with appropriate dimensions. $F(t_k)$ is unknown time-varying matrix satisfying $F^T(t_k)F(t_k) \leq I$. It is assumed that the matrix $[I - GF(t_k)]^{-1}$ is invertible for any $F(t_k)$ and $I - G^TG > 0$.

Remark 4.1. The model (4.2) describes a wider class of parameter uncertainties than norm-bounded parameter uncertainties. It is easy to see that the linear fractional parameter uncertainties can be reduced to norm-bounded parameter uncertainties when $G = 0$.

Before ending this section, the following lemmas will be used to prove our main results in the next section.

Lemma 4.2. *[266] The property of delta operator: for any time function $x(t_k)$ and $y(t_k)$*

$$\delta(x(t_k)y(t_k)) = \delta(x(t_k))y(t_k) + x(t_k)\delta(y(t_k)) + T\delta(x(t_k))\delta(y(t_k)),$$

where T is a sampling period.

Lemma 4.3. *[124] For any constant positive semi-definite symmetric matrix W, two positive integers r and r_0 satisfying $r \geq r_0 \geq 1$, the following inequality holds*

$$\left(\sum_{i=r_0}^{r} x(i)\right)^T W \left(\sum_{i=r_0}^{r} x(i)\right) \leq (r - r_0 + 1)\sum_{i=r_0}^{r} x^T(i)Wx(i).$$

Lemma 4.4. *[269] For some given matrices Υ, D and E of appropriate dimension and with Υ symmetric, then*

$$\Upsilon + D\hat{F}(t_k)E + E^T\hat{F}^T(t_k)D^T \leq 0,$$

where $\hat{F}(t_k)$ is given as in (4.2), if and only if there exists a scalar $\varepsilon > 0$ such that

$$\Upsilon + \begin{bmatrix} \varepsilon^{-1}E^T & \varepsilon D \end{bmatrix} \begin{bmatrix} I & -G \\ -G^T & I \end{bmatrix}^{-1} \begin{bmatrix} \varepsilon^{-1}E \\ \varepsilon D^T \end{bmatrix} \leq 0.$$

4.3 Main Results

The following control law is employed to deal with the problem of stabilization via state feedback:

$$u(t_k) = Kx(t_k), \tag{4.3}$$

where K is the state feedback controller gain to be determined such that the closed-loop system is asymptotically stable for any $d_m \leq d(k) \leq d_M$. Consider system (4.1) associated with the control law (4.3), then the resulting closed-loop system can be expressed as follows:

$$\delta x(t_k) = (A + BK + \Delta A(t_k))x(t_k) + (A_d + \Delta A_d(t_k))x(t_k - d(k)). \tag{4.4}$$

In this section, we let $A_K(t_k) = A+BK+\Delta A(t_k)$ and $A_d(t_k) = A_d+\Delta A_d(t_k)$. An LMI approach is developed for designing the desired state feedback controller. The following theorem presents a delay-range-dependent result given in terms of LMIs.

Theorem 4.5. *Considering the uncertain closed-loop system (4.4), if there exist matrices $X > 0$, $Z > 0$, $M > 0$, $N > 0$ and Y, as well as a positive scalar α such that the following LMI holds*

$$\Sigma = \begin{bmatrix} \Sigma(1,1) & AX + BY & A_dX & 0 & 0 & \alpha D \\ * & \Sigma(2,2) & A_dX & \frac{1}{d_M}N & XE_1^T & \alpha D \\ * & * & -M & 0 & XE_2^T & 0 \\ * & * & * & -Z - \frac{1}{d_M}N & 0 & 0 \\ * & * & * & * & -\alpha I & \alpha G \\ * & * & * & * & * & -\alpha I \end{bmatrix} < 0, \tag{4.5}$$

with

$$\Sigma(1,1) = (T - 2)X + d_M N$$
$$\Sigma(2,2) = (AX + BY)^T + AX + BY + (d_M - d_m + T)M + M - \frac{1}{d_M}N + Z$$

Then system (4.4) is robustly asymptotically stable. Moreover, a suitable stabilizing state feedback controller for system (4.1) can be chosen by $u(t_k) = YX^{-1}x(t_k)$.

Proof. Construct the Lyapunov-Krasovskii functional in delta domain as follows:

$$V(t_k) = V_1(t_k) + V_2(t_k) + V_3(t_k) + V_4(t_k) + V_5(t_k),$$

with

$$\begin{aligned}
V_1(t_k) &= x^T(t_k)Px(t_k), \quad V_2(t_k) = \mathtt{T}\sum_{i=1}^{n} x^T(t_k - i\mathtt{T})Qx(t_k - i\mathtt{T}),\\
V_3(t_k) &= \mathtt{T}\sum_{i=1}^{n_M} x^T(t_k - i\mathtt{T})Sx(t_k - i\mathtt{T}),\\
V_4(t_k) &= \mathtt{T}^2 \sum_{i=n_m}^{n_M}\sum_{j=1}^{i} x^T(t_k - j\mathtt{T})Qx(t_k - j\mathtt{T}),\\
V_5(t_k) &= \sum_{i=1}^{n_M}\sum_{j=1}^{i} e^T(t_k - j\mathtt{T})Re(t_k - j\mathtt{T}),
\end{aligned}$$

where $e(j) = x(j) - x(j + \mathtt{T})$, so there exist $\delta x(j) = -e(j)/\mathtt{T}$ and $e(t - i\mathtt{T}) = x(t - i\mathtt{T}) - x(t - (i-1)\mathtt{T})$.

Taking the delta operator manipulations of $V_1(t_k)$ along the trajectory of system (4.4), using Lemma 4.2, we can obtain:

$$\begin{aligned}
\delta V_1(t_k) &= \delta^T(x(t_k))Px(t_k) + x^T(t_k)P\delta(x(t_k)) + \mathtt{T}\delta^T(x(t_k))P\delta(x(t_k))\\
&= x^T(t_k)A_K^T(t_k)Px(t_k) + x^T(t - d(k))A_d^T(t_k)Px(t_k)\\
&\quad + x^T(t_k)PA_d(t_k)x(t - d(k)) + x^T(t_k)PA_K(t_k)x(t_k)\\
&\quad + \mathtt{T}\delta^T(x(t_k))P\delta(x(t_k)).
\end{aligned} \tag{4.6}$$

Taking the delta operator manipulations of $V_2(t_k)$, $V_3(t_k)$ and $V_4(t_k)$, it can be obtained that

$$\begin{aligned}
\delta V_2(t_k) &= \frac{1}{\mathtt{T}}\left[\mathtt{T}\sum_{i=1}^{n} x^T(t_k - (i-1)\mathtt{T})Qx(t_k - (i-1)\mathtt{T})\right.\\
&\quad \left. - \mathtt{T}\sum_{i=1}^{n} x^T(t_k - i\mathtt{T})Qx(t_k - i\mathtt{T})\right]\\
&\le x^T(t_k)Qx(t_k) - x^T(t_k - d(k))Qx(t_k - d(k))\\
&\quad + \mathtt{T}\sum_{i=n_m}^{n_M} x^T(t_k - i\mathtt{T})Qx(t_k - i\mathtt{T}).
\end{aligned} \tag{4.7}$$

$$\delta V_3(t_k) = x^T(t_k)Sx(t_k) - x^T(t_k - d_M)Sx(t_k - d_M). \tag{4.8}$$

$$\begin{aligned}\delta V_4(t_k) &= \mathrm{T}\sum_{i=n_m}^{n_M}\left(\sum_{j=1}^{i} x^T(t_k-(j-1)\mathrm{T})Qx(t_k-(j-1)\mathrm{T})\right.\\ &\qquad \left.-\sum_{j=1}^{i} x^T(t_k-j\mathrm{T})Qx(t_k-j\mathrm{T})\right)\\ &= (d_M - d_m + \mathrm{T})x^T(t_k)Qx(t_k)\\ &\qquad -\mathrm{T}\sum_{i=n_m}^{n_M} x^T(t_k-i\mathrm{T})Qx(t_k-i\mathrm{T}).\end{aligned} \tag{4.9}$$

Taking the delta operator manipulations of $V_5(t_k)$, and using Lemma 4.3, results in

$$\begin{aligned}\delta V_5(t_k) &= \frac{1}{\mathrm{T}}\left[\sum_{i=1}^{n_M}\sum_{j=1}^{i} e^T(t_k-(j-1)\mathrm{T})Re(t_k-(j-1)\mathrm{T})\right.\\ &\qquad\left.-\sum_{i=1}^{n_M}\sum_{j=1}^{i} e^T(t_k-j\mathrm{T})Re(t_k-j\mathrm{T})\right]\\ &= \frac{1}{\mathrm{T}}\left[\sum_{i=1}^{n_M} e^T(t_k)Re(t_k) - \sum_{i=1}^{n_M} e^T(t_k-i\mathrm{T})Re(t_k-i\mathrm{T})\right]\\ &\le \frac{n_M}{\mathrm{T}}e^T(t_k)Re(t_k) - \frac{1}{n_M\mathrm{T}}\left[\sum_{i=1}^{n_M} e(t_k-i\mathrm{T})\right]^T R\left[\sum_{i=1}^{n_M} e(t_k-i\mathrm{T})\right]\\ &= -\frac{1}{d_M}\left[x(t_k-d_M)-x(t_k)\right]^T R\left[x(t_k-d_M)-x(t_k)\right]\\ &\qquad + d_M\delta^T(x(t_k))R\delta(x(t_k)).\end{aligned} \tag{4.10}$$

For the positive definite real matrix P, one has that

$$\begin{aligned}0 &= -2\delta^T(x(t_k))P[\delta(x(t_k)) - A_K(t_k)x(t_k) - A_d(t_k)x(t-d(k))]\\ &= -2\delta^T(x(t_k))P\delta(x(t_k)) + 2\delta^T(x(t_k))PA_K(t_k)x(t_k)\\ &\quad +2\delta^T(x(t_k))PA_d(t_k)x(t-d(k)).\end{aligned} \tag{4.11}$$

Finally, it follows from (4.6)-(4.11) that

$$\delta V(t_k) \le \xi^T(t_k)\Sigma_1\xi(t_k) < 0, \tag{4.12}$$

where

$$\xi^T(t_k) = \left[\,\delta^T(x(t_k))\ x^T(t_k)\ x^T(t_k-d(k))\ x^T(t_k-d_M)\,\right],$$

and

$$\Sigma_1 = \begin{bmatrix} (\mathtt{T}-2)P + d_M R & PA_K(t_k) & PA_d(t_k) & 0 \\ * & \Sigma_1(2,2) & PA_d(t_k) & \frac{1}{d_M}R \\ * & * & -Q & 0 \\ * & * & * & -S - \frac{1}{d_M}R \end{bmatrix}$$

$$= \Upsilon + \begin{bmatrix} PD \\ PD \\ 0 \\ 0 \end{bmatrix} \hat{F}(t_k) \begin{bmatrix} 0 \\ E_1^T \\ E_2^T \\ 0 \end{bmatrix}^T + \begin{bmatrix} 0 \\ E_1^T \\ E_2^T \\ 0 \end{bmatrix} \hat{F}^T(t_k) \begin{bmatrix} PD \\ PD \\ 0 \\ 0 \end{bmatrix}^T < 0,$$

where

$$\Sigma_1(2,2) = A_K^T(t_k)P + PA_K(t_k) + (d_M - d_m + \mathtt{T} + 1)Q - \tfrac{1}{d_M}R + S$$

$$\Upsilon = \begin{bmatrix} (\mathtt{T}-2)P + d_M R & P(A+BK) & PA_d & 0 \\ * & \Upsilon(2,2) & PA_d & \frac{1}{d_M}R \\ * & * & -Q & 0 \\ * & * & * & -S - \frac{1}{d_M}R \end{bmatrix}.$$

with

$$\Upsilon(2,2) = (A+BK)^T P + P(A+BK) + \beta Q - \frac{1}{d_M}R + S$$

By the Lemma 4.4, from the inequality $\Sigma_1 < 0$, there exists a scalar $\varepsilon > 0$ such that $\Sigma_2 = \Upsilon + \Theta < 0$ where

$$\Theta = \left[\varepsilon \begin{bmatrix} 0 \\ E_1^T \\ E_2^T \\ 0 \end{bmatrix} \ \varepsilon^{-1} \begin{bmatrix} PD \\ PD \\ 0 \\ 0 \end{bmatrix} \right] \begin{bmatrix} -I & G \\ G^T & -I \end{bmatrix}^{-1} \begin{bmatrix} \varepsilon \ \begin{bmatrix} 0 & E_1 & E_2 & 0 \end{bmatrix} \\ \varepsilon^{-1} \begin{bmatrix} D^T P & D^T P & 0 & 0 \end{bmatrix} \end{bmatrix}$$

Using Schur's complement, $\Sigma_2 < 0$ can be changed to $\Sigma_3 < 0$, where

$$\Sigma_3 = \begin{bmatrix} (\mathtt{T}-2)P + d_M R & P(A+BK) & PA_d & 0 & 0 & \varepsilon^{-1}PD \\ * & \Sigma_3(2,2) & PA_d & \frac{1}{d_M}R & \varepsilon E_1^T & \varepsilon^{-1}PD \\ * & * & -Q & 0 & \varepsilon E_2^T & 0 \\ * & * & * & -S - \frac{1}{d_M}R & 0 & 0 \\ * & * & * & * & -I & G \\ * & * & * & * & * & -I \end{bmatrix}$$

with

$$\Sigma_3(2,2) = (A+BK)^T P + P(A+BK) + (d_M - d_m + \mathtt{T} + 1)Q - \frac{1}{d_M}R + S$$

Pre-multiplying and post-multiplying $\Sigma_3 < 0$ by the diagonal matrix

$$\mathrm{diag}\{P^{-1},\ P^{-1},\ P^{-1},\ P^{-1},\ \varepsilon I,\ \varepsilon I\}$$

and letting $\varepsilon^2 = \alpha$, $X = P^{-1}$, $Y = KX = KP^{-1}$, $Z = P^{-1}SP^{-1}$, $M = P^{-1}QP^{-1}$ and $N = P^{-1}RP^{-1}$. The inequality $\Sigma_3 < 0$ is equivalent to $\Sigma < 0$. From Definition 1.2, it can be easily concluded that the DOS (4.1) with $u(t_k) = Kx(t_k)$ is asymptotically stabilizable if (4.5) holds. Furthermore, the explicit expression of the state feedback controller is given by $u(t_k) = YX^{-1}x(t_k)$.

From the proof of Theorem 4.5, we can get the following results easily.

Corollary 4.6. *Uncertain DOS (4.1) with $u(t_k) = 0$ is asymptotically stable if there exist matrices $P > 0$, $Q > 0$, $R > 0$ and $S > 0$ such that the following LMI holds:*

$$\begin{bmatrix} (T-2)P + d_M R & PA & PA_d & 0 & 0 & \varepsilon^{-1}PD \\ * & \Phi & PA_d & \frac{1}{d_M}R & \varepsilon E_1^T & \varepsilon^{-1}PD \\ * & * & -Q & 0 & \varepsilon E_2^T & 0 \\ * & * & * & -S - \frac{1}{d_M}R & 0 & 0 \\ * & * & * & * & -I & G \\ * & * & * & * & * & -I \end{bmatrix} < 0$$

with

$$\Phi = A^T P + PA + (d_M - d_m + T)Q + Q - \frac{1}{d_M}R + S.$$

Corollary 4.7. *The DOS (4.1) with $\hat{F}(t_k) \equiv 0$ is stabilizable if there exist matrices $X > 0$, $Z > 0$, Y, $M > 0$ and $N > 0$ such that the following LMI holds:*

$$\begin{bmatrix} (T-2)X + d_M N & AX + BY & A_d X & 0 \\ * & \Omega & A_d X & \frac{1}{d_M}N \\ * & * & -M & 0 \\ * & * & * & -Z - \frac{1}{d_M}N \end{bmatrix} < 0,$$

with

$$\Omega = (AX + BY)^T + AX + BY + (d_M - d_m + T)M + M - \frac{1}{d_M}N + Z.$$

In this case, a suitable stabilizing state feedback controller can be chosen by $u(t_k) = YX^{-1}x(t_k)$.

Remark 4.8. It is noted that if letting $\mathtt{T} = 0$, the system (4.1) can be changed to the ones in s-domain; when $\mathtt{T} = 1$ the system (4.1) can be reduced into the one in z-domain, which was considered in [65, 70].

4.4 Numerical Example

In the following, we give an example on truck-trailer model shown in Fig. 3.1 to demonstrate the results obtained in previous sections.

Example 4.9. The proposed method will be applied to design a robust state feedback controller for truck trailer system with time-delay proposed in [227] and [23]. The time-delay model with uncertainties is given by:

$$\begin{aligned}
\dot{x}_1(t_k) &= -\frac{a\Gamma(t_k)}{t_0}x_1(t_k) - \frac{(1-a)\Gamma(t_k)}{t_0}x_1(t-d(k)) + \frac{v\bar{t}}{lt_0}u(t_k)\\
\dot{x}_2(t_k) &= \frac{a\Gamma(t_k)}{t_0}x_1(t_k) + \frac{(1-a)\Gamma(t_k)}{t_0}x_1(t-d(k))\\
\dot{x}_3(t_k) &= \frac{v\bar{t}}{t_0}\sin\left(x_2(t_k) + \frac{a\Gamma(t_k)}{2}x_1(t_k) + \frac{(1-a)\Gamma(t_k)}{2}x_1(t-d(k))\right)
\end{aligned}$$

where $\Gamma(t_k) = v\bar{t}/(L+\Delta L(t_k))$, $a = 0.7$, $v = -1.0$, $\bar{t} = 2.0$, $t_0 = 0.5$, $L = 5.5$, $l = 2.8$ and $-0.2519 \leq \Delta L \leq 0.2891$. When

$$x_2(t_k) + \frac{a\Gamma(t_k)}{2}x_1(t_k) + \frac{(1-a)\Gamma(t_k)}{2}x_1(t-d(k)) \approx 0$$

it is easy to see that the truck trailer system can be sampled to the class of DOS when $\mathtt{T} = 0.01$s as follows:

$$\begin{aligned}
\delta x(t_k) = &\left(A + DF(t_k)\left[I - GF(t_k)\right]^{-1}E_1\right)x(t_k) + Bu(t_k)\\
&+ \left(A_d + DF(t_k)\left[I - GF(t_k)\right]^{-1}E_2\right)x(t-d(k)) \qquad (4.13)
\end{aligned}$$

with

$$A = \begin{bmatrix} 0.5109 & 0 & 0\\ -0.5109 & 0 & 0\\ 0.5212 & -4 & 0 \end{bmatrix},\ A_d = \begin{bmatrix} 0.219 & 0 & 0\\ -0.219 & 0 & 0\\ 0.2234 & 0 & 0 \end{bmatrix},\ B = \begin{bmatrix} -1.4338\\ 0.0052\\ -0.0053 \end{bmatrix},$$

$$F(t_k) = \begin{bmatrix} \sin(t_k) & 0 & 0\\ 0 & \cos(t_k) & 0\\ 0 & 0 & \sin(t_k) \end{bmatrix},\ D = \begin{bmatrix} 0.05 & 0 & 0\\ 0 & 0.05 & 0\\ 0 & 0 & 0.05 \end{bmatrix},$$

$$G = \begin{bmatrix} 0.01 & 0 & 0\\ 0 & 0.01 & 0\\ 0 & 0 & 0.01 \end{bmatrix},\ E_1 = \begin{bmatrix} 0.5091 & 0 & 0\\ -0.5091 & 0 & 0\\ -0.5091 & 0 & 0 \end{bmatrix},\ E_2 = \begin{bmatrix} 0.2182 & 0 & 0\\ -0.2182 & 0 & 0\\ -0.2182 & 0 & 0 \end{bmatrix},$$

In order to design a state feedback controller for DOS (4.13), we first choose $d_m = 1$ and $d_M = 2$. Then, using the Matlab LMI Control Toolbox to solve the LMIs in (3.3), we obtain the solutions as follows:

$$X = \begin{bmatrix} 2.6533 & 0.7365 & -0.6091 \\ 0.7365 & 0.5698 & 2.7557 \\ -0.6091 & 2.7557 & 64.5982 \end{bmatrix}, \ M = \begin{bmatrix} 1.9562 & 0.5470 & 0.1126 \\ 0.5470 & 0.1875 & 0.3841 \\ 0.1126 & 0.3841 & 5.6783 \end{bmatrix},$$
$$N = \begin{bmatrix} 0.1101 & 0.0407 & -0.1649 \\ 0.0407 & 0.0684 & 0.7767 \\ -0.1649 & 0.7767 & 20.7341 \end{bmatrix}, \ Z = \begin{bmatrix} 0.3608 & 0.0898 & 0.5517 \\ 0.0898 & 0.0674 & 0.5531 \\ 0.5517 & 0.5531 & 6.7294 \end{bmatrix},$$
$$Y = \begin{bmatrix} 4.3841 & 0.2786 & -1.4729 \end{bmatrix}, \ \alpha = 11.044.$$

Therefore, by Theorem 4.5, a desired state feedback controller can be given by $K = YX^{-1}$ as follows:

$$u(t_k) = \begin{bmatrix} 2.7862 & -3.9426 & 0.1717 \end{bmatrix} x(t_k).$$

The state response of the class of DOS in this example is given in Fig. 4.1, where $x_1(t_k)$, $x_2(t_k)$, $x_3(t_k)$ are denoted by $x(1)$, $x(2)$, $x(3)$, respectively.

Moreover, Fig. 4.1 shows the simulation results of the input $u(t_k)$.

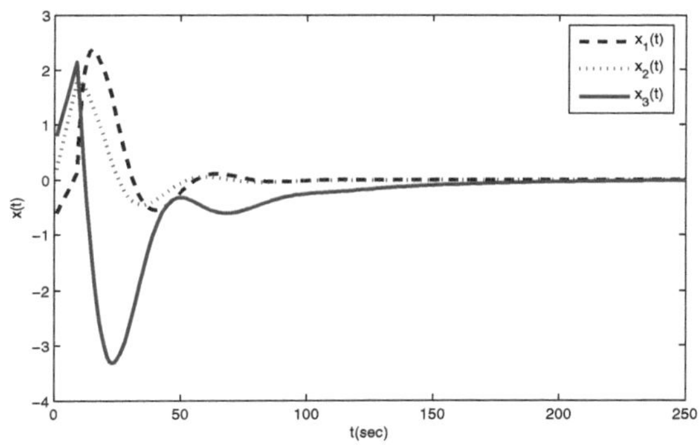

Fig. 4.1 The state in the closed-loop system

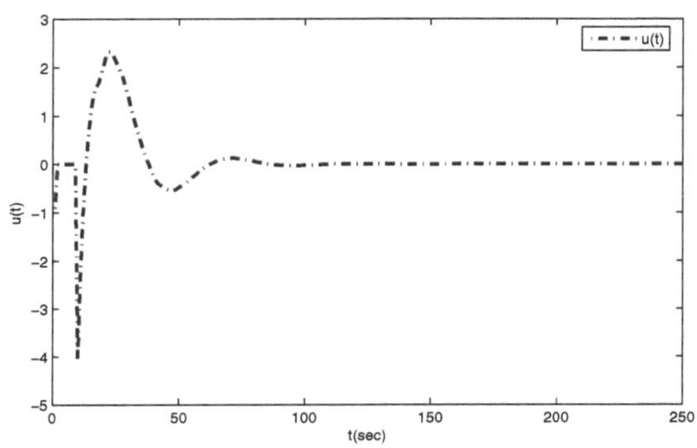

Fig. 4.2 The input in the closed-loop system

Furthermore, if we choose $d_m = 1$, the largest time-delay is $d_M = 5.33$. Then the state feedback controller is $K = \begin{bmatrix} 2.7022 & -1.1204 & 0.0001 \end{bmatrix}$. From these simulation results, it can be seen that the designed state feedback controller ensures the robust asymptotic stability of the closed-loop DOS (4.13).

4.5 Summary

In this chapter, The stabilization of DOSs with time-varying delays has been considered. A state feedback controller for linear fractional uncertainties has been designed such that the closed-loop systems is robustly asymptotically stable. A delay-dependent sufficient conditions have been developed by using novel Lyapunov-Krasovskii functions in delta domain and all results are presented in terms of LMIs form. The proposed method could unify some previous related results of delayed continuous and discrete systems into delayed DOSs framework. Numerical example has been given to illustrate the effectiveness of the theoretic results obtained.

Chapter 5
Robust H_∞ Control for T-S Fuzzy DOSs with Time Delays

5.1 Introduction

Fuzzy technique has been widely and successfully used in nonlinear system modeling and control for more than two decades. T-S fuzzy model was introduced in [223]. The T-S fuzzy system has emerged as one of active and fruitful areas of fuzzy control in recent years, see for example [44, 47, 148, 210, 216] and the references therein. Fuzzy automata and syntactic analysis approach have been used for fault diagnosis [199]. Two different methods on simulate fuzzy random variables have been presented in [81]. Fuzzy H_∞ control has been investigated in [76, 152, 179]. By using a fuzzy linear model, an MIMO adaptive sliding mode controller has been proposed in systems [37]. Fuzzy filter has been established by using Lyapuonv function approach in [181, 297]. A stabilization problem for T-S fuzzy systems with nonuniform uncertain sampling was investigated in [65]. By a fuzzy output feedback controller, a stabilizing problem of nonlinear systems has been addressed in [180]. Stabilization analysis for a T-S discrete fuzzy model has been presented based on a non-PDC control law [85]. Using a non-PDC control law, an extensive result has been presented for T-S discrete fuzzy systems in [45].

Delta operator approach was introduced in [82] and [170]. The transformations between delta operator and shift operator transfer function models have been highlighted in [177]. The computational formulation, properties and applications of DOSs have been illustrated in [176]. The relationships between optimal realization sets in shift operator and delta operator have been established in [142]. A structure in shift operator and delta operator has been derived based on a polynomial-operator approach [141]. A direct model reference adaptive control algorithm has been described in [123]. A tabular method has been presented for DOS in [193]. A technique to directly obtain an approximate DOS has been presented by a given continuous system in [194]. Problems of controller structures realized have been addressed in [33, 222]. An approach inspired from delta-operator has been proposed to

H. Yang et al.: Analysis and Synthesis of Delta Operator Systems, LNCIS 430, pp. 61–81.
springerlink.com © Springer-Verlag Berlin Heidelberg 2012

implement distributed delay [307]. A design method of 2-D digital filters in δ-domain has been presented in [134]. To improve the state matching performance, a refined LMI-based digital redesign has been proposed by using delta operator [137]. The robustness problem for DOSs has been firstly presented in [215]. However, to the best of our knowledge, there have been few papers on the H_∞ control for T-S fuzzy systems with delays via delta operator approach, which motivates us to make an effort in this chapter.

5.2 Problem Statement

In the chapter, we consider the following delayed uncertain fuzzy DOS, which is represented by T-S fuzzy model composed of a set of fuzzy implications, and each implication is expressed by a linear system model. The ith rule of this Takagi-Sugeno model is of the following form:

Rule i: If $\theta_1(t)$ is M_{i1} and $\cdots$ and $\theta_p(t)$ is M_{ip} then

$$\begin{aligned} \delta x(t) &= (A_{1i} + \Delta A_{1i}(t_k))x(t_k) + (A_{2i} + \Delta A_{2i}(t_k))x(t_k - n\mathrm{T}) \\ &\quad +(B_{1i} + \Delta B_{1i}(t_k))\omega(t_k) + (B_{2i} + \Delta B_{2i}(t_k))u(t_k) \qquad (5.1) \\ z(t_k) &= (C_{1i} + \Delta C_{1i}(t_k))x(t_k) + (C_{2i} + \Delta C_{2i}(t_k))x(t_k - n\mathrm{T}) \\ &\quad +(D_{1i} + \Delta D_{1i}(t_k))\omega(t_k) + (D_{2i} + \Delta D_{2i}(t_k))u(t_k) \qquad (5.2) \\ x(t_k) &= \phi(t_k) \qquad t \in [-h,\ 0] \quad i = 1, 2, \cdots, r, \end{aligned}$$

where $x(t_k) \in R^q$ is the state variable; $u(t_k) \in R^m$ is control input; $z(t_k) \in R^p$ is control output; $\omega(t_k) \in R^l$ is the disturbance input which belongs to $L_2[0, \infty)$. $n\mathrm{T}$ is the bounded time delay in the state, and n is a positive integer, where $n\mathrm{T} \le h$; $\phi(t_k)$ is a quasi-continuous vector-valued initial function; $\theta_1(t_k), \theta_2(t_k), \cdots, \theta_p(t_k)$ are the premise variables, each M_{ij} $(j = 1, 2, \cdots, p)$ are the fuzzy sets, r is the number of if-then rules. The parametric uncertainties $\Delta A_{1i}(t_k)$, $\Delta A_{2i}(t_k)$, $\Delta B_{1i}(t_k)$, $\Delta B_{2i}(t_k)$, $\Delta C_{1i}(t_k)$, $\Delta C_{2i}(t_k)$, $\Delta D_{1i}(t_k)$ and $\Delta D_{2i}(t_k)$ are time-varying matrices with appropriate dimensions. They are defined as follows:

$$\begin{aligned} &\begin{bmatrix} \Delta A_{1i}(t_k) \ \Delta A_{2i}(t_k) \ \Delta B_{1i}(t_k) \ \Delta B_{2i}(t_k) \\ \Delta C_{1i}(t_k) \ \Delta C_{2i}(t_k) \ \Delta D_{1i}(t_k) \ \Delta D_{2i}(t_k) \end{bmatrix} \\ &= \begin{bmatrix} G_{1i} \\ G_{2i} \end{bmatrix} F_i(t_k) \begin{bmatrix} E_{1i} \ E_{2i} \ E_{3i} \ E_{4i} \end{bmatrix} \end{aligned} \qquad (5.3)$$

where G_{1i}, G_{2i}, E_{1i}, E_{2i}, E_{3i} and E_{4i} are known constant real matrices with appropriate dimensions. $F_i(t_k)$ is unknown time-varying matrices with Lebesgue measurable elements bounded by:

$$F_i^T(t_k)F_i(t_k) \le I, \quad \forall\, t_k \quad i = 1, 2, \cdots, r. \qquad (5.4)$$

Through the use of fuzzy 'blending', the final output of the fuzzy DOS is inferred as follows:

$$\delta x(t_k) = \sum_{i=1}^{r} \mu_i(\theta(t_k))[(A_i + \Delta A_i(t_k))x(t_k) + (A_{2i} + \Delta A_{2i}(t_k))x(t_k - n\mathrm{T}) \\ +(B_{1i} + \Delta B_{1i}(t_k))\omega(t_k) + (B_{2i} + \Delta B_{2i}(t_k))u(t_k)], \tag{5.5}$$

$$z(t_k) = \sum_{i=1}^{r} \mu_i(\theta(t_k))[(C_i + \Delta C_i(t_k))x(t_k) + (C_{2i} + \Delta C_{2i}(t_k))x(t_k - n\mathrm{T}) \\ +(D_{1i} + \Delta D_{1i}(t_k))\omega(t_k) + (D_{2i} + \Delta D_{2i}(t_k))u(t_k)], \tag{5.6}$$

with

$$\mu_i(\theta(t_k)) = \frac{\omega_i(\theta(t_k))}{\sum_{i=1}^{r} \omega_i(\theta(t_k))}, \ \omega_i(\theta(t_k)) = \prod_{l=1}^{p} M_{il}(\theta_l(t_k)),$$

in which $M_{il}(\theta_l(t_k))$ is the grade of membership of $\theta_l(t_k)$ in M_{il}. In this chapter, we assume $\omega_i(\theta(t_k)) \geq 0$, $i = 1, 2, \cdots, r$, and $\sum_{i=1}^{r} \omega_i(\theta(t_k)) > 0$ for all $t_k \geq 0$. Therefore, $\mu_i(\theta(t_k)) \geq 0$, and $\sum_{i=1}^{r} \mu_i(\theta(t_k)) = 1$ for all $t_k \geq 0$.

The objective of the following definition is to give H_∞ control conditions for DOSs.

Definition 5.1. Consider the fuzzy uncertain system (5.1)-(5.2) and the state feedback $u(t_k) = K_i x(t_k)$. If the following conditions are satisfied:

(i) With $\omega(t_k) = 0$, the closed-loop system (5.1)-(5.2) and $u(t_k) = K_i x(t_k)$ is asymptotically stable,

(ii) With zero initial condition (i.e.$\phi = 0$), the following condition is satisfied:

$$J = \sum_{t=0}^{\infty} [\gamma^{-1} z^T(t_k) z(t_k) - \gamma \omega^T(t_k)\omega(t_k)] \leq 0,$$

for some $\gamma > 0$. Then the control input $u(t_k) = K_i x(t_k)$ is said to be the H_∞ control of system (5.1)-(5.2) with disturbance attenuation γ. The parameter γ is called the H_∞ norm bound of the control.

Before ending this section, we recall the following lemmas which will be used to prove our main results in this chapter.

Lemma 5.2. *[196] The property of delta operator: for any time function $x(t_k)$ and $y(t_k)$*

$$\delta(x(t_k)y(t_k)) = \delta(x(t_k))y(t_k) + x(t_k)\delta(y(t_k)) + T\delta(x(t_k))\delta(y(t_k)),$$

where T is a sampling period.

Lemma 5.3. *[124] For any constant positive semi-definite symmetric matrix W, two positive integers r and r_0 satisfying $r \geq r_0 \geq 1$, the following inequality holds*

$$\left[\sum_{i=r_0}^{r} x(i)\right]^T W \left[\sum_{i=r_0}^{r} x(i)\right] \leq (r - r_0 + 1) \sum_{i=r_0}^{r} x^T(i) W x(i).$$

Lemma 5.4. *[269] Given constant matrices G and H and a symmetric constant matrix Υ of appropriate dimensions, the following inequality holds:*

$$\Upsilon + GF(t_k)H + H^T F^T(t_k) G^T \leq 0,$$

where $F(t_k)$ satisfies $F^T(t_k)F(t_k) \leq I$, if and only if for $\varepsilon > 0$

$$\Upsilon + \varepsilon^{-1} H^T H + \varepsilon G G^T \leq 0.$$

5.3 Main Results

We consider the design problem of robust fuzzy H_∞ state feedback controller for system (5.1)-(5.2). Based on parallel distributed compensation, the following fuzzy control law is employed to deal with the problem of stabilization via state feedback control rule:

Rule i: If $\theta_1(t_k)$ is M_{i1} and $\cdots$ and $\theta_p(t_k)$ is M_{ip} then

$$u(t_k) = K_i x(t_k). \tag{5.7}$$

where K_i $(i = 1, 2, \cdots r)$ are the local control gain K_i $(i = 1, 2, \cdots r)$ such that the closed-loop system (5.1)-(5.2) is asymptotically stable for any $0 \leq n\mathrm{T} \leq h$. Hence, the overall fuzzy control law is represented by

$$u(t_k) = \sum_{i=1}^{r} \mu_i(\theta(t_k)) K_i x(t_k). \tag{5.8}$$

Consider the fuzzy DOS (5.5)-(5.6) associated with fuzzy control law (5.8), then the resulting closed-loop fuzzy system can be expressed as follows:

$$\begin{aligned}
\delta x(t_k) &= \sum_{i=1}^{r} \sum_{j=1}^{r} \mu_i(\theta(t_k)) \mu_j(\theta(t_k)) [(A_{1i} + B_{2i} K_j + \Delta A_{1i}(t_k) \\
&\quad + \Delta B_{2i}(t_k) K_j) x(t_k) + (A_{2i} + \Delta A_{2i}(t_k)) x(t_k - n\mathrm{T}) \\
&\quad + (B_{1i} + \Delta B_{1i}(t_k)) \omega(t_k)], \\
&= \sum_{i=1}^{r} \mu_i^2(\theta(t_k)) [(A_{1i} + B_{2i} K_i + \Delta A_{1i}(t_k) + \Delta B_{2i}(t_k) K_i) x(t_k)
\end{aligned}$$

$$
\begin{aligned}
&+(A_{2i}+\Delta A_{2i}(t_k))x(t_k-n\mathrm{T})+(B_{1i}+\Delta B_{1i}(t_k))\omega(t_k)] \\
&+\sum_{i<j}^{r}\mu_i(\theta(t_k))\mu_j(\theta(t_k))[(A_{1i}+B_{2i}K_j+\Delta A_{1i}(t_k)+\Delta B_{2i}(t_k)K_j \\
&+A_{1i}+B_{2j}K_i+\Delta A_{1j}(t_k)+\Delta B_{2j}(t_k)K_i)x(t_k) \\
&+(A_{2i}+\Delta A_{2i}(t_k)+A_{2j}+\Delta A_{2j}(t_k))x(t_k-n\mathrm{T}) \\
&+(B_{1i}+\Delta B_{1i}(t_k)+B_{1j}+\Delta B_{1j}(t_k))\omega(t_k)], \qquad (5.9)
\end{aligned}
$$

$$
\begin{aligned}
z(t_k) &= \sum_{i=1}^{r}\sum_{j=1}^{r}\mu_i(\theta(t_k))\mu_j(\theta(t_k))[(C_{1i}+D_{2i}K_j \\
&+\Delta C_{1i}(t_k)+\Delta D_{2i}(t_k)K_j)x(t_k)+(C_{2i}+\Delta C_{2i}(t_k))x(t_k-n\mathrm{T}) \\
&+(D_{1i}+\Delta D_{1i}(t_k))\omega(t_k)], \\
&= \sum_{i=1}^{r}\mu_i^2(\theta(t_k))[(C_{1i}+D_{2i}K_i+\Delta C_{1i}(t_k)+\Delta D_{2i}(t_k)K_i)x(t_k) \\
&+(C_{2i}+\Delta C_{2i}(t_k))x(t_k-n\mathrm{T})+(D_{1i}+\Delta D_{1i}(t_k))\omega(t_k)] \\
&+\sum_{i<j}^{r}\mu_i(\theta(t_k))\mu_j(\theta(t_k))[(C_{1i}+D_{2i}K_j+\Delta C_{1i}(t_k)+\Delta D_{2i}(t_k)K_j \\
&+C_{1i}+D_{2j}K_i+\Delta C_{1j}(t_k)+\Delta D_{2j}(t_k)K_i)x(t_k) \\
&+(C_{2i}+\Delta C_{2i}(t_k)+C_{2j}+\Delta C_{2j}(t_k))x(t_k-n\mathrm{T}) \\
&+(D_{1i}+\Delta D_{1i}(t_k)+D_{1j}+\Delta D_{1j}(t_k))\omega(t_k)]. \qquad (5.10)
\end{aligned}
$$

Let $\delta x(t_k)=X_1$ and $z(t_k)=Z_1$ for $1\le i\le r$, $\delta x(t_k)=X_2$ and $z(t_k)=Z_2$ for $1\le i<j\le r$, where

$$
\begin{aligned}
X_1 &= \sum_{i=1}^{r}\mu_i^2(\theta(t_k))[(A_{ki}+\Delta A_{ki}(t_k))x(t_k)+(B_{1i}+\Delta B_{1i}(t_k))x(t_k-n\mathrm{T}) \\
&+(B_{1i}+\Delta B_{1i}(t_k))\omega(t_k)], \qquad (5.11) \\
Z_1 &= \sum_{i=1}^{r}\mu_i^2(\theta(t_k))[(C_{ki}+\Delta C_{ki}(t_k))x(t_k)+(C_{2i}+\Delta C_{2i}(t_k))x(t_k-n\mathrm{T}) \\
&+(D_{1i}+\Delta D_{1i}(t_k))\omega(t_k)], \qquad (5.12)
\end{aligned}
$$

with

$$
\begin{aligned}
A_{ki}&=A_{1i}+B_{2i}K_i, \quad \Delta A_{ki}(t_k)=\Delta A_{1i}(t_k)+\Delta B_{2i}(t_k)K_i, \\
C_{ki}&=C_{1i}+D_{2i}K_i, \quad \Delta C_{ki}(t_k)=\Delta C_{1i}(t_k)+\Delta D_{2i}(t_k)K_i.
\end{aligned}
$$

And

$$
X_2=\sum_{i<j}^{r}\mu_i(\theta(t_k))\mu_j(\theta(t_k))[(A_{1i}+B_{2i}K_j+A_{1i}+\Delta A_{1i}(t_k)+B_{2j}K_i
$$

$$
\begin{aligned}
&+\Delta B_{2i}(t_k)K_j + \Delta A_{1j}(t_k) + \Delta B_{2j}(t_k)K_i)x(t_k) \\
&+(A_{2i} + \Delta A_{2i}(t_k) + A_{2j} + \Delta A_{2j}(t_k))x(t_k - n\mathrm{T}) \\
&+(B_{1i} + \Delta B_{1i}(t_k) + B_{1j} + \Delta B_{1j}(t_k))\omega(t_k)],
\end{aligned} \tag{5.13}
$$

$$
\begin{aligned}
Z_2 = \sum_{i<j}^{r} \mu_i(\theta(t_k))\mu_j(\theta(t_k))&[(C_{1i} + D_{2i}K_j + C_{1i} + \Delta C_{1i}(t_k) + D_{2j}K_i \\
&+\Delta D_{2i}(t_k)K_j + \Delta C_{1j}(t_k) + \Delta D_{2j}(t_k)K_i)x(t_k) \\
&+(C_{2i} + \Delta C_{2i}(t_k) + C_{2j} + \Delta C_{2j}(t_k))x(t_k - n\mathrm{T}) \\
&+(D_{1i} + \Delta D_{1i}(t_k) + D_{1j} + \Delta D_{1j}(t_k))\omega(t_k)].
\end{aligned} \tag{5.14}
$$

For the sake of convenience, we define

$$
\begin{aligned}
A_k(t_k) &= \sum_{i=1}^{r} \mu_i^2(\theta(t_k))(A_{ki} + \Delta A_{ki}(t_k)), \\
A_2(t_k) &= \sum_{i=1}^{r} \mu_i^2(\theta(t_k))(A_{2i} + \Delta A_{2i}(t_k)), \\
C_k(t_k) &= \sum_{i=1}^{r} \mu_i^2(\theta(t_k))(C_{ki} + \Delta C_{ki}(t_k)), \\
C_2(t_k) &= \sum_{i=1}^{r} \mu_i^2(\theta(t_k))(C_{2i} + \Delta C_{2i}(t_k)), \\
B_1(t_k) &= \sum_{i=1}^{r} \mu_i^2(\theta(t_k))(B_{1i} + \Delta B_{1i}(t_k)), \\
D_1(t_k) &= \sum_{i=1}^{r} \mu_i^2(\theta(t_k))(D_{1i} + \Delta D_{1i}(t_k)).
\end{aligned}
$$

In this section, a sufficient condition for the solvability of the robust H_∞ control problem is proposed and an LMI approach for designing the desired fuzzy state feedback controller is developed. Our main result in this chapter is given as follows.

Theorem 5.5. *Given a scalar $\gamma > 0$, then uncertain fuzzy DOS (5.9)-(5.10) is robustly asymptotically stable with disturbance attenuation γ, if there exist matrices $X > 0$, Y_i $(i = 1, 2, \cdots, r)$, $M > 0$ and $N > 0$, some positive scalars α_i, β_i and β_j such that the following LMIs hold:*

$$
\begin{bmatrix}
\Phi_{11} & \Phi_{12} & \Phi_{13} & B_{1i} & \Phi_{15} & \Phi_{16} \\
* & \Phi_{22} & A_{2i}X & B_{1i} & \alpha_i G_{1i} G_{2i}^T & 0 \\
* & * & \Phi_{33} & 0 & XC_{2i}^T & XE_{2i}^T \\
* & * & * & -\gamma I & D_{1i}^T & E_{3i}^T \\
* & * & * & * & \Phi_{55} & 0 \\
* & * & * & * & * & -\alpha_i I
\end{bmatrix} < 0, \quad (1 \le i \le r) \tag{5.15}
$$

and

$$\begin{bmatrix} \Psi_{11} & \Psi_{12} & \Psi_{13} & \Psi_{14} & \Psi_{15} & B_{1i} & B_{1j} & \Psi_{18} & \Psi_{19} & \Psi_{110} & \Psi_{111} \\ * & \Psi_{22} & 0 & A_{2i}X & 0 & B_{1i} & 0 & \beta_i G_{1i}G_{2i}^T & 0 & 0 & 0 \\ * & * & \Psi_{33} & 0 & A_{2j}X & 0 & B_{1j} & 0 & \beta_j G_{1j}G_{2j}^T & 0 & 0 \\ * & * & * & \Psi_{44} & 0 & 0 & 0 & XC_{2i}^T & 0 & XE_{2i}^T & 0 \\ * & * & * & * & \Psi_{55} & 0 & 0 & 0 & XC_{2j}^T & 0 & XE_{2j}^T \\ * & * & * & * & * & -\gamma I & 0 & D_{1i}^T & 0 & E_{3i}^T & 0 \\ * & * & * & * & * & * & -\gamma I & 0 & D_{1j}^T & 0 & E_{3j}^T \\ * & * & * & * & * & * & * & \Psi_{88} & 0 & 0 & 0 \\ * & * & * & * & * & * & * & * & \Psi_{99} & 0 & 0 \\ * & * & * & * & * & * & * & * & * & -\beta_i I & 0 \\ * & * & * & * & * & * & * & * & * & * & -\beta_j I \end{bmatrix} < 0,$$

$$(1 \le i < j \le r) \tag{5.16}$$

with

$$\begin{aligned}
&\Phi_{11} = XA_{1i}^T + Y_i^T B_{2i}^T + A_{1i}X + B_{2i}Y_i - \frac{1}{nT}N + M + \alpha_i G_{1i}G_{1i}^T, \\
&\Phi_{13} = A_{2i}X + \frac{1}{nT}N,\ \Phi_{12} = XA_{1i}^T + Y_i^T B_{2i}^T + \alpha_i G_{1i}G_{1i}^T, \\
&\Phi_{15} = XC_{1i}^T + Y_i^T D_{2i}^T + \alpha_i G_{1i}G_{2i}^T,\ \Phi_{16} = XE_{1i}^T + Y_i^T E_{4i}^T, \\
&\Phi_{22} = TX + nTN - 2X + \alpha_i G_{1i}G_{1i}^T,\ \Phi_{33} = -M - \frac{1}{nT}N, \\
&\Phi_{55} = -\gamma I + \alpha_i G_{2i}G_{2i}^T,\ \Psi_{11} = XA_{1i}^T + Y_j^T B_{2i}^T + A_{1i}X + B_{2i}Y_j + 2M \\
&\qquad + XA_{1j}^T + Y_i^T B_{2j}^T + A_{1j}X + B_{2j}Y_i - \frac{2}{nT}N + \beta_i G_{1i}G_{1i}^T + \beta_j G_{1j}G_{1j}^T, \\
&\Psi_{12} = XA_{1i}^T + Y_j^T B_{2i}^T + \beta_i G_{1i}G_{1i}^T,\ \Psi_{13} = XA_{1j}^T + Y_i^T B_{2j}^T + \beta_j G_{1j}G_{1j}^T, \\
&\Psi_{14} = A_{2i}X + \frac{1}{nT}N,\ \Psi_{15} = A_{2j}X + \frac{1}{nT}N, \\
&\Psi_{18} = XC_{1i}^T + Y_j^T D_{2i}^T + \beta_i G_{1i}G_{2i}^T,\ \Psi_{19} = XC_{1j}^T + Y_i^T D_{2j}^T + \beta_j G_{1j}G_{2j}^T, \\
&\Psi_{110} = XE_{1i}^T + Y_j^T E_{4i}^T,\quad \Psi_{111} = XE_{1j}^T + Y_i^T E_{4j}^T, \\
&\Psi_{22} = TX + nTN - 2X + \beta_i G_{1i}G_{1i}^T, \Psi_{33} = TX + nTN - 2X + \beta_j G_{1j}G_{1j}^T, \\
&\Psi_{44} = -M - \frac{1}{nT}N,\ \Psi_{55} = -M - \frac{1}{nT}N, \\
&\Psi_{88} = -\gamma I + \beta_i G_{2i}G_{2i}^T,\quad \Psi_{99} = -\gamma I + \beta_j G_{2j}G_{2j}^T.
\end{aligned}$$

In this case, a suitable stabilizing fuzzy robust H_∞ state feedback controller is given as follows

$$u(t_k) = Y_i X^{-1} x(t_k).$$

for $i = 1, 2, \cdots, r$.

Proof. Firstly, considering DOS (5.9) with $\omega(t_k)=0$, LMIs (5.15) and (5.16) are changed to

$$\begin{bmatrix} \Phi_{11} & \Phi_{12} & \Phi_{13} & XE_{1i}^T+Y_i^TE_{4i}^T \\ * & \Phi_{22} & A_{2i}X & 0 \\ * & * & \Phi_{33} & XE_{2i}^T \\ * & * & * & -\alpha_i I \end{bmatrix} < 0, \ (1\le i\le r) \tag{5.17}$$

and

$$\begin{bmatrix} \Psi_{11} & \Psi_{12} & \Psi_{13} & \Psi_{14} & \Psi_{15} & \Psi_{16} & \Psi_{17} \\ * & \Psi_{22} & 0 & A_{2i}X & 0 & 0 & 0 \\ * & * & \Psi_{33} & 0 & A_{2j}X & 0 & 0 \\ * & * & * & \Psi_{44} & 0 & XE_{2i}^T & 0 \\ * & * & * & * & \Psi_{55} & 0 & XE_{2j}^T \\ * & * & * & * & * & -\beta_i I & 0 \\ * & * & * & * & * & * & -\beta_j I \end{bmatrix} < 0, \ (1\le i<j\le r) \tag{5.18}$$

with

$$\begin{aligned}
\Phi_{11} &= XA_{1i}^T+Y_i^TB_{2i}^T+A_{1i}X+B_{2i}Y_i-\frac{1}{n\mathrm{T}}N+M+\alpha_iG_{1i}G_{1i}^T,\\
\Phi_{13} &= A_{2i}X+\frac{1}{n\mathrm{T}}N,\ \Phi_{12}=XA_{1i}^T+Y_i^TB_{2i}^T+\alpha_iG_{1i}G_{1i}^T,\\
\Phi_{33} &= -M-\frac{1}{n\mathrm{T}}N,\ \Phi_{22}=\mathrm{T}X+n\mathrm{T}N-2X+\alpha_iG_{1i}G_{1i}^T,\\
\Psi_{11} &= XA_{1i}^T+Y_j^TB_{2i}^T+A_{1i}X+B_{2i}Y_j+2M+XA_{1j}^T+Y_iB_{2j}^T\\
&\quad +A_{1j}X+B_{2j}Y_i-\frac{2}{n\mathrm{T}}N+\beta_iG_{1i}G_{1i}^T+\beta_jG_{1j}G_{1j}^T,\\
\Psi_{12} &= XA_{1i}^T+Y_j^TB_{2i}^T+\beta_iG_{1i}G_{1i}^T+\beta_jG_{1j}G_{1j}^T,\\
\Psi_{13} &= XA_{1j}^T+Y_i^TB_{2j}^T,\ \Psi_{14}=A_{2i}X+\frac{1}{n\mathrm{T}}N,\\
\Psi_{15} &= A_{2j}X+\frac{1}{n\mathrm{T}}N,\ \Psi_{16}=XE_{1i}^T+Y_j^TE_{4i}^T,\\
\Psi_{17} &= XE_{1j}^T+Y_i^TE_{4j}^T,\ \Psi_{22}=\mathrm{T}X+n\mathrm{T}N-2X+\beta_iG_{1i}G_{1i}^T,\\
\Psi_{33} &= \mathrm{T}X+n\mathrm{T}N-2X+\beta_jG_{1j}G_{1j}^T,\quad \Psi_{44}=-M-\frac{1}{n\mathrm{T}}N,\\
\Psi_{55} &= -M-\frac{1}{n\mathrm{T}}N.
\end{aligned}$$

In the following, this theorem will be proved in two steps. In the first step, fuzzy system (3.5)-(3.6) is considered as $1\le i\le r$. Construct Lyapunov-Krasovskii functional in delta domain as follows:

$$V(t_k)=V_1(t_k)+V_2(t_k)+V_3(t_k)$$

with

$$
\begin{aligned}
V_1(t_k) &= x^T(t_k)Px(t_k),\\
V_2(t_k) &= \mathrm{T}\sum_{i=1}^{n} x^T(t_k - i\mathrm{T})Qx(t_k - i\mathrm{T}),\\
V_3(t_k) &= \sum_{i=1}^{n}\sum_{j=1}^{i} e^T(t_k - j\mathrm{T})Re(t_k - j\mathrm{T}),
\end{aligned}
$$

where $e(j) = x(j) - x(j+\mathrm{T})$, so there exist $\delta x(j) = -e(j)/\mathrm{T}$ and $e(t_k - i\mathrm{T}) = x(t_k - i\mathrm{T}) - x(t_k - (i-1)\mathrm{T})$.

By using Lemma 5.2 and taking the delta operator manipulations of $V_1(t_k)$ along the trajectory of system (5.11) and (5.12), we can obtain:

$$
\begin{aligned}
\delta V_1(t_k) &= \delta^T(x(t_k))Px(t_k) + x^T(t_k)P\delta(x(t_k)) + \mathrm{T}\delta^T(x(t_k))P\delta(x(t_k))\\
&= x^T(t_k)A_k^T(t_k)Px(t_k) + x^T(t_k - n\mathrm{T})A_2^T(t_k)Px(t_k)\\
&\quad +\omega^T(t_k)B_1^T(t_k)Px(t_k) + x^T(t_k)PA_k(t_k)x(t_k)\\
&\quad +x^T(t_k)PA_2(t_k)x(t_k - n\mathrm{T}) + x^T(t_k)PB_1(t_k)\omega(t_k)\\
&\quad +\mathrm{T}\delta^T(x(t_k))P\delta(x(t_k)).
\end{aligned}
$$

Taking the delta operator manipulations of $V_2(t_k)$, it is given that

$$
\begin{aligned}
\delta V_2(t_k) &= \frac{1}{\mathrm{T}}\Bigg[\mathrm{T}\sum_{i=1}^{n} x^T(t_k - (i-1)\mathrm{T})Qx(t_k - (i-1)\mathrm{T})\\
&\qquad -\mathrm{T}\sum_{i=1}^{n} x^T(t_k - i\mathrm{T})Qx(t_k - i\mathrm{T})\Bigg]\\
&= x^T(t_k)Qx(t_k) - x^T(t_k - n\mathrm{T})Qx(t_k - n\mathrm{T}).
\end{aligned}
$$

Taking the delta operator manipulations of $V_3(t_k)$ and using Lemma 5.3, there exists

$$
\begin{aligned}
\delta V_3(t_k) &= \frac{1}{\mathrm{T}}\Bigg[\sum_{i=1}^{n}\sum_{j=1}^{i} e^T(t_k - (i-1)\mathrm{T})Re(t_k - (i-1)\mathrm{T})\\
&\qquad -\sum_{i=1}^{n}\sum_{j=1}^{i} e^T(t_k - i\mathrm{T})Re(t_k - i\mathrm{T})\Bigg]\\
&= \frac{1}{\mathrm{T}}\left[\sum_{i=1}^{n} e^T(t_k)Re(t_k) - \sum_{i=1}^{n} e^T(t_k - i\mathrm{T})Re(t_k - i\mathrm{T})\right]\\
&\le \frac{n}{\mathrm{T}}e^T(t_k)Re(t_k) - \frac{1}{n\mathrm{T}}\left[\sum_{i=1}^{n} e(t_k - i\mathrm{T})\right]^T R\left[\sum_{i=1}^{n} e(t_k - i\mathrm{T})\right]
\end{aligned}
$$

$$
\begin{aligned}
&= -\frac{1}{n\mathrm{T}}\left[x(t_k - n\mathrm{T}) - x(t_k)\right]^T R\left[x(t_k - n\mathrm{T}) - x(t_k)\right] \\
&\quad + n\mathrm{T}\delta^T(x(t_k))R\delta(x(t_k))
\end{aligned}
$$

For the positive definite real matrix P, one has that

$$
\begin{aligned}
0 &= -2\delta^T(x(t_k))P[\delta(x(t_k)) - A_k(t_k)x(t_k) - A_2(t_k)x(t_k - n\mathrm{T}) \\
&\quad - B_1(t_k)\omega(t_k)] \\
&= -2\delta^T(x(t_k))P\delta(x(t_k)) + 2\delta^T(x(t_k))PA_k(t_k)x(t_k) \\
&\quad + 2\delta^T(x(t_k))PA_2(t_k)x(t_k - n\mathrm{T}) + 2\delta^T(x(t_k))PB_1(t_k)\omega(t_k).
\end{aligned}
$$

And there exists

$$
\begin{aligned}
&\gamma^{-1}z^T(t_k)z(t_k) - \gamma\omega^T(t_k)\omega(t_k) \\
&= \gamma^{-1}[x^T(t_k)C_k^T(t_k)C_k(t_k)x(t_k) + x^T(t_k)C_k^T(t_k)C_2(t_k)x(t_k - n\mathrm{T}) \\
&\quad + x^T(t_k)C_k^T(t_k)D_1(t_k)\omega(t_k) + x^T(x - n\mathrm{T})C_2^T(t_k)C_k(t_k)x(t_k) \\
&\quad + x^T(x - n\mathrm{T})C_2^T(t_k)C_2(t_k)x(t_k - n\mathrm{T}) + x^T(x - n\mathrm{T})C_2^T(t_k)D_1(t_k)\omega(t_k) \\
&\quad + \omega^T(t_k)D_1^T(t_k)C_k(t_k)x(t_k) + \omega^T(t_k)D_1^T(t_k)C_2(t_k)x(t_k - n\mathrm{T}) \\
&\quad + \omega^T(t_k)D_1^T(t_k)D_1(t_k)\omega(t_k)] - \gamma\omega^T(t_k)\omega(t_k).
\end{aligned}
$$

Lastly, it can be easily shown that

$$
\gamma^{-1}z^T(t_k)z(t_k) - \gamma\omega^T(t_k)\omega(t_k) + \delta V(t_k) \leq \xi^T(t_k)\Sigma_1\xi(t_k) < 0
$$

where

$$
\xi^T(t_k) = \left[x^T(t_k)\ \delta^T(x(t_k))\ x^T(t_k - n\mathrm{T})\ \omega^T(t_k)\right]
$$

and

$$
\Sigma_1 = \begin{bmatrix}
\Sigma_1(1,1) & A_k^T(t_k)P & \Sigma_1(1,3) & PB_1(t_k) + \gamma^{-1}C_k^T(t_k)D_1(t_k) \\
* & \Sigma_1(2,2) & PA_2(t_k) & PB_1(t_k) \\
* & * & \Sigma_1(3,3) & \gamma^{-1}C_2^T(t_k)D_1(t_k) \\
* & * & * & -\gamma I + \gamma^{-1}D_1^T(t_k)D_1(t_k)
\end{bmatrix} < 0
$$

with

$$
\begin{aligned}
&\Sigma_1(1,1) = A_k^T(t_k)P + PA_k(t_k) + Q - \frac{1}{n\mathrm{T}}R + \gamma^{-1}C_k^T(t_k)C_k(t_k), \\
&\Sigma_1(1,3) = PA_2(t_k) + \frac{1}{n\mathrm{T}}R + \gamma^{-1}C_k^T(t_k)C_2(t_k), \\
&\Sigma_1(2,2) = \mathrm{T}P + n\mathrm{T}R - 2P,\ \Sigma_1(3,3) = -Q - \frac{1}{n\mathrm{T}}R + \gamma^{-1}C_2^T(t_k)C_2(t_k).
\end{aligned}
$$

Using Schur's complement, $\Sigma_1 < 0$ can be changed to $\Sigma_2 < 0$ where

$$\Sigma_2 = \begin{bmatrix} \Sigma_2(1,1) & A_k^T(t_k)P & PA_2(t_k) + \frac{1}{n\mathrm{T}}R & PB_1(t_k) & C_k^T(t_k) \\ * & \Sigma_2(2,2) & PA_2(t_k) & PB_1(t_k) & 0 \\ * & * & -Q - \frac{1}{n\mathrm{T}}R & 0 & C_2^T(t_k) \\ * & * & * & -\gamma I & D_1^T(t_k) \\ * & * & * & * & -\gamma I \end{bmatrix},$$

with

$$\Sigma_2(1,1) = A_k^T(t_k)P + PA_k(t_k) + Q - \frac{1}{n\mathrm{T}}R, \ \Sigma_2(2,2) = \mathrm{T}P + n\mathrm{T}R - 2P.$$

It is easy to see that $\Sigma_2 = \sum_{i=1}^{r} \mu_i^2(\theta(t_k))\Sigma_3$, where

$$\Sigma_3 = \Upsilon + \begin{bmatrix} PG_{1i} \\ PG_{1i} \\ 0 \\ 0 \\ G_{2i} \end{bmatrix} F_i(t_k) \begin{bmatrix} \Sigma_3(1) \\ 0 \\ E_{2i}^T \\ E_{3i}^T \\ 0 \end{bmatrix}^T + \begin{bmatrix} \Sigma_3(1) \\ 0 \\ E_{2i}^T \\ E_{3i}^T \\ 0 \end{bmatrix} F_i^T(t_k) \begin{bmatrix} PG_{1i} \\ PG_{1i} \\ 0 \\ 0 \\ G_{2i} \end{bmatrix}^T,$$

where $\Sigma_3(1) = (E_{1i} + E_{4i}K_i)^T$,

$$\Upsilon = \begin{bmatrix} \Upsilon(1,1) & (A_{1i} + B_{2i}K_i)^T P & PA_{2i} + \frac{1}{n\mathrm{T}}R & PB_{1i} & C_{1i}^T + K_i^T D_{2i}^T \\ * & \mathrm{T}P + n\mathrm{T}R - 2P & PA_{2i} & PB_{1i} & 0 \\ * & * & -Q - \frac{1}{n\mathrm{T}}R & 0 & C_{2i}^T \\ * & * & * & -\gamma I & D_{1i}^T \\ * & * & * & * & -\gamma I \end{bmatrix},$$

with

$$\Upsilon(1,1) = (A_{1i} + B_{2i}K_i)^T P + P(A_{1i} + B_{2i}K_i) + Q - \frac{1}{n\mathrm{T}}R.$$

Since $\mu_i(\theta(t_k)) \geq 0$ and $\Sigma_3 < 0$ holds for any $1 \leq r$, there exists $\Sigma_2 < 0$. By using Lemma 5.4, $\Sigma_3 < 0$ can be changed to

$$\Sigma_3 \leq \Upsilon + \alpha_i \begin{bmatrix} PG_{1i} \\ PG_{1i} \\ 0 \\ 0 \\ G_{2i} \end{bmatrix} \begin{bmatrix} PG_{1i} \\ PG_{1i} \\ 0 \\ 0 \\ G_{2i} \end{bmatrix}^T + \alpha_i^{-1} \begin{bmatrix} \Sigma_3(1) \\ 0 \\ E_{2i}^T \\ E_{3i}^T \\ 0 \end{bmatrix} \begin{bmatrix} \Sigma_3(1) \\ 0 \\ E_{2i}^T \\ E_{3i}^T \\ 0 \end{bmatrix}^T < 0. \quad (5.19)$$

From Schur's complement, it can be seen that (5.19) is equivalent to $\Sigma_4 < 0$, where

$$\Sigma_4 = \begin{bmatrix} \Sigma_4(1,1) & \Sigma_4(1,2) & PA_{2i} + \frac{1}{n\mathrm{T}}R & PB_{1i} & \Sigma_4(1,5) & \Sigma_3(1) \\ * & \Sigma_4(2,2) & PA_{2i} & PB_{1i} & \alpha_i PG_{1i}G_{2i}^T & 0 \\ * & * & -Q - \frac{1}{n\mathrm{T}}R & 0 & C_{2i}^T & E_{2i}^T \\ * & * & * & -\gamma I & D_{1i}^T & E_{3i}^T \\ * & * & * & * & -\gamma I + \alpha_i G_{2i}G_{2i}^T & 0 \\ * & * & * & * & * & -\alpha_i I \end{bmatrix},$$

with

$$\Sigma_4(1,1) = (A_{1i} + B_{2i}K_i)^T P + P(A_{1i} + B_{2i}K_i) - \frac{1}{n\mathrm{T}}R + Q + \alpha_i PG_{1i}G_{1i}^T P,$$
$$\Sigma_4(1,2) = (A_{1i} + B_{2i}K_i)^T P + \alpha_i PG_{1i}G_{1i}^T P,$$
$$\Sigma_4(1,5) = (C_{1i} + D_{2i}K_i)^T + \alpha_i PG_{1i}G_{2i}^T,$$
$$\Sigma_4(2,2) = \mathrm{T}P + n\mathrm{T}R - 2P + \alpha_i PG_{1i}G_{1i}^T P.$$

Pre-multiplying and post-multiplying $\Sigma_4 < 0$ by diagonal matrix

$$\mathrm{diag}\{P^{-1}, P^{-1}, P^{-1}, I, I\}$$

and letting $X = P^{-1}$, $Y_i = K_i X = K_i P^{-1}$, $M = P^{-1}QP^{-1}$ and $N = P^{-1}RP^{-1}$, the inequality $\Sigma_4 < 0$ is equivalent to $\Sigma_5 < 0$, where Σ_5 can be written as

$$\Sigma_5 = \begin{bmatrix} \Sigma_5(1,1) & \Sigma_5(1,2) & A_{2i}X + \frac{1}{n\mathrm{T}}N & B_{1i} & \Sigma_5(1,5) & \Sigma_5(1,6) \\ * & \Sigma_5(2,2) & A_{2i}X & B_{1i} & \alpha_i G_{1i}G_{2i}^T & 0 \\ * & * & -M - \frac{1}{n\mathrm{T}}N & 0 & XC_{2i}^T & XE_{2i}^T \\ * & * & * & -\gamma I & D_{1i}^T & E_{3i}^T \\ * & * & * & * & -\gamma I + \alpha_i G_{2i}G_{2i}^T & 0 \\ * & * & * & * & * & -\alpha_i I \end{bmatrix},$$

with

$$\Sigma_5(1,1) = (A_{1i}X + B_{2i}Y_i)^T + A_{1i}X + B_{2i}Y_i - \frac{1}{n\mathrm{T}}N + M + \alpha_i G_{1i}G_{1i}^T,$$
$$\Sigma_5(1,2) = (A_{1i}X + B_{2i}Y_i)^T + \alpha_i G_{1i}G_{1i}^T,$$
$$\Sigma_5(1,5) = (C_{1i}X + D_{2i}Y_i)^T + \alpha_i G_{1i}G_{2i}^T, \ \Sigma_5(1,6) = (E_{1i}X + E_{4i}Y_i)^T$$
$$\Sigma_5(2,2) = \mathrm{T}X + n\mathrm{T}N - 2X + \alpha_i G_{1i}G_{1i}^T.$$

Based on Definition 1.2, the fuzzy closed-loop state feedback system (5.9) with $\omega(t_k) = 0$ is asymptotically stable for inequalities (5.17) and (5.18). The solution to the robust H_∞ state feedback controller design problem is presented. Considering the H_∞ performance index in Definition 5.1,

$$J = \sum_{t=0}^{\infty}[\gamma^{-1}z^T(t_k)z(t_k) - \gamma\omega^T(t_k)\omega(t_k)],$$

and assuming, without loss of generality, system (5.9) is stable with zero initial condition, it implies $V(\cdot)|_{t=0} = 0$ and $V(\cdot)|_{t\to\infty} \to \epsilon$ with $\epsilon \to 0$, if $\omega(t_k) = 0$ or $\epsilon < \infty$ if $\omega(t_k) \neq 0$. In this way, the above index can be rewritten as

$$J = \sum_{t=0}^{\infty}[\gamma^{-1}z^T(t_k)z(t_k) - \gamma\omega^T(t_k)\omega(t_k) + \delta V(t_k)].$$

And it can be obtained that

$$J = \sum_{t=0}^{\infty}[\gamma^{-1}z^T(t_k)z(t_k) - \gamma\omega^T(t_k)\omega(t_k)] \leq \sum_{t=0}^{\infty}[\zeta^T(t_k)\Sigma_5\zeta(t_k)] \leq 0, \quad (5.20)$$

with

$$\zeta^T(t_k) = \begin{bmatrix} x^T(t_k) \ \delta^T(x(t_k)) \ x^T(t_k - n\mathrm{T}) \ \omega^T(t_k) \ y_1^T(t_k) \ y_2^T(t_k) \end{bmatrix}.$$

where $y_1(t_k)$ and $y_2(t_k)$ are arbitrary vectors with appropriate dimensions.

In the second step, consider the fuzzy DOS (5.13)-(5.14) when $1 \leq i < j \leq r$, following the similar line of above process a matrix Σ_6, which is similar to Σ_5, can be given as follows:

$$\Sigma_6 = \begin{bmatrix} \Omega_{11} & \Omega_{12} \\ \Omega_{12}^T & \Omega_{22} \end{bmatrix} < 0, \quad (5.21)$$

with

$$\Omega_{11} = \begin{bmatrix} \Sigma_6(1,1) & \Sigma_6(1,2) & \Sigma_6(1,3) & \Sigma_6(1,4) & A_{2j}X + \frac{1}{n\mathrm{T}}N & B_{1i} & B_{1j} \\ \Sigma_6(1,2)^T & \Sigma_6(2,2) & 0 & A_{2i}X & 0 & B_{1i} & 0 \\ \Sigma_6(1,3)^T & 0 & \Sigma_6(3,3) & 0 & A_{2j}X & 0 & B_{1j} \\ \Sigma_6(1,4)^T & XA_{2i}^T & 0 & -M - \frac{1}{n\mathrm{T}}N & 0 & 0 & 0 \\ \Sigma_6(1,5)^T & 0 & XA_{2j}^T & 0 & -M - \frac{1}{n\mathrm{T}}N & 0 & 0 \\ B_{2i}^T & B_{2i}^T & 0 & 0 & 0 & -\gamma I & 0 \\ B_{2j}^T & 0 & B_{2j}^T & 0 & 0 & 0 & -\gamma I \end{bmatrix},$$

$$\Omega_{12} = \begin{bmatrix} \Sigma_6(1,8) & \Sigma_6(1,9) & (E_{1i}X + E_{4i}Y_j)^T & (E_{1j}X + E_{4j}Y_i)^T \\ 0 & 0 & 0 & 0 \\ 0 & 0 & 0 & 0 \\ (C_{2i}X)^T & 0 & (E_{2i}X)^T & 0 \\ 0 & (C_{2j}X)^T & 0 & (E_{2j}X)^T \\ D_{1i}^T & 0 & E_{3i}^T & 0 \\ 0 & D_{1j}^T & 0 & E_{3j}^T \end{bmatrix},$$

$$\Omega_{22} = diag\{-\gamma I + \beta_i G_{2i}G_{2i}^T \quad -\gamma I + \beta_j G_{2j}G_{2j}^T \quad -\beta_i I \quad -\beta_j I\},$$

and

$$
\begin{aligned}
\Sigma_6(1,1) &= (A_{1i}X + B_{2i}Y_j)^T + (A_{1i}X + B_{2i}Y_j) + (A_{1j}X + B_{2j}Y_i)^T \\
&\quad + (A_{1j}X + B_{2j}Y_i) + 2M - \frac{2}{n\mathrm{T}}N + \beta_i G_{1i}G_{1i}^T + \beta_j G_{1j}G_{1j}^T, \\
\Sigma_6(1,4) &= A_{2i}X + \frac{1}{n\mathrm{T}}N, \ \Sigma_6(1,2) = (A_{1i}X + B_{2i}Y_j)^T + \beta_i G_{1i}G_{1i}^T, \\
\Sigma_6(1,3) &= (A_{1j}X + B_{2j}Y_i)^T + \beta_j G_{1j}G_{1j}^T, \\
\Sigma_6(1,8) &= (C_{1i}X + D_{2i}Y_j)^T + \beta_i G_{1i}G_{1i}^T, \\
\Sigma_6(1,9) &= (C_{1j}X + D_{2j}Y_i)^T + \beta_j G_{1j}G_{2j}^T \\
\Sigma_6(2,2) &= \mathrm{T}X + n\mathrm{T}N - 2X + \beta_i G_{1i}G_{1i}^T, \\
\Sigma_6(3,3) &= \mathrm{T}X + n\mathrm{T}N - 2X + \beta_j G_{1j}G_{1j}^T.
\end{aligned}
$$

From Definition 5.1, the fuzzy DOS (5.1)-(5.2) with $u(t_k) = K_i x(t_k)$ is asymptotically stable and has an H_∞ norm bound γ. Furthermore, the state feedback controller is given by $K_i = Y_i X^{-1}$.

From the proof of Theorem 5.5, it is easy to see that if there are no uncertainties in the fuzzy DOS (5.1)-(5.2), i.e. $F_i(t_k) \equiv 0 \quad (i = 1, 2, \cdots, r)$, we will have the following corollary.

Corollary 5.6. *Given a scalar $\gamma > 0$, then the uncertain fuzzy DOS in (5.9)-(5.10) with $F_i(t_k) \equiv 0 \quad (i = 1, 2, \cdots, r)$ is stabilization with disturbance attenuation γ if there exist matrices $X > 0$, Y_i $(i = 1, 2, \cdots, r)$, $M > 0$ and N such that the following LMIs hold:*

$$
\begin{bmatrix}
\Phi_{11} & \Phi_{12} & \Phi_{13} & B_{1i} & \Phi_{15} \\
* & \Phi_{22} & A_{2i}X & B_{1i} & 0 \\
* & * & \Phi_{33} & 0 & XC_{2i}^T \\
* & * & * & -\gamma I & D_{1i}^T \\
* & * & * & * & -\gamma I
\end{bmatrix} < 0, \ (1 \le i \le r) \tag{5.22}
$$

and

$$
\begin{bmatrix}
\Psi_{11} & \Psi_{12} & \Psi_{13} & \Psi_{14} & \Psi_{15} & B_{1i} & B_{1j} & \Psi_{18} & \Psi_{19} \\
* & \Psi_{22} & 0 & A_{2i}X & 0 & B_{1i} & 0 & 0 & 0 \\
* & * & \Psi_{33} & 0 & A_{2j}X & 0 & B_{1j} & 0 & 0 \\
* & * & * & \Psi_{44} & 0 & 0 & 0 & XC_{2i}^T & 0 \\
* & * & * & * & \Psi_{55} & 0 & 0 & 0 & XC_{2j}^T \\
* & * & * & * & * & -\gamma I & 0 & D_{1i}^T & 0 \\
* & * & * & * & * & * & -\gamma I & 0 & D_{1j}^T \\
* & * & * & * & * & * & * & -\gamma I & 0 \\
* & * & * & * & * & * & * & * & -\gamma I
\end{bmatrix} < 0,
$$

$$(1 \le i < j \le r) \tag{5.23}$$

with

$$\begin{aligned}
\Phi_{11} &= XA_{1i}^T + Y_i^T B_{2i}^T + A_{1i}X + B_{2i}Y_i - \frac{1}{nT}N + M, \\
\Phi_{12} &= XA_{1i}^T + Y_i^T B_{2i}^T,\ \Phi_{13} = A_{2i}X + \frac{1}{nT}N,\ \Phi_{15} = XC_{1i}^T + Y_i^T D_{2i}^T, \\
\Phi_{22} &= TX + nTN - 2X,\ \Phi_{33} = -M - \frac{1}{nT}N, \\
\Psi_{11} &= XA_{1i}^T + Y_j^T B_{2i}^T + A_{1i}X + B_{2i}Y_j + 2M + XA_{1j}^T + Y_i^T B_{2j}^T + A_{1j}X \\
&\quad + B_{2j}Y_i - \frac{2}{nT}N,\ \Psi_{12} = XA_{1i}^T + Y_j^T B_{2i}^T, \\
\Psi_{13} &= XA_{1j}^T + Y_i^T B_{2j}^T,\ \Psi_{14} = A_{2i}X + \frac{1}{nT}N,\ \Psi_{15} = A_{2j}X + \frac{1}{nT}N, \\
\Psi_{18} &= XC_{1i}^T + Y_j^T D_{2i}^T,\ \Psi_{19} = XC_{1j}^T + Y_i^T D_{2j}^T,\ \Psi_{22} = TX + nTN - 2X, \\
\Psi_{33} &= TX + nTN - 2X,\ \Psi_{44} = -M - \frac{1}{nT}N,\ \Psi_{55} = -M - \frac{1}{nT}N.
\end{aligned}$$

In this case, a suitable stabilizing fuzzy state feedback controller can be chosen by

$$u(t_k) = Y_i X^{-1} x(t_k).$$

for $i = 1, 2, \cdots, r$.

Remark 5.7. For every fuzzy rule, if the fuzzy DOS (5.1)-(5.2) have the same input matrices, which means that $B_{2i} = B_2$, $\Delta B_{2i}(t_k) = \Delta B_2(t_k)$, $D_{2i} = D_2$ and $\Delta D_{2i}(t_k) = \Delta D_2(t_k)$. The closed-loop fuzzy DOS (5.9)-(5.10) can be rewritten as:

$$\begin{aligned}
\delta x(t_k) &= \sum_{i=1}^{r} \mu_i(\theta(t_k))[(A_{1i} + B_2 K_j + \Delta A_{1i}(t_k) + \Delta B_2(t_k)K_j)x(t_k) \\
&\quad + (B_{1i} + \Delta B_{1i}(t_k))x(t_k - n\mathrm{T}) + (B_{1i} + \Delta B_{1i}(t_k))\omega(t_k)], \quad (5.24) \\
z(t_k) &= \sum_{i=1}^{r} \mu_i(\theta(t_k))[(C_{1i} + D_2 K_j + \Delta C_{1i}(t_k) + \Delta D_2(t_k)K_j)x(t_k) \\
&\quad + (C_{2i} + \Delta C_{2i}(t_k))x(t_k - n\mathrm{T}) + (D_{1i} + \Delta D_{1i}(t_k))\omega(t_k)]. \quad (5.25)
\end{aligned}$$

In this case, investigating the fuzzy DOS (5.24)-(5.25), we get the same state feedback controller K_j. That is a suitable stabilizing fuzzy state feedback controller can be chosen by $u(t_k) = K_j x(t_k)$.

5.4 Numerical Examples

In the following, the proposed method will be applied to design a fuzzy robust H_∞ state feedback controller for the class of fuzzy DOSs.

Example 5.8. A fuzzy DOS is described as follows:
Rule 1: If $x_1(t_k)$ is M_1, then

$$\begin{aligned}\delta x(t_k) &= (A_{11}+\Delta A_{11}(t_k))x(t_k)+(A_{21}+\Delta A_{21}(t_k))x(t_k-n\mathrm{T})\\&\quad+(B_{11}+\Delta B_{11}(t_k))\omega(t_k)+(B_{21}+\Delta B_{21}(t_k))u(t_k),\\ z(t_k) &= (C_{11}+\Delta C_{11}(t_k))x(t_k)+(C_{21}+\Delta C_{21}(t_k))x(t_k-n\mathrm{T})\\&\quad+(D_{11}+\Delta D_{11}(t_k))\omega(t_k)+(D_{21}+\Delta D_{21}(t_k))u(t_k).\end{aligned}$$

Rule 2: If $x_1(t_k)$ is M_2, then

$$\begin{aligned}\delta x(t_k) &= (A_{21}+\Delta A_{21}(t_k))x(t_k)+(A_{22}+\Delta A_{22}(t_k))x(t_k-n\mathrm{T})\\&\quad+(B_{12}+\Delta B_{12}(t_k))\omega(t_k)+(B_{22}+\Delta B_{22}(t_k))u(t_k),\\ z(t_k) &= (C_{21}+\Delta C_{21}(t_k))x(t_k)+(C_{22}+\Delta C_{22}(t_k))x(t_k-n\mathrm{T})\\&\quad+(D_{12}+\Delta D_{12}(t_k))\omega(t_k)+(D_{22}+\Delta D_{22}(t_k))u(t_k).\end{aligned}$$

and the membership function for rule 1 and rule 2 are

$$M_1(x_1(t_k)) = \frac{1}{1+exp(-2x_1(t_k))},\quad M_2(x_1(t_k)) = 1-M_1(x_1(t_k)),$$

with

$$\begin{aligned}
&A_{11}=\begin{bmatrix}0.6 & -0.4\\ 0.2 & 2\end{bmatrix},\ A_{21}=\begin{bmatrix}-0.04 & 0.2\\ 0 & 0\end{bmatrix},\ A_{12}=\begin{bmatrix}0.8 & -0.5\\ 0.5 & 1.8\end{bmatrix},\\
&A_{22}=\begin{bmatrix}-0.4 & 0.2\\ 0 & -0.1\end{bmatrix},\ B_{11}=\begin{bmatrix}-0.5 & 0\\ 0 & 0\end{bmatrix},\ B_{21}=\begin{bmatrix}0.8 & 0\\ 0 & 0.6\end{bmatrix},\\
&B_{12}=\begin{bmatrix}-0.4 & 0\\ 0.2 & 0.1\end{bmatrix},\ B_{22}=\begin{bmatrix}0.7 & 0\\ 0 & 0.5\end{bmatrix},\\
&C_{11}=\begin{bmatrix}1 & 0\end{bmatrix},\ C_{21}=\begin{bmatrix}0.1 & 0\end{bmatrix},\ C_{12}=\begin{bmatrix}0.8 & 0\end{bmatrix},\ C_{22}=\begin{bmatrix}0.08 & 0\end{bmatrix},\\
&D_{11}=\begin{bmatrix}0 & 0.01\end{bmatrix},\mathrm{d}_{21}=\begin{bmatrix}0 & 0.1\end{bmatrix},\ D_{12}=\begin{bmatrix}0 & 0.02\end{bmatrix},\ D_{22}=\begin{bmatrix}0 & -0.1\end{bmatrix},\\
&G_{21}=\begin{bmatrix}1 & 0\end{bmatrix},\ G_{22}=\begin{bmatrix}2 & 0\end{bmatrix},\ G_{11}=1.0\cdot I,\ G_{12}=1.1\cdot I,\\
&E_{11}=E_{21}=E_{31}=E_{41}=0.05\cdot I,\ F_1(t_k)=\mathrm{diag}\{\cos(t_k),\sin(t_k)\},\\
&E_{12}=E_{22}=E_{32}=E_{42}=0.08\cdot I,\ F_2(t_k)=\mathrm{diag}\{\sin(t_k),\cos(t_k)\}.
\end{aligned}$$

In this example, we choose the performance level $\gamma = 0.75$. In order to design a fuzzy state feedback controller for the T-S model, we first choose $\mathrm{T} = 0.001$s. Then, using the Matlab LMI Control Toolbox to solve the LMIs in (5.15) and (5.16), we obtain the solution as follows:

$$\begin{aligned}
&X=\begin{bmatrix}0.5552 & -0.1067\\ -0.1067 & 0.3919\end{bmatrix},\ Y_1=\begin{bmatrix}-1.8679 & 0.6903\\ 0.5010 & -2.1422\end{bmatrix},\\
&Y_2=\begin{bmatrix}-2.8377 & 1.0125\\ 0.6796 & -2.2186\end{bmatrix},\ M=\begin{bmatrix}0.2244 & -0.1615\\ -0.1615 & 0.1254\end{bmatrix},\\
&N=\begin{bmatrix}0.0263 & 0.0580\\ 0.0580 & 0.1278\end{bmatrix},\ \begin{matrix}\alpha_1=0.1047,\ \alpha_2=0.0925,\\ \beta_1=0.1001,\ \beta_2=0.0944.\end{matrix}
\end{aligned}$$

And the largest n can be get to 1691. Therefore, the largest time delay in this example is 1.691. By Theorem 5.5, a desired fuzzy H_∞ state feedback controller can be given by $K_i = Y_i X^{-1}$ $(i = 1, 2)$ as follows:

$$K_1 = \begin{bmatrix} -3.1929 & 0.8918 \\ -0.1567 & -5.5089 \end{bmatrix}, \quad K_2 = \begin{bmatrix} -4.8692 & 1.2574 \\ 0.1431 & -5.6222 \end{bmatrix}.$$

The simulation results of the state response of the class of fuzzy DOSs in this example is given in Fig. 5.1, where $x(t_k)$ is denoted by $x_1(t_k)$, $x_2(t_k)$. Fig. 5.2 shows the simulation results of the input $u(t_k)$. From these simulation results, it can be seen the designed fuzzy state feedback controller ensures the robust asymptotic stability of the closed-loop fuzzy DOS (5.9)-(5.10) and guarantees a prescribed performance level.

In the following, a practical application simulation example on truck-trailer model is given.

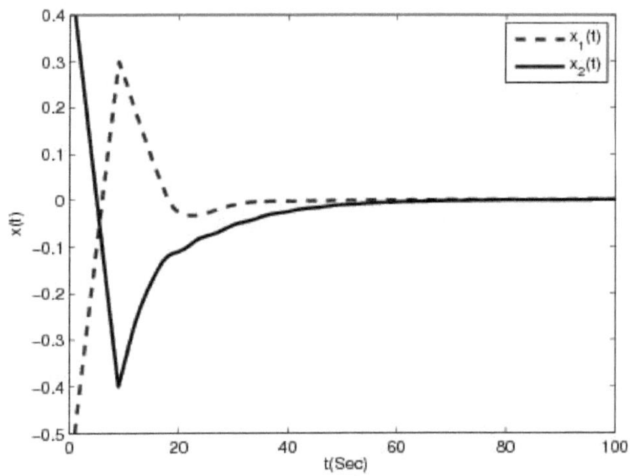

Fig. 5.1 The trajectories of the state in the closed-loop system of this example

Example 5.9. The proposed method will be applied to design a robust state feedback controller for truck trailer system with time-delay proposed in Chapter 4. The truck-trailer model is shown in Fig. 3.1. The following time-delay model with uncertainties is given by:

$$\dot{x}_1(t_k) = -\frac{a\Gamma(t_k)}{t_0}x_1(t_k) - \frac{(1-a)\Gamma(t_k)}{t_0}x_1(t-d(k)) + \frac{v\bar{t}}{lt_0}u(t_k)$$

$$\dot{x}_2(t_k) = \frac{a\Gamma(t_k)}{t_0}x_1(t_k) + \frac{(1-a)\Gamma(t_k)}{t_0}x_1(t-d(k))$$

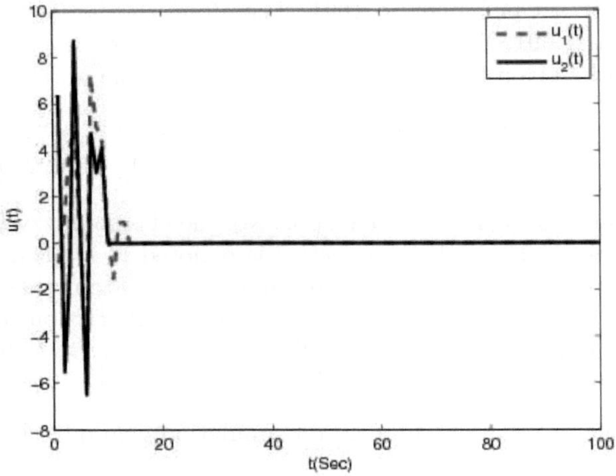

Fig. 5.2 The trajectories of the input in the system of this example

$$\dot{x}_3(t_k) = \frac{v\bar{t}}{t_0}\sin\left(x_2(t_k) + \frac{a\Gamma(t_k)}{2}x_1(t_k) + \frac{(1-a)\Gamma(t_k)}{2}x_1(t-d(k))\right)$$

where $\Gamma(t_k) = v\bar{t}/(L+\Delta L(t_k))$, $v=-1.0$, $\bar{t}=2.0$, $t_0=0.5$, $L=5.5$, $l=2.8$, $-0.2519 \le \Delta L \le 0.2891$. The constant a is the retarded coefficient, which satisfies the conditions: $a \in [0,1]$. The limits 1 and 0 correspond to no delay term and to a completed delay term, respectively. In this example, we assume $a=0.7$. Let $\mathtt{T}=0.01$s, the above truck trailer system can be sampled to the following fuzzy delta operator models to design the fuzzy controller:

Rule 1

IF $\theta(t_k) = x_2(t_k) + a\frac{v\bar{t}}{2(L+\Delta L(t_k))}x_1(t_k) + (1-a)\frac{v\bar{t}}{2(L+\Delta L(t_k))}x_1(t_k - n\mathrm{T})$ is about 0.

THEN

$$\begin{aligned}\delta x(t_k) &= (A_{11}+\Delta A_{11}(t_k))x(t_k) + (A_{21}+\Delta A_{21}(t_k))x(t_k-n\mathrm{T})\\ &\quad +(B_{21}+\Delta B_{21}(t_k))u(t_k),\\ z(t_k) &= (C_{11}+\Delta C_{11}(t_k))x(t_k) + (C_{21}+\Delta C_{21}(t_k))x(t_k-n\mathrm{T}).\end{aligned}$$

Rule 2

IF $\theta(t_k) = x_2(t_k) + a\frac{v\bar{t}}{2(L+\Delta L(t_k))}x_1(t_k) + (1-a)\frac{v\bar{t}}{2(L+\Delta L(t_k))}x_1(t_k - n\mathrm{T})$ is about π and $-\pi$.

THEN

$$\begin{aligned}\delta x(t_k) &= (A_{21}+\Delta A_{21}(t_k))x(t_k) + (A_{22}+\Delta A_{22}(t_k))x(t_k-n\mathrm{T})\\ &\quad +(B_{22}+\Delta B_{22}(t_k))u(t_k),\end{aligned}$$

$$z(t_k) = (C_{21} + \Delta C_{21}(t_k))x(t_k) + (C_{22} + \Delta C_{22}(t_k))x(t_k - n\mathtt{T}).$$

The membership function for Rule 1 and Rule 2 are

$$\begin{aligned} h_1(\theta(t_k)) &= \left(1 - \frac{1}{1 + e^{-3(\theta(t_k) - 0.5\pi)}}\right)\left(\frac{1}{1 + e^{-3(\theta(t_k) - 0.5\pi)}}\right), \\ h_2(\theta(t_k)) &= 1 - h_1(\theta(t_k)), \end{aligned}$$

with

$$\begin{aligned}
A_{11} &= \begin{bmatrix} 0.5109 & 0 & 0 \\ -0.5109 & 0 & 0 \\ 0.5212 & -4 & 0 \end{bmatrix}, \; B_{21} = \begin{bmatrix} -1.4338 \\ 0.0052 \\ -0.0053 \end{bmatrix}, \\
A_{12} &= \begin{bmatrix} 0.5109 & 0 & 0 \\ -0.5109 & 0 & 0 \\ 0.8234 & -4 & 0 \end{bmatrix}, \; B_{22} = \begin{bmatrix} -1.4338 \\ 0.0052 \\ -0.0084 \end{bmatrix}, \\
A_{21} &= [a_{21} \; 0_{3\times 2}], \; a_{21} = [0.2190 \; -0.2190 \; 0.2234]^T, \\
A_{22} &= [a_{22} \; 0_{3\times 2}], \; a_{22} = [0.2190 \; -0.2190 \; 0.3529]^T, \\
C_{11} &= C_{12} = [4.90 \; -2.00 \; 0.03], \; C_{21} = C_{22} = [2.1 \; 0 \; 0], \\
G_{11} &= G_{12} = [h \; h \; h], \; G_{21} = G_{22} = h^T, \\
E_{31} &= E_{32} = E_{41} = E_{42} = 0.05 \cdot h, \quad h = [1 \; 1 \; 1]^T, \\
F_1(t_k) &= F_2(t_k) = \mathrm{diag}\{\sin(t_k), \; \cos(t_k), \; \sin(t_k)\}, \\
E_{11} &= E_{12} = [e_1 \; 0_{3\times 2}], \; e_1 = [0.5091 \; -0.5091 \; -0.5091]^T, \\
E_{21} &= E_{22} = [e_2 \; 0_{3\times 2}], \; e_2 = [0.2182 \; -0.2182 \; -0.2182]^T.
\end{aligned}$$

Firstly, we choose the performance level $\gamma = 0.1$ and sampling period $\mathtt{T} = 0.01$s. Then, using the Matlab LMI Control Toolbox to solve the LMIs in (5.15) and (5.16), the following solution is obtained as:

$$\begin{aligned}
X &= \begin{bmatrix} 0.0047 & 0.0044 & -0.1208 \\ 0.0044 & 0.0167 & 0.4848 \\ -0.1208 & 0.4848 & 61.3076 \end{bmatrix}, \; M = \begin{bmatrix} 0.0023 & 0.0016 & -0.0469 \\ 0.0016 & 0.0013 & -0.0366 \\ -0.0469 & -0.0366 & 1.0700 \end{bmatrix}, \\
N &= \begin{bmatrix} 0.0005 & 0.0010 & -0.0246 \\ 0.0010 & 0.0021 & -0.0403 \\ -0.0246 & -0.0403 & 1.8589 \end{bmatrix}, \; \begin{matrix} Y_1 = [0.0076 \; 0.0029 \; -0.0525], \\ Y_2 = [0.0076 \; 0.0029 \; -0.0519], \end{matrix} \\
\alpha_1 &= 0.0829, \quad \alpha_2 = 0.0829, \; \beta_1 = 0.0826, \quad \beta_2 = 0.0830.
\end{aligned}$$

Therefore, the largest time delay in this example is 6.18. By Theorem 3.1, a desired fuzzy H_∞ state feedback controller can be given by $K_i = Y_i X^{-1}$ $(i = 1, 2)$ as follows:

$$K_1 = \begin{bmatrix} 2.7625 & -0.8947 & 0.0117 \end{bmatrix}, \; K_2 = \begin{bmatrix} 2.7669 & -0.8967 & 0.0117 \end{bmatrix}.$$

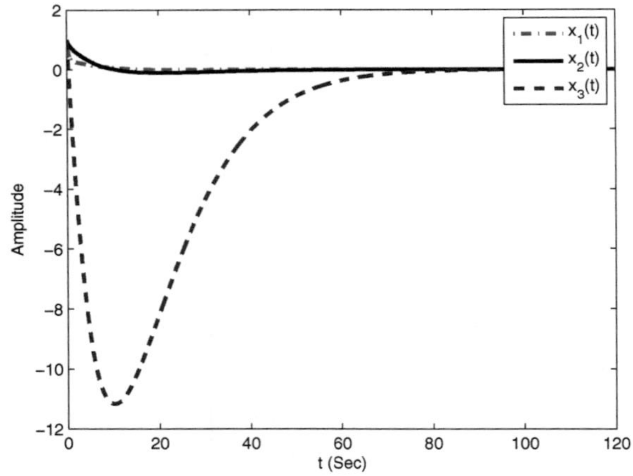

Fig. 5.3 The trajectories of the state in the closed-loop system

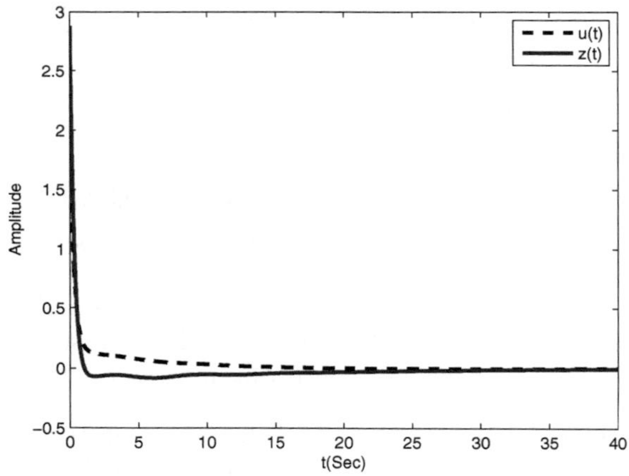

Fig. 5.4 The trajectories of the input and output

The simulation results of the state response of the class of fuzzy DOSs in this example is given in Fig. 5.3, where $x(t_k)$ is denoted by $x_1(t_k)$, $x_2(t_k)$ and $x_3(t_k)$. Fig. 5.4 shows the simulation results of the input $u(t_k)$ and output $z(t_k)$. From the practical application simulation results, it can be seen the designed fuzzy state feedback controller ensures the robust asymptotic stability of closed-loop fuzzy delta operator truck trailer system and guarantees a prescribed performance level.

5.5 Summary

A bridge between robust fuzzy control and DOS framework has been made in this chapter. Based on some Lyapunov-Krasovskii functional method in delta domain and LMI technique, the problem of robust H_∞ control for uncertain fuzzy DOSs with time delays has been considered. A new fuzzy state feedback controller has been designed such that the closed-loop uncertain fuzzy DOS with time delays is robustly asymptotically stable and satisfies a prescribed H_∞ performance level. Numerical examples have been given to illustrate the feasibility and effectiveness of the developed technique.

Part III
Filter Design for DOSs

Chapter 6
Robust H_∞ Filter for Markovian Jump DOSs with Time-Varying Delays

6.1 Introduction

Recently, the problems of the filtering has been studied by a number of authors, see for example [9, 71, 72, 178] and the references therein. Both a Kalman-filter-based iterative algorithm and a recursive frequency estimator using linear prediction have been considered in [305]. In [122], the synthesis of two observer-based nonlinear control algorithms was performed within the globally linearizing control framework. A problem of optimal linear Kalman filtering with packet losses has been considered in [109]. When parameter uncertainty arises in a system model, the robust H_∞ filtering problem has been studied in [75]. Because H_∞ filtering provides a bound for the worst-case estimation error without the need for knowledge of noise statistics, it has been demonstrated that H_∞ filtering has the advantage of being less sensitive to uncertainties of the underlying system. The class of Markovian jump systems is an important family of systems subject to abrupt variations. The Markovian jump systems have finite modes, which may jump from one to another at different times and between different modes. A system with such a "jumping" character may be modeled as a hybrid system, and the parameter jumps among different modes can be seen as discrete events. Note that the concept of Markovian jump filter has already been used in some papers, for example [6, 20, 209, 275]. Robust Kalman filtering for continuous systems with Markovian jumping parameters was given in [160, 208].

The so called delta operator instead of traditional shift operator was constructed to study sampling continuous time systems in [82]. The delta operator approach can unify the related results of the continuous and discrete systems into a delta operator framework. The robustness problem for some DOSs with parametric uncertainties has been investigated in [215]. An optimal predictive control technique using delta operator approach was considered in [49]. An H_∞ control approach on DOSs has also been reported in [52]. Delayed systems have been found many applications in a variety of areas and have been extensively studied for example [19, 29]. However, to the best of our knowledge, very few results are available on delta operator H_∞ filtering

H. Yang et al.: Analysis and Synthesis of Delta Operator Systems, LNCIS 430, pp. 85–98.
springerlink.com © Springer-Verlag Berlin Heidelberg 2012

problem for Markovian jump DOS with linear fractional parameter uncertainties and time-varying delays. This problem is important and challenging in both theory and practice, which motivated us for this study.

6.2 Problem Statement

Given a probability space $(\Omega, \mathbf{F}, \mathbf{P})$, where Ω is the sample space, $\mathbf{F}$ is the algebra of events and $\mathbf{P}$ is the probability measure defined on $\mathbf{F}$, $\{\eta_{t_k}, t_k \geq 0\}$ is a homogeneous, finite-state Markovian process with right continuous trajectories and taking values in a finite set $S = \{1, 2, \cdots, s\}$ with generator $\Lambda = (\lambda_{ij})$. The transition probability from mode i at time t_k to the mode j at time $t_k + \mathtt{T}$, $i, j \in S$, is

$$Pr(\eta_{t_k+\mathtt{T}} = j|\ \eta_{t_k} = i) = \begin{cases} \lambda_{ij}\mathtt{T}, & i \neq j, \\ 1 + \lambda_{ij}\mathtt{T}, & i = j, \end{cases} \tag{6.1}$$

where the transition probability rates satisfy $\lambda_{ij} \geq 0$ for $i, j \in S$, $i \neq j$ and $\lambda_{ii} = -\sum_{j=1,\ j\neq i}^{s} \lambda_{ij}$. The Markov process transition rate matrix Λ_δ is defined by

$$\Lambda_\delta = \frac{\Pi_s(\mathtt{T}) - \Pi_s(0)}{\mathtt{T} - 0} = \frac{e^{\Lambda_s \mathtt{T}} - I}{\mathtt{T}} = \begin{bmatrix} \lambda_{11} & \lambda_{12} & \cdots & \lambda_{1s} \\ \lambda_{21} & \lambda_{22} & \cdots & \lambda_{2s} \\ \vdots & \vdots & \ddots & \vdots \\ \lambda_{s1} & \lambda_{s2} & \cdots & \lambda_{ss} \end{bmatrix}.$$

where Π_z is the Markov process transition probability matrix in z-domain and Λ_s is the continuous Markov process transition rate matrix in s-domain.

Remark 6.1. The Markov process transition rate matrix Λ_s is a special case of Λ_δ, since there exists

$$\lim_{\mathtt{T}\to 0} \Lambda_\delta = \lim_{\mathtt{T}\to 0} \frac{e^{\Lambda_s \mathtt{T}} - I}{\mathtt{T}} = \lim_{\mathtt{T}\to 0} \Lambda_s e^{\Lambda_s \mathtt{T}} = \Lambda_s.$$

Furthermore, the transition rate matrix Λ_δ also covers the transition probability matrix Π_z which is relative to Π_s for $\Pi_z = \Pi_s(1)$. Therefore, it can be said that the continuous-time Markov process transition rate matrix Λ_s and the Markov process transition probability matrix Π_z are the special case of the Markov process transition rate matrix Λ_δ in δ-domain when sampling period $\mathtt{T} \to 0$ and $\mathtt{T} = 1$, respectively.

In this chapter, the following uncertain jump DOS with time-varying delays is considered:

$$\begin{aligned} \delta x(t_k) = {} & (A(\eta_{t_k}) + \Delta A(\eta_{t_k}, t_k))x(t_k) + B(\eta_{t_k})\omega(t_k) \\ & +(A_d(\eta_{t_k}) + \Delta A_d(\eta_{t_k}, t_k))x(t_k - d(k)), \end{aligned} \tag{6.2}$$

$$y(t_k) = C(\eta_{t_k})x(t_k) + D(\eta_{t_k})\omega(t_k), \tag{6.3}$$
$$z(t_k) = L(\eta_{t_k})x(t_k), \tag{6.4}$$
$$x(t_k) = \phi(t_k), \qquad t_k \in [-d_M,\ 0]. \tag{6.5}$$

where $x(t_k) \in R^n$ is the state variable; $z(t_k) \in R^p$ is control output; $\omega(t_k) \in R^l$ is the disturbance input which belongs to $L_2[0, \infty)$. The time-delay $d(k)$ is a time-varying function that satisfies $0 \le d_m \le d(k) \le d_M$, with $d_m = n_m\mathrm{T}$ and $d_M = n_M\mathrm{T}$, n_m and n_M are two known positive integers, from which we let $n_m \le n \le n_M$. It is easy to get the relationship $x(t_k - n\mathrm{T}) = x(t_k - d(k))$. For notational simplicity, in the sequel, for $\eta_k = i \in S$, we denote $A(\eta_{t_k})$ by A_i, $\Delta A(\eta_{t_k}, t_k)$ by $\Delta A_i(t_k)$, $A_i(t_k) = A_i + \Delta A_i(t_k)$ and so on. The linear fractional parametric uncertainties $\Delta A_i(t_k)$ and $\Delta A_{di}(t_k)$ are time-varying matrices with appropriate dimensions, which are defined as follows:

$$\left[\, \Delta A_i(t_k)\ \Delta A_{di}(t_k) \,\right] = M_i \hat{F}_i(t_k) \left[\, N_{1i}\ N_{2i} \,\right], \tag{6.6}$$
$$\hat{F}_i(t_k) = F_i(t_k)\left[I - H_i F_i(t_k)\right]^{-1}, \tag{6.7}$$

where M_i, N_{1i} and N_{2i} are known constant real matrices with appropriate dimensions, and $F_i(t_k)$ is unknown time-varying matrix satisfying $F_i^T(t_k)F_i(t_k) \le I$, $\forall t_k \ge 0$. It is assumed that the matrix $[I - H_i F_i(t_k)]^{-1}$ is invertible for any $F_i(t_k)$ and $I - H_i^T H_i > 0$.

Remark 6.2. The model (6.6)-(6.7) describes a wider class of parameter uncertainties than norm-bounded parameter uncertainties. It is easy to see that the linear fractional parameter uncertainties can be reduced to norm-bounded parameter uncertainties when $H_i = 0$.

We focus attention to the design of a linear Markovian jump nth-order filter for which the jumping process is available for $t_k \ge 0$ and has the following state-space model:

$$\delta\hat{x}(t_k) = G_i\hat{x}(t_k) + K_i y(t_k), \quad \hat{z}(t_k) = L_i\hat{x}(t_k), \tag{6.8}$$

where $\hat{x}(t_k) \in R^q$, G_i and K_i are the matrices to be determined. In terms of the state error $e(t_k) = x(t_k) - \hat{x}(t_k)$, the following error dynamics systems can be obtained:

$$\begin{aligned} \delta e(t_k) &= G_i e(t_k) + (A_i - K_iC_i - G_i)x(t_k) + A_{di}x(t_k - d(k)) \\ &\quad + (B_i + K_iD_i)\omega(t_k), \end{aligned} \tag{6.9}$$
$$\tilde{z}(t_k) = L_i e(t_k), \tag{6.10}$$

Let $\tilde{x}(t_k) = \left[x^T(t_k)\ e^T(t_k)\right]^T$, then the filtering error dynamics systems can be described by

$$\delta\tilde{x}(t_k) = A_{fi}(t_k)\tilde{x}(t_k) + A_{dfi}(t_k)\tilde{x}(t_k - d(k)) + D_{fi}\omega(t_k), \tag{6.11}$$
$$\tilde{z}(t_k) = L_{fi}(t_k)\tilde{x}(t_k), \tag{6.12}$$

where

$$A_{fi}(t_k) = A_{fi} + \Delta A_{fi}(t_k), \quad A_{dfi}(t_k) = A_{dfi} + \Delta A_{dfi}(t_k), \quad L_{fi} = \begin{bmatrix} 0 & L_i \end{bmatrix},$$
$$A_{fi} = \begin{bmatrix} A_i & 0 \\ A_i - K_iC_i - G_i & G_i \end{bmatrix}, \quad A_{dfi} = \begin{bmatrix} A_{di} & 0 \\ A_{di} & 0 \end{bmatrix}, \quad D_{fi} = \begin{bmatrix} B_i \\ B_i - K_iD_i \end{bmatrix},$$
$$\Delta A_{fi}(t_k) = \begin{bmatrix} M_i \\ M_i \end{bmatrix} \hat{F}_i(t_k) \begin{bmatrix} N_{1i} & 0 \end{bmatrix}, \quad \Delta A_{dfi}(t_k) = \begin{bmatrix} M_i \\ M_i \end{bmatrix} \hat{F}_i(t_k) \begin{bmatrix} N_{2i} & 0 \end{bmatrix}.$$

Definition 6.3. A jump DOS is stochastic asymptotically stable in delta domain, if the following conditions hold:
(i) $V(x(t_k)) \geq 0$, with equality if and only if $x(t_k) = 0$;
(ii) $\sigma V(x(t_k)) = [\mathbf{E}[V(x(t_k + \mathtt{T}))] - V(x(t_k))]/\mathtt{T} < 0$;
where $V(x(t_k))$ is a Lyapunov function in delta domain.

Definition 6.4. The robust H_∞ filtering problem to be addressed is formulated as follows: consider the uncertain DOS (6.2)-(6.5) and a prescribed level of noise attenuation $\gamma > 0$, determine an asymptotically stable filter in the form of (6.8) such that the the following requirements are satisfied:
(i) The filtering error system (6.11)-(6.12) is robustly asymptotically stable;
(ii) With zero initial condition (i.e.$\phi = 0$), the filtering error system (6.11)-(6.12) satisfies:

$$J = \sum_{t=0}^{\infty} [\tilde{z}^T(t_k)\tilde{z}(t_k) - \gamma^2 \omega^T(t_k)\omega(t_k)] \leq 0,$$

for with any nonzero $\omega(t_k) \in L_2[0, \infty)$ and all admissible uncertainties. The parameter γ is called the H_∞ norm bound of the control.

Before ending this section, the following lemmas will be recalled to prove our main results in the next section.

Lemma 6.5. [124] For any constant positive semi-definite symmetric matrix W, two positive integers r and r_0 satisfying $r \geq r_0 \geq 1$, the following inequality holds

$$\left(\sum_{i=r_0}^{r} x(i)\right)^T W \left(\sum_{i=r_0}^{r} x(i)\right) \leq (r - r_0 + 1) \sum_{i=r_0}^{r} x^T(i) W x(i).$$

Lemma 6.6. [269] For some given matrices Υ, D and E of appropriate dimension and with Υ symmetric, then

$$\Upsilon + D\hat{F}(t_k)E + E^T\hat{F}^T(t_k)D^T \leq 0,$$

where $\hat{F}(t_k)$ is given as in (6.6), if and only if there exists a scalar $\varepsilon > 0$ such that

$$\Upsilon + \begin{bmatrix} \varepsilon^{-1}E^T & \varepsilon D \end{bmatrix} \begin{bmatrix} I & -G \\ -G^T & I \end{bmatrix} \begin{bmatrix} \varepsilon^{-1}E \\ \varepsilon D^T \end{bmatrix} \leq 0.$$

6.3 Main Results

In this section, we will develop a LMI approach to solve the robust H_∞ filtering problem for the uncertain time-delay DOSs (6.2)-(6.5). Our main result is given as follows.

Theorem 6.7. *Given a scalar $\gamma > 0$, then the uncertain error dynamics DOS in (6.11)-(6.12) is robustly asymptotically stable with the disturbance attenuation γ, if there exist matrices $P_i > 0$, $Q > 0$ and $R > 0$, positive scalars α_i such that the following LMI holds:*

$$\Sigma = \begin{bmatrix} \Sigma(1,1) & \Sigma(1,2) & P_iA_{dfi} & P_iD_{fi} & 0 & P_iM_{fi} \\ * & \Sigma(2,2) & P_iA_{dfi} + \frac{1}{d_M}R & P_iD_{fi} & \alpha_iN_{1fi}^T & P_iM_{fi} \\ * & * & -Q - \frac{1}{d_M}R & 0 & \alpha_iN_{2fi}^T & 0 \\ * & * & * & -\gamma^2I & 0 & 0 \\ * & * & * & * & -\alpha_iI & \alpha_iH_i \\ * & * & * & * & * & -\alpha_iI \end{bmatrix} < 0, \quad (6.13)$$

with

$$\Sigma(1,1) = T^2\sum_{j=1}^{s}\lambda_{ij}P_j + (T-2)P_i + d_MR_i, \;\; \Sigma(1,2) = T\sum_{j=1}^{s}\lambda_{ij}P_j + P_iA_{fi},$$

$$\Sigma(2,2) = \sum_{j=1}^{s}\lambda_{ij}P_j + PA_{fi} + A_{fi}^TP_i + (d_M - d_n + 1)Q - \frac{1}{d_M}R + L_{fi}^TL_{fi},$$

$$N_{1fi} = \begin{bmatrix} N_{1i} & 0 \end{bmatrix}, \; N_{2fi} = \begin{bmatrix} N_{2i} & 0 \end{bmatrix}, \; M_{fi} = \begin{bmatrix} M_i \\ M_i \end{bmatrix}.$$

Proof. Consider the uncertain time-delay DOS (6.11) with $\omega(t_k) = 0$. The LMI (6.13) will be changed to

$$\Sigma_0 = \begin{bmatrix} \Sigma_0(1,1) & \Sigma_0(1,2) & P_iA_{dfi} & 0 & P_iM_{fi} \\ * & \Sigma_0(2,2) & P_iA_{dfi} + \frac{1}{d_M}R & \alpha_iN_{1fi}^T & P_iM_{fi} \\ * & * & -Q - \frac{1}{d_M}R & \alpha_iN_{2fi}^T & 0 \\ * & * & * & -\alpha_iI & \alpha_iH_i \\ * & * & * & * & -\alpha_iI \end{bmatrix} < 0 \quad (6.14)$$

with

$$\Sigma_0(1,1)=\mathtt{T}^2\sum_{j=1}^{s}\lambda_{ij}P_j+(\mathtt{T}-2)P_i+d_MR_i,\ \Sigma_0(1,2)=\mathtt{T}\sum_{j=1}^{s}\lambda_{ij}P_j+P_iA_{fi},$$

$$\Sigma_0(2,2)=\sum_{j=1}^{s}\lambda_{ij}P_j+PA_{fi}+A_{fi}^TP_i+(d_M-d_n+1)Q-\frac{1}{d_M}R+L_{fi}^TL_{fi}.$$

From Definition 6.3, we can conclude that the DOS (6.11) with $\omega(t_k)=0$ is robustly asymptotically stable if inequality (6.14) holds. Construct Lyapunov-Krasovskii functional in delta domain as follows:

$$V(\tilde{x},t_k,\eta_{t_k})=V_1(\tilde{x},t_k,\eta_{t_k})+V_2(\tilde{x},t_k,\eta_{t_k})+V_3(\tilde{x},t_k,\eta_{t_k})+V_4(\tilde{x},t_k,\eta_{t_k}),$$

with

$$\begin{aligned}
V_1(\tilde{x},t_k,\eta_{t_k})&=\tilde{x}^T(t_k)P(\eta_{t_k})\tilde{x}(t_k),\\
V_2(\tilde{x},t_k,\eta_{t_k})&=\mathtt{T}\sum_{i=1}^{n}\tilde{x}^T(t_k-i\mathtt{T})Q\tilde{x}(t_k-i\mathtt{T}),\\
V_3(\tilde{x},t_k,\eta_{t_k})&=\mathtt{T}^2\sum_{i=n_m+1}^{n_M}\sum_{j=1}^{i}\tilde{x}^T(t_k-j\mathtt{T})Q\tilde{x}(t_k-j\mathtt{T}),\\
V_4(\tilde{x},t_k,\eta_{t_k})&=\sum_{i=1}^{n}\sum_{j=1}^{i}e^T(t_k-j\mathtt{T})Re(t_k-j\mathtt{T}),
\end{aligned}$$

where $e(j)=\tilde{x}(j)-\tilde{x}(j+\mathtt{T})$, so there exist $\delta\tilde{x}(j)=-e(j)/\mathtt{T}$ and $e(t_k-i\mathtt{T})=\tilde{x}(t_k-i\mathtt{T})-\tilde{x}(t_k-(i-1)\mathtt{T})$. The delta operator of the stochastic process $\{\tilde{x}(t_k),\eta_{t_k},t_k\geq 0\}$, acting on $V(\tilde{x},t_k,\eta_{t_k})$ at the point $\{\tilde{x},t_k,\eta_{t_k}=i\}$, can be expressed as follows:

$$\sigma(V(\tilde{x},t_k,\eta_{t_k}))=\frac{\mathbf{E}[V(\tilde{x},\ t_k+\mathtt{T},\ \eta_{t_k+\mathtt{T}})]-V(\tilde{x},t_k,\eta_{t_k})}{\mathtt{T}}. \tag{6.15}$$

Let the mode at time t be i, that is $\eta_{t_k}=i$. Recall that at time $t_k+\mathtt{T}$, the system may jump to any mode $\eta_{t_k+\mathtt{T}}=j$. By letting $\mathbf{E}(P_j)=\mathtt{T}\sum_{j=1}^{s}\lambda_{ij}P_j+P_i$ and taking the stochastic delta operator manipulations of along the trajectory of system (6.11) we can obtain:

$$\begin{aligned}
\sigma V_1(\tilde{x},t_k,\eta_{t_k})&=\frac{1}{\mathtt{T}}\left\{\mathbf{E}\left[\tilde{x}^T(t_k+\mathtt{T})P(\eta_{t_k+\mathtt{T}})\tilde{x}(t_k+\mathtt{T})\right]\right.\\
&\qquad\left.-\tilde{x}^T(t_k)P(\eta(t_k))\tilde{x}(t_k)\right\}\\
&=\sum_{j=1}^{s}\lambda_{ij}\tilde{x}^T(t_k+\mathtt{T})P_j\tilde{x}(t_k+\mathtt{T})\\
&\quad+\frac{1}{\mathtt{T}}\left[\tilde{x}^T(t_k+\mathtt{T})P_i\tilde{x}(t_k+\mathtt{T})-\tilde{x}^T(t_k)P_i\tilde{x}(t_k)\right]
\end{aligned}$$

$$
\begin{aligned}
&= \mathtt{T}^2 \sum_{j=1}^{s} \lambda_{ij} \delta^T(\tilde{x}(t_k)) P_j \delta(\tilde{x}(t_k)) + \mathtt{T} \sum_{j=1}^{s} \lambda_{ij} \delta^T(\tilde{x}(t_k)) P_j \tilde{x}(t_k) \\
&+ \mathtt{T} \sum_{j=1}^{s} \lambda_{ij} \tilde{x}^T(t_k) P_j \delta(\tilde{x}(t_k)) \\
&+ \sum_{j=1}^{s} \lambda_{ij} \tilde{x}^T(t_k) P_j \tilde{x}(t_k) + \mathtt{T} \delta^T(\tilde{x}(t_k)) P_i \delta(\tilde{x}(t_k)) \\
&+ \delta^T(\tilde{x}(t_k)) P_i \tilde{x}(t_k) + \tilde{x}^T(t_k) P_i \delta(\tilde{x}(t_k))
\end{aligned} \tag{6.16}
$$

where $\tilde{x}^T(t_k)P_i\delta(\tilde{x}(t_k)) = \tilde{x}^T(t_k)P_i[A_{fi}(t_k)\tilde{x}(t_k) + A_{dfi}(t_k)\tilde{x}(t_k - d(k)) + D_{fi}\omega(t_k)]$.

Remark 6.8. If we let $\mathtt{T} = 0$, the manipulation result using stochastic delta operator for $V_1(x, t_k, \eta_{t_k})$ is equal to the one using weak infinitesimal operator in [209].

Taking the stochastic delta operator manipulations of $V_2(\tilde{x}, t_k, \eta_{t_k})$ and $V_3(\tilde{x}, t_k, \eta_{t_k})$, it can be obtained that

$$
\begin{aligned}
\sigma V_2(\tilde{x}, t_k, \eta_{t_k}) &= \frac{1}{\mathtt{T}} \left[\mathtt{T} \sum_{i=1}^{n} \tilde{x}^T(t_k - (i-1)\mathtt{T}) Q \tilde{x}(t_k - (i-1)\mathtt{T}) \right. \\
&\qquad \left. - \mathtt{T} \sum_{i=1}^{n} \tilde{x}^T(t_k - i\mathtt{T}) Q \tilde{x}(t_k - i\mathtt{T}) \right] \\
&\leq \tilde{x}^T(t_k) Q \tilde{x}(t_k) - \tilde{x}^T(t_k - d(k)) Q \tilde{x}(t_k - d(k)) \\
&\qquad + \mathtt{T} \sum_{i=n_m+1}^{n_M} \tilde{x}^T(t_k - i\mathtt{T}) Q \tilde{x}(t_k - i\mathtt{T}).
\end{aligned} \tag{6.17}
$$

$$
\begin{aligned}
\sigma V_3(\tilde{x}, t_k, \eta_{t_k}) &= \mathtt{T} \sum_{i=n_m+1}^{n_M} \left(\sum_{j=1}^{i} \tilde{x}^T(t_k - (j-1)\mathtt{T}) Q \tilde{x}(t_k - (j-1)\mathtt{T}) \right. \\
&\qquad \left. - \sum_{j=1}^{i} \tilde{x}^T(t_k - j\mathtt{T}) Q \tilde{x}(t_k - j\mathtt{T}) \right) \\
&= (d_M - d_m) \tilde{x}^T(t_k) Q \tilde{x}(t_k) \\
&\qquad - \mathtt{T} \sum_{i=n_m+1}^{n_M} \tilde{x}^T(t_k - i\mathtt{T}) Q \tilde{x}(t_k - i\mathtt{T}).
\end{aligned} \tag{6.18}
$$

Using Lemma 6.5 and taking the stochastic delta operator manipulations of $V_4(\tilde{x}, t_k, \eta_{t_k})$, there exists

$$\begin{aligned}
\sigma V_4(\tilde{x}, t_k, \eta_{t_k}) &= \frac{1}{\mathrm{T}}\left[\sum_{i=1}^{n}\sum_{j=1}^{i} e^T(t_k-(j-1)\mathrm{T})Re(t_k-(j-1)\mathrm{T})\right.\\
&\qquad\left.-\sum_{i=1}^{n}\sum_{j=1}^{i} e^T(t_k-j\mathrm{T})Re(t_k-j\mathrm{T})\right]\\
&= \frac{1}{\mathrm{T}}\left[\sum_{i=1}^{n} e^T(t_k)Re(t_k)-\sum_{i=1}^{n} e^T(t_k-i\mathrm{T})Re(t_k-i\mathrm{T})\right]\\
&\le \frac{n}{\mathrm{T}}e^T(t_k)Re(t_k)-\frac{1}{n\mathrm{T}}\left[\sum_{i=1}^{n} e(t_k-i\mathrm{T})\right]^T R\left[\sum_{i=1}^{n} e(t_k-i\mathrm{T})\right]\\
&\le -\frac{1}{d_M}\left[\tilde{x}(t_k-d(k))-\tilde{x}(t_k)\right]^T R\left[\tilde{x}(t_k-d(k))-\tilde{x}(t_k)\right]\\
&\quad +d_M\delta^T(\tilde{x}(t_k))R\delta(\tilde{x}(t_k)). \qquad (6.19)
\end{aligned}$$

For any real matrix P_i, one has that

$$\begin{aligned}
0 &= -2\delta^T(\tilde{x}(t_k))P_i[\delta(\tilde{x}(t_k))-A_{fi}(t_k)\tilde{x}(t_k)\\
&\quad -A_{dfi}(t_k)\tilde{x}(t_k-d(k))-D_{fi}\omega(t_k)]\\
&= -2\delta^T(\tilde{x}(t_k))P_i\delta(\tilde{x}(t_k))+2\delta^T(\tilde{x}(t_k))P_iA_{fi}(t_k)\tilde{x}(t_k)\\
&\quad +2\delta^T(\tilde{x}(t_k))P_iA_{dfi}(t_k)\tilde{x}(t_k-d(k))+2\delta^T(\tilde{x}(t_k))P_iB_{di}\omega(t_k). \qquad (6.20)
\end{aligned}$$

In addition

$$\tilde{z}^T(t_k)\tilde{z}(t_k)-\gamma^2\omega^T(t_k)\omega(t_k)=\tilde{x}^T(t_k)L_{fi}^TL_{fi}\tilde{x}(t_k)-\gamma^2\omega^T(t_k)\omega(t_k). \quad (6.21)$$

Finally, it follows from (6.16)-(6.21) that

$$\sigma V(\tilde{x}, t_k, \eta_{t_k}) \le \varphi^T(t_k)\Sigma_1\varphi(t_k) < 0, \qquad (6.22)$$

where

$$\eta^T(t_k)=\begin{bmatrix}\delta^T(\tilde{x}(t_k)) & \tilde{x}^T(t_k) & \tilde{x}^T(t_k-d(k)) & \omega^T(t_k)\end{bmatrix},$$

and

$$\Sigma_1=\begin{bmatrix}\Sigma_1(1,1) & \Sigma_1(1,2) & P_iA_{dfi}(t_k) & P_iD_{fi}\\ * & \Sigma_1(2,2) & P_iA_{dfi}(t_k)+\frac{1}{d_M}R_i & P_iD_{fi}\\ * & * & -Q_i-\frac{1}{d_M}R_i & 0\\ * & * & * & -\gamma^2 I\end{bmatrix}<0, \qquad (6.23)$$

with

$$\Sigma_1(1,1)=\mathrm{T}^2\sum_{j=1}^{s}\lambda_{ij}P_j+(\mathrm{T}-2)P_i+d_MR_i,$$

$$\Sigma_1(1,2) = \mathrm{T}\sum_{j=1}^{s}\lambda_{ij}P_j + P_iA_{fi}(t_k),$$

$$\Sigma_1(2,2) = \sum_{j=1}^{s}\lambda_{ij}P_j + PA_{fi}(t_k) + A_{fi}^T(t_k)P_i + (d_M - d_n + 1)Q_i$$
$$-\frac{1}{d_M}R_i + L_{fi}^TL_{fi}.$$

From (6.6) and (6.23), we get that

$$\Sigma_1 = \Upsilon + \xi_i\hat{F}(t_k)\zeta_i + \zeta_i^T\hat{F}^T(t_k)\xi_i^T < 0, \tag{6.24}$$

where

$$\Upsilon = \begin{bmatrix} \Upsilon(1,1) & \Upsilon(1,2) & P_iA_{dfi} & P_iD_{fi} \\ * & \Upsilon(2,2) & P_iA_{dfi} + \frac{1}{d_M}R_i & P_iD_{fi} \\ * & * & -Q_i - \frac{1}{d_M}R_i & 0 \\ * & * & * & -\gamma^2 I \end{bmatrix},$$

$$\xi_i = \begin{bmatrix} P_iM_{fi} \\ P_iM_{fi} \\ 0 \\ 0 \end{bmatrix}, \quad \zeta_i = \begin{bmatrix} 0 \\ N_{1fi}^T \\ N_{2fi}^T \\ 0 \end{bmatrix}^T,$$

with

$$\Upsilon(1,1) = \mathrm{T}^2\sum_{j=1}^{s}\lambda_{ij}P_j + (\mathrm{T}-2)P_i + d_MR_i, \ \Upsilon(1,2) = \mathrm{T}\sum_{j=1}^{s}\lambda_{ij}P_j + P_iA_{fi},$$

$$\Upsilon(2,2) = \sum_{j=1}^{s}\lambda_{ij}P_j + PA_{fi} + A_{fi}^TP_i + (d_M - d_n + 1)Q_i - \frac{1}{d_M}R_i + L_{fi}^TL_{fi}.$$

By Lemma 6.6, there exists a scalar $\varepsilon_i > 0$ such that the (6.24) equal to the following inequality

$$\Sigma_1 = \Upsilon + \begin{bmatrix} \varepsilon_i\zeta_i^T & \varepsilon_i^{-1}\xi_i \end{bmatrix} \begin{bmatrix} -I & H_i \\ H_i^T & -I \end{bmatrix}^{-1} \begin{bmatrix} \varepsilon_i\,\zeta_i \\ \varepsilon_i^{-1}\xi_i^T \end{bmatrix} < 0. \tag{6.25}$$

Using Schur's complement, $\Sigma_1 < 0$ can be changed to $\Sigma_2 < 0$, where

$$\Sigma_2 = \begin{bmatrix} \Sigma_2(1,1) & \Sigma_2(1,2) & P_iA_{dfi} & P_iD_{fi} & 0 & \varepsilon_i^{-1}P_iM_{fi} \\ * & \Sigma_2(2,2) & P_iA_{dfi} + \frac{1}{d_M}R_i & P_iD_{fi} & \varepsilon_iN_{1fi}^T & \varepsilon_i^{-1}P_iM_{fi} \\ * & * & -Q_i - \frac{1}{d_M}R_i & 0 & \varepsilon_iN_{2fi}^T & 0 \\ * & * & * & -\gamma^2 I & 0 & 0 \\ * & * & * & * & -I & H_i \\ * & * & * & * & * & -I \end{bmatrix},$$

with

$$\Sigma_2(1,1) = \mathtt{T}^2\sum_{j=1}^{s}\lambda_{ij}P_j + (\mathtt{T}-2)P_i + d_M R_i,\ \Sigma_2(1,2) = \mathtt{T}\sum_{j=1}^{s}\lambda_{ij}P_j + P_i A_{fi},$$

$$\Sigma_2(2,2) = \sum_{j=1}^{s}\lambda_{ij}P_j + PA_{fi} + A_{fi}^T P_i + (d_M - d_n + 1)Q_i - \frac{1}{d_M}R_i + L_{fi}^T L_{fi}.$$

Then pre- and post-multiplying $\Sigma_2 < 0$ by diag$\{I,\ I,\ I,\ I,\ \varepsilon_i I,\ \varepsilon_i I\}$, respectively, and letting $\varepsilon_i^2 = \alpha_i$, the inequality $\Sigma_2 < 0$ is equivalent to $\Sigma < 0$. From Definition 6.3, we conclude that the DOS (6.11) with $\omega(t_k) = 0$ is robustly asymptotically stable if inequality (6.14) holds. Next, considering the H_∞ performance index

$$J = \sum_{t=0}^{\infty}[\tilde{z}^T(t_k)\tilde{z}(t_k) - \gamma^2\omega^T(t_k)\omega(t_k)],$$

and assuming, without loss of generality, system (6.11) is stable with zero initial condition, it implies $V(\cdot)|_{t_k=0} = 0$ and $V(\cdot)|_{t_k\to\infty} \to \epsilon$ with $\epsilon \to 0$, if $\omega(t_k) = 0$ or $\epsilon < \infty$ if $\omega(t_k) \neq 0$. In this way, the above index can be rewritten as

$$J = \sum_{t=0}^{\infty}[\tilde{z}^T(t_k)\tilde{z}(t_k) - \gamma^2\omega^T(t_k)\omega(t_k) + \delta V(t_k)].$$

Then, it can be obtained that

$$J = \sum_{t=0}^{\infty}[\tilde{z}^T(t_k)\tilde{z}(t_k) - \gamma^2\omega^T(t_k)\omega(t_k)] \le \sum_{t=0}^{\infty}[\zeta^T(t_k)\Sigma\zeta(t_k)] \le 0, \quad (6.26)$$

with

$$\zeta^T(t_k) = \left[\,\delta^T(x(t_k))\ x^T(t_k)\ x^T(t_k - d(k))\ \omega^T(t_k)\ y_1^T(t_k)\ y_2^T(t_k)\,\right].$$

where $y_1(t_k)$ and $y_2(t_k)$ are arbitrary vectors with appropriate dimensions. From Definition 6.4, the delay dependent sufficient conditions for the solvability of the robust H_∞ filtering problem of the jump DOS (6.2)-(6.5) is presented in terms of LMI (6.13).

Remark 6.9. It should be noted that when $\mathtt{T} = 0$ the delta domain Lyapunov function $V(\tilde{x}, t_k, \eta_{t_k})$ and all of the DOS in this chapter can be changed to the ones in s-domain. On the other hand, when $\mathtt{T} = 1$ they can be changed to the ones in z-domain.

Next, the main result on the solvability of the robust H_∞ filtering problem is given as follows:

Theorem 6.10. *Consider the uncertain time-delay DOS (6.2)-(6.5), then the robust H_∞ filtering problem is solvable if there exist scalars $\alpha_i > 0$, some matrices $P_{1i} > 0$, $P_{2i} > 0$, $Q_1 > 0$, $Q_2 > 0$, $Q_3 > 0$, $R_1 > 0$, $R_2 > 0$, $R_3 > 0$, Ψ_i and Φ_i such that the following LMIs hold:*

$$\begin{bmatrix} \Omega_{11} & d_M R_2 & \Omega_{13} & 0 & P_{1i}A_{di} & 0 & P_{1i}B_i & 0 & P_{1i}M_i \\ * & \Omega_{22} & \Omega_{23} & \Omega_{24} & P_{2i}A_{di} & 0 & \Omega_{27} & 0 & P_{2i}M_i \\ * & * & \Omega_{33} & \Omega_{34} & \Omega_{35} & \frac{1}{d_M}R_2 & P_{1i}B_i & \alpha_i N_{1i}^T & P_{1i}M_i \\ * & * & * & \Omega_{44} & \Omega_{45} & \frac{1}{d_M}R_3 & \Omega_{47} & 0 & P_{2i}M_i \\ * & * & * & * & \Omega_{55} & \Omega_{56} & 0 & \alpha_i N_{2i}^T & 0 \\ * & * & * & * & * & \Omega_{66} & 0 & 0 & 0 \\ * & * & * & * & * & * & -\gamma^2 I & 0 & 0 \\ * & * & * & * & * & * & * & -\alpha_i I & \alpha_i H_i \\ * & * & * & * & * & * & * & * & -\alpha_i I \end{bmatrix} < 0, \quad (6.27)$$

with

$$\Omega_{11} = T^2\sum_{j=1}^{s}\lambda_{ij}P_{1j} + (T-2)P_{1i} + d_M R_1, \; \Omega_{13} = T\sum_{j=1}^{s}\lambda_{ij}P_{1j} + P_{1i}A_i,$$

$$\Omega_{22} = T^2\sum_{j=1}^{s}\lambda_{ij}P_{2j} + (T-2)P_{2i} + d_M R_3, \; \Omega_{23} = P_{2i}A_i + \Phi_i C + \Psi_i,$$

$$\Omega_{24} = T\sum_{j=1}^{s}\lambda_{ij}P_{2j} + \Psi_i, \; \Omega_{27} = P_{2i}B_i + \Phi_i D_i, \; \Omega_{35} = P_{1i}A_{di} + \frac{1}{d_M}R_1,$$

$$\Omega_{45} = P_{2i}A_{di} + \frac{1}{d_M}R_2, \; \Omega_{47} = P_{2i}B_i + \Phi_i D_i,$$

$$\Omega_{33} = \sum_{j=1}^{s}\lambda_{ij}P_{1j} + P_{1i}A_i + A_i^T P_{1i} + (d_M - d_m + 1)Q_1 - \frac{1}{d_M}R_1,$$

$$\Omega_{34} = A_i^T P_{1i} + C_i^T \Phi_i^T + \Psi_i^T + (d_M - d_m + 1)Q_2 - \frac{1}{d_M}R_2,$$

$$\Omega_{44} = \sum_{j=1}^{s}\lambda_{ij}P_{2j} + P_{2i}A_i + A_i^T P_{2i} + (d_M - d_m + 1)Q_3 - \frac{1}{d_M}R_3,$$

$$\Omega_{55} = -Q_1 - \frac{1}{d_M}R_1, \; \Omega_{56} = -Q_2 - \frac{1}{d_M}R_2, \; \Omega_{66} = -Q_3 - \frac{1}{d_M}R_3.$$

In this case, a desired robustly asymptotically stable delta operator H_∞ filter is given in the form of (6.8) with parameters as follows: $G_i = P_{2i}^{-1}\Psi_i$, $K_i = P_{2i}^{-1}\Phi_i$.

Proof. *To facilitate further development, the following matrix expressions are introduced:*

$$P_i = \begin{bmatrix} P_{1i} & 0 \\ 0 & P_{2i} \end{bmatrix}, \quad Q = \begin{bmatrix} Q_1 & Q_2 \\ Q_2 & Q_3 \end{bmatrix}, \quad R = \begin{bmatrix} R_1 & R_2 \\ R_2 & R_3 \end{bmatrix},$$

and let $\Psi_i = P_{2i}G_i$ and $\Phi_i = P_{2i}K_i$, the LMIs (6.13) is changed to (6.27).

If the system mode set $S = \{1\}$, the jump DOS (6.2)-(6.5) is simplified into a general linear systems as follows:

$$\delta x(t_k) = (A + \Delta A(t_k))x(t_k) + (A_d + \Delta A_d(t_k))x(t_k - d(k)) + B\omega(t_k) \quad (6.28)$$

$$y(t_k) = Cx(t_k) + D\omega(t_k), \quad z(t_k) = Lx(t_k). \quad (6.29)$$

The following H_∞ filter will be designed:

$$\delta\hat{x}(t_k) = G\hat{x}(t_k) + Ky(t_k), \quad \hat{z}(t_k) = L\hat{x}(t_k), \quad (6.30)$$

The following error dynamics DOSs which are similar with (6.11)-(6.12) can be obtained:

$$\delta\tilde{x}(t_k) = A_f(t_k)\tilde{x}(t_k) + A_{df}(t_k)\tilde{x}(t_k - d(k)) + D_f\omega(t_k), \quad (6.31)$$

$$\tilde{z}(t_k) = L_f(t_k)\tilde{x}(t_k), \quad (6.32)$$

Corollary 6.11. *Given a scalar $\gamma > 0$, then the uncertain error dynamics DOS in (6.31) is robustly asymptotically stable with the disturbance attenuation γ, if there exist matrices $P > 0$, $Q > 0$ and $R > 0$, positive a scalar α such that the following LMI holds:*

$$\begin{bmatrix} (T-2)P + d_M R & PA_f & PA_{df} & PD_f & 0 & PM_f \\ * & \Xi(2,2) & PA_{df} + \frac{1}{d_M}R & PD_f & \alpha N_{1f}^T & PM_f \\ * & * & -Q - \frac{1}{d_M}R & 0 & \alpha N_{2f}^T & 0 \\ * & * & * & -\gamma^2 I & 0 & 0 \\ * & * & * & * & -\alpha I & \alpha H \\ * & * & * & * & * & -\alpha I \end{bmatrix} < 0, \quad (6.33)$$

with

$$\Xi(2,2) = PA_f + A_f^T P + (d_M - d_n + 1)Q - d_M^{-1}R + L_f^T L_f,$$

$$N_{1f} = \begin{bmatrix} N_1 & 0 \end{bmatrix}, \quad N_{2f} = \begin{bmatrix} N_2 & 0 \end{bmatrix}, \quad M_f = \begin{bmatrix} M^T & M^T \end{bmatrix}^T.$$

Corollary 6.12. *Consider the uncertain time-delay DOS (6.28)-(6.29), then the robust H_∞ filtering problem is solvable if there exist a scalar $\alpha > 0$, some matrices $P_1 > 0$, $P_2 > 0$, $Q_1 > 0$, $Q_2 > 0$, $Q_3 > 0$, $R_1 > 0$, $R_2 > 0$, $R_3 > 0$, Ψ and Φ such that the following LMI holds:*

$$\begin{bmatrix} \Gamma_{11} & d_M R_2 & P_1 A & 0 & P_1 A_d & 0 & P_1 B & 0 & P_1 M \\ * & \Gamma_{22} & \Gamma_{23} & \Psi & P_2 A_d & 0 & P_2 B + \Phi D & 0 & P_2 M \\ * & * & \Gamma_{33} & \Gamma_{34} & \Gamma_{35} & \frac{1}{d_M} R_2 & P_1 B & \alpha N_1^T & P_1 M \\ * & * & * & \Gamma_{44} & \Gamma_{45} & \frac{1}{d_M} R_3 & P_2 B + \Phi D & 0 & P_2 M \\ * & * & * & * & \Gamma_{55} & \Gamma_{56} & 0 & \alpha N_2^T & 0 \\ * & * & * & * & * & \Gamma_{66} & 0 & 0 & 0 \\ * & * & * & * & * & * & -\gamma^2 I & 0 & 0 \\ * & * & * & * & * & * & * & -\alpha I & \alpha H \\ * & * & * & * & * & * & * & * & -\alpha I \end{bmatrix} < 0, \quad (6.34)$$

with

$$\begin{aligned}
&\Gamma_{11} = (T-2)P_1 + d_M R_1, \ \Gamma_{35} = P_1 A_d + \frac{1}{d_M} R_1, \ \Gamma_{45} = P_2 A_d + \frac{1}{d_M} R_2 \\
&\Gamma_{33} = P_1 A + A^T P_1 + (d_M - d_m + 1)Q_1 - d_M^{-1} R_1, \ \Gamma_{56} = -Q_2 - \frac{1}{d_M} R_2 \\
&\Gamma_{22} = (T-2)P_2 + d_M R_3, \ \Gamma_{55} = -Q_1 - \frac{1}{d_M} R_1, \ \Gamma_{66} = -Q_3 - \frac{1}{d_M} R_3, \\
&\Gamma_{34} = A^T P_1 + C^T \Phi^T + \Psi^T + (d_M - d_m + 1)Q_2 - d_M^{-1} R_2, \\
&\Gamma_{23} = P_2 A + \Phi C + \Psi, \ \Gamma_{44} = P_2 A + A^T P_2 + (d_M - d_m + 1)Q_3 - d_M^{-1} R_3.
\end{aligned}$$

In this case, a desired robustly asymptotically stable delta operator H_∞ filter is given in the form of (6.30) with parameters as follows: $G = P_2^{-1}\Psi$ and $K = P_2^{-1}\Phi$.

6.4 Numerical Example

Example 6.13. The time-varying delayed uncertain DOS (6.2)-(6.5) with $i = 1, 2$ is considered. The parameters are described by

$$\begin{aligned}
&A_1 = \begin{bmatrix} -0.1 & -0.5 \\ 0.1 & -0.5 \end{bmatrix}, \ A_{d1} = \begin{bmatrix} 0.1 & 0.1 \\ 0.1 & -0.3 \end{bmatrix}, \ M_1 = \begin{bmatrix} 0.01 \\ 0.01 \end{bmatrix}, \\
&C_1 = \begin{bmatrix} 0.1 \ 0 \end{bmatrix}, \ D_1 = \begin{bmatrix} 0.1 \end{bmatrix}, \ H_1 = \begin{bmatrix} 0.02 \end{bmatrix}, \\
&L_1 = \begin{bmatrix} 0.1 \ 0.1 \end{bmatrix}, \ N_{11} = \begin{bmatrix} 0.03 \ 0.01 \end{bmatrix}, \ N_{21} = \begin{bmatrix} 0.02 \ 0.02 \end{bmatrix}, \\
&A_2 = \begin{bmatrix} -0.2 & 0.4 \\ 0.2 & -0.4 \end{bmatrix}, \ A_{d2} = \begin{bmatrix} -0.2 & 0.2 \\ 0.2 & -0.4 \end{bmatrix}, \ M_2 = \begin{bmatrix} 0.02 \\ 0.02 \end{bmatrix} \\
&C_2 = \begin{bmatrix} 0.2 \ 0 \end{bmatrix}, \ D_2 = \begin{bmatrix} 0.2 \end{bmatrix}, \ H_2 = \begin{bmatrix} 0.01 \end{bmatrix}, \\
&L_2 = \begin{bmatrix} 0.2 \ 0.2 \end{bmatrix},, \ N_{12} = \begin{bmatrix} 0.01 \ 0.02 \end{bmatrix}, \ N_{22} = \begin{bmatrix} 0.02 \ 0.01 \end{bmatrix}.
\end{aligned}$$

In order to design a robust stable delta operator filter such that the filtering error system is stable and satisfies a prescribed H_∞ performance level. First, we choose T $= 0.01$ and $d_m = 1$. Then, using the Matlab LMI Control Toolbox to solve the LMIs in (6.27), the solution is obtained as follows:

$$P_{11} = \begin{bmatrix} 0.1883 & -0.0625 \\ -0.0625 & 0.3334 \end{bmatrix}, \ P_{21} = \begin{bmatrix} 0.1858 & -0.0517 \\ -0.0517 & 0.3328 \end{bmatrix},$$

$$P_{12} = \begin{bmatrix} 0.1858 & -0.0517 \\ -0.0517 & 0.3328 \end{bmatrix}, \quad P_{22} = \begin{bmatrix} 0.0281 & 0.0159 \\ 0.0159 & 0.0108 \end{bmatrix},$$

$$Q_1 = \begin{bmatrix} 0.0072 & -0.0062 \\ -0.0062 & 0.024 \end{bmatrix}, \ Q_2 = 10^{-5} \begin{bmatrix} 0.4689 & -0.0637 \\ -0.0637 & 0.1447 \end{bmatrix},$$

$$Q_3 = 10^{-3} \begin{bmatrix} 0.1344 & -0.0163 \\ -0.0163 & 0.006 \end{bmatrix}, \ R_1 = \begin{bmatrix} 0.0285 & -0.0285 \\ -0.0285 & 0.0502 \end{bmatrix},$$

$$R_2 = 10^{-4} \begin{bmatrix} 0.135 & 0.0479 \\ 0.0479 & 0.0961 \end{bmatrix}, \ R_3 = 10^{-4} \begin{bmatrix} 0.2346 & 0.0048 \\ 0.0048 & 0.2069 \end{bmatrix},$$

$$\Psi_1 = \begin{bmatrix} -0.0664 & -0.0378 \\ -0.0427 & -0.0278 \end{bmatrix}, \ \Psi_2 = \begin{bmatrix} -0.0037 & -0.0053 \\ 0.0039 & -0.0036 \end{bmatrix},$$

$$\Phi_1 = \begin{bmatrix} 0.6106 \\ 0.3816 \end{bmatrix}, \ \Phi_2 = \begin{bmatrix} 0.0104 \\ -0.023 \end{bmatrix}, \ \alpha_1 = 0.0338, \ \alpha_2 = 0.2064.$$

And the largest d_M can be obtained as 6.12 and the smallest $\gamma = 0.7$. By Theorem 6.10, a desired H_∞ filter can be constructed as:

$$\delta\hat{x}(t_k) = G_1\hat{x}(t_k) + K_1 y(t_k), \qquad \delta\hat{x}(t_k) = G_2\hat{x}(t_k) + K_2 y(t_k).$$

where

$$G_1 = \begin{bmatrix} -0.4112 & -0.2371 \\ -0.1920 & -0.1203 \end{bmatrix}, \ K_1 = \begin{bmatrix} 3.7690 \\ 1.7322 \end{bmatrix},$$

$$G_2 = \begin{bmatrix} -1.9396 & 0.0023 \\ 3.2044 & -0.3353 \end{bmatrix}, \ K_2 = \begin{bmatrix} 9.0480 \\ -15.3729 \end{bmatrix}.$$

From this results, it is easy can be seen that the designed delta operator H_∞ filter meet the specified requirement.

6.5 Summary

In this chapter, A robust H_∞ filtering problem for DOSs with time-varying delays and Markovian jump parameters has been considered. A wider class of parameter uncertainties than norm-bounded parameter uncertainties has been described in this model. Sufficient conditions for the filter to satisfy prescribed H_∞ performance have been given in terms of LMIs, which can stabilize the system and guarantee a prescribed H_∞ performance on attenuation of all admissible parameter uncertainties. Numerical example has been given to illustrate the feasibility and effectiveness of the developed technique.

Chapter 7
A Novel DOKF Design and Convergence Analysis

7.1 Introduction

Kalman filter is a very powerful tool in control system design since 1960 when Kalman offered his original paper [132]. Kalman filter design problems for control systems have been given in [130, 183, 225, 256] and the references therein. The convergence analysis of an extended Kalman filter for nonlinear deterministic systems has been presented in [21]. Kalman filter was also used in GPS/SINS integrated navigation fusion algorithm in [36]. Kalman filter was also used for networked data fusion [262]. Kalman filter has also been widely used in many engineering and scientific areas, such as [7, 8, 92, 131, 229]. The control of an inverted pendulum has been studied and demonstrated in control laboratories in connection with the control of launching the rocket for decades. There is a long list of references related to inverted pendulum control for example [80, 184, 217, 224]. In this chapter, we use the inverted pendulum model in an experiment to illustrate the effectiveness of the designed Kalman filter and its implementation.

Goodwin constructed delta operator instead of traditional shift operator for sampling continuous systems with short sampling periods [82]. Many results on DOSs have been obtained, see for example [33, 137, 170, 254] A delta operator Riccati differential equation has been gained for discrete-time Kalman filter in [202], however, a detailed DOKF algorithm was not given in the results. A generalization structure of the transposed direct-form II filter structures in shift operator and delta operator was derived in [141]. A least square error criterion based procedure to design 2-D digital filters in delta-domain was presented in [134]. However, to the best of our knowledge, there are few papers considering the convergence of DOKFs up to now. Therefore, it is an interesting idea and stimulating to apply a DOKF to estimate the full-state of DOSs.

H. Yang et al.: Analysis and Synthesis of Delta Operator Systems, LNCIS 430, pp. 99–118.
springerlink.com © Springer-Verlag Berlin Heidelberg 2012

7.2 Problem Formulation

The following continuous system is considered as:

$$\dot{x}(t) = A_s x(t) + B_s u(t) + \omega(t) \tag{7.1}$$
$$y(t) = C_s x(t) + \upsilon(t) \tag{7.2}$$

where $x(t) \in R^n$ is the plant state, $u(t) \in R^p$ is the control input, $y(t) \in R^m$ is the measurement output, $A_s \in R^{n\times n}$ is the system matrix, $B_s \in R^{n\times p}$ is the input matrix, $C_s \in R^{m\times n}$ is the measurement matrix, $\omega(t) \in R^n$ is the input noise, $\upsilon(t) \in R^m$ is the measurement noise. The two noises are given as $\omega(t_k) \sim (0, Q_s)$ and $\upsilon(t_k) \sim (0, R_s)$, where Q_s and R_s are ordinary covariance matrices.

In this chapter, we consider a fast reaction case for system (7.1)-(7.2). Hence, a small sampling period should be used to meet the fast reaction case such as in an inverted-pendulum model. By using delta operator approach, the following discrete system is obtained:

$$\delta x(t_k) = Ax(t_k) + Bu(t_k) + D\omega(t_k) \tag{7.3}$$
$$y(t_k) = Cx(t_k) + \upsilon(t_k) \tag{7.4}$$

where A, B, C and D are constant matrices given as below:

$$A = \frac{e^{A_s \mathrm{T}} - I}{\mathrm{T}}, \quad B = \frac{1}{\mathrm{T}} \int_0^{\mathrm{T}} e^{A_s(\mathrm{T}-\tau)} B_s d\tau,$$
$$C = C_s, \quad D = \frac{1}{\mathrm{T}} \int_0^{\mathrm{T}} e^{A_s(\mathrm{T}-\tau)} d\tau.$$

The uncorrelated white noises $\omega(t_k) \in R^n$ and $\upsilon(t_k) \in R^m$ are given as follows:

$$\omega(t_k) \sim (0, Q), \qquad Q = Q_s \tag{7.5}$$
$$\upsilon(t_k) \sim (0, R), \qquad R = R_s \tag{7.6}$$

where Q and R are ordinary covariance matrices.

Before ending this section, the following lemmas will be used to develop our main results in the sequel.

Lemma 7.1. *[212] The continuous-time Kalman filter for the continuous system (7.1)-(7.2) is summarized as follows.*

i For two constants $t > 0$, $\ell > 0$ and the Kronecker delta function $\delta(t)$, the uncorrelated white noises $\omega(t) \in R^n$ and $\upsilon(t) \in R^m$ with zero mean and two ordinary covariance matrices Q_s and R_s satisfy

$$\boldsymbol{E}[\omega(t)\omega^T(\ell)] = Q_s \delta(t - \ell)$$

$$\begin{aligned} \boldsymbol{E}[\upsilon(t)\upsilon^T(\ell)] &= R_s\delta(t-\ell) \\ \boldsymbol{E}[\omega(t)\upsilon^T(\ell)] &= 0 \end{aligned}$$

ii *The initial state $x(0)$ is independent of $\omega(t) \in R^n$ and $\upsilon(t) \in R^m$*

$$\begin{aligned} \hat{x}(0) &= \boldsymbol{E}[x(0)] \\ P(0) &= \boldsymbol{E}[(\hat{x}(0) - x(0))(\hat{x}(0) - x(0))^T]. \end{aligned}$$

iii *The continuous-time Kalman filter equations for the continuous system (7.1)-(7.2) are given as*

$$\begin{aligned} K_s &= P(t)C_s^T R_s^{-1} \\ \dot{\hat{x}}(t) &= A_s\hat{x}(t) + B_s u(t) + K_s(y(t) - C_s\hat{x}(t)) \\ \dot{P}(t) &= -P(t)C_s^T R_s C_s P(t) + P(t)A_s^T + A_s P(t) + Q_s \end{aligned}$$

Lemma 7.2. *[212] The discrete-time Kalman filter is summarized as follows.*

i *For the ordinary covariance matrices Q_z and R_z, the dynamic discrete-time system is given by the following equations:*

$$\begin{aligned} x_k &= Fx_{k-1} + Gu_{k-1} + \omega_{k-1} \\ y_k &= Hx_k + \upsilon_k \\ \boldsymbol{E}[\omega_k\omega_j^T] &= Q_z\delta_{k-j} \\ \boldsymbol{E}[\upsilon_k\upsilon_j^T] &= R_z\delta_{k-j} \\ \boldsymbol{E}[\omega_k\upsilon_j^T] &= 0 \end{aligned}$$

ii *The Kalman filter is initialized as follows*

$$\begin{aligned} \hat{x}_0^+ &= \boldsymbol{E}[x_0] \\ P_0^+ &= \boldsymbol{E}[(x_0 - \hat{x}_0^+)(x_0 - \hat{x}_0^+)^T] \end{aligned}$$

iii *The discrete-time Kalman filter is given by the following equations, which are computed for each time step $k = 1, 2, \cdots$:*

$$\begin{aligned} P_k^- &= FP_{k-1}^+F^T + Q_z \\ K_k &= P_k^- H^T(HP_k^- H^T + R_z)^{-1} \\ &= P_k^+ H^T R_z^{-1} \\ \hat{x}_k^- &= F\hat{x}_{k-1}^+ + Gu_{k-1} \\ &= \textit{a priori state estimate} \\ \hat{x}_k^+ &= \hat{x}_k^- + K_k(y_k - H\hat{x}_k^-) \\ &= \textit{a posteriori state estimate} \\ P_k^+ &= (I - K_kH)P_k^-(I - K_kH)^T + K_kR_zK_k^T \\ &= [(P_k^-)^{-1} + H^T R_z^{-1} H]^{-1} \\ &= (I - K_kH)P_k^- \end{aligned}$$

7.3 Main Results

7.3.1 DOKF Design Approach

In this section, a discrete-time Kalman filter called DOKF for system (7.3)-(7.4) is introduced.

Theorem 7.3. *A single algorithm of Kalman filter named DOKF is given for the DOS (7.3)-(7.4) with white noise processes (7.5)-(7.6) and initial conditions:*

$$\hat{x}(0) = E[x(0)]$$
$$P(0) = E[(\hat{x}(0) - x(0))(\hat{x}(0) - x(0))^T]$$

which operate recursively performing a single cycle each time and a new set of measurements becomes available. The DOKF is given by the following equations:

$$K(t_k) = P(t_k)C^T(CP(t_k)C^T\mathrm{T} + R)^{-1} \tag{7.7}$$

$$\begin{aligned} \delta\hat{x}(t_k) &= A\hat{x}(t_k) + Bu(t_k) + K(t_k)[y(t_k) - C\hat{x}(t_k)] \\ &\quad - \mathrm{T}K(t_k)C[A\hat{x}(t_k) + Bu(t_k)] \end{aligned} \tag{7.8}$$

$$\begin{aligned} \delta P(t_k) &= -K(t_k)CP(t_k) + AP(t_k) + P(t_k)A^T \\ &\quad + Q + \mathrm{T}A(I - K(t_k)C\mathrm{T})P(t_k)A^T \\ &\quad - \mathrm{T}[AK(t_k)CP(t_k) + K(t_k)CP(t_k)A^T] \end{aligned} \tag{7.9}$$

Each iteration propagates the estimate from the time of last measurement is obtained to the current time.

Proof. Transforming the DOS (7.3)-(7.4), we have the following discrete system

$$x_k = Fx_{k-1} + Gu_{k-1} + \Lambda\omega_{k-1} \tag{7.10}$$

$$y_k = Hx_k + \upsilon_k \tag{7.11}$$

where $x_k = x(t_k)$, $\omega_{k-1} = \omega(t_{k-1})$, $\upsilon_k = \upsilon(t_k)$, $F = (I + A\mathrm{T})$, $G = B\mathrm{T}$, $\Lambda = D\mathrm{T}$ and $H = C$. The uncorrelated white noises $\omega_k \in R^n$ and $\upsilon_k \in R^m$ are given as follows:

$$\omega_k \sim (0, Q_z), \quad Q_z = \mathrm{T}Q$$
$$\upsilon_k \sim (0, R_z), \quad R_z = R/\mathrm{T}$$

By Lemma 7.2, the following discrete-time Kalman filter gain for system (7.10)-(7.11) is given as

$$K_k = P_k^- H^T (H P_k^- H^T + R_z)^{-1}$$

Letting $K(t_k) = K_k/\mathtt{T}$, it can be derived that

$$\begin{aligned} K_k &= P_k^- C^T (C P_k^- C^T + R/\mathtt{T})^{-1} \\ K(t_k) &= P_k^- C^T (C P_k^- C^T \mathtt{T} + R)^{-1} \end{aligned}$$

Using Lemma 7.2 again, the estimation-error covariances are

$$\begin{aligned} P_k^+ &= (I - K_k H) P_k^- \\ P_{k+1}^- &= F P_k^+ F^T + Q_z \end{aligned}$$

Note that P_{k+1}^- can be rewritten as

$$\begin{aligned} P_{k+1}^- &= (I + A\mathtt{T}) P_k^+ (I + A\mathtt{T})^T + \mathtt{T} Q \\ &= P_k^+ + \mathtt{T}(A P_k^+ + P_k^+ A^T + Q) + \mathtt{T}^2 A P_k^+ A^T \end{aligned}$$

Substituting P_k^- gives

$$\begin{aligned} P_{k+1}^- = {} & (I - K_k C) P_k^- + \mathtt{T} A (I - K_k C) P_k^- + \mathtt{T}[(I - K_k C) P_k^- A^T + Q] \\ & + \mathtt{T}^2 A (I - K_k H) P_k^- A^T \end{aligned}$$

Subtracting P_k^- from both sides and divided by $\mathtt{T}$, we have

$$\begin{aligned} \frac{P_{k+1}^- - P_k^-}{\mathtt{T}} = {} & -\frac{K_k C P_k^-}{\mathtt{T}} + A P_k^- - A K_k C P_k^- + P_k^- A^T - K_k C P_k^- A^T + Q \\ & + \mathtt{T} A (I - K_k H) P_k^- A^T \end{aligned}$$

Letting $P(t_k) = P_k^-$ and $\delta P(t_k) = (P_{k+1}^- - P_k^-)/\mathtt{T}$, there exists

$$\delta P(t_k) = -K(t_k) C P(t_k) + A P(t_k) + P(t_k) A^T + Q + \gamma(\mathtt{T}) \tag{7.12}$$

with

$$\gamma(\mathtt{T}) = \mathtt{T}[A(I - K(t_k) C \mathtt{T}) P(t_k) A^T - A K(t_k) C P(t_k) - K(t_k) C P(t_k) A^T])$$

From Lemma 7.2, the Kalman filter for $\hat{x}(t_k)$ is

$$\begin{aligned} \hat{x}_k^- &= F \hat{x}_{k-1}^+ + G u_{k-1} \\ \hat{x}_k^+ &= \hat{x}_k^- + K_k (y_k - H \hat{x}_k^-) \end{aligned}$$

The above measurement update equation is rewritten as

$$\begin{aligned}\hat{x}_k^+ &= F\hat{x}_{k-1}^+ + Gu_{k-1} + K_k(y_k - HF\hat{x}_{k-1}^+ - HGu_{k-1}) \\ &= K_k(y_k - C\hat{x}_{k-1}^+ - \mathtt{T}CA\hat{x}_{k-1}^+ - \mathtt{T}CBu_{k-1}) \\ &\quad +(I + A\mathtt{T})\hat{x}_{k-1}^+ + B\mathtt{T}u_{k-1}\end{aligned}$$

Subtracting x_{k-1}^+ from both sides, divided by $\mathtt{T}$, we obtain

$$\begin{aligned}\frac{\hat{x}_k^+ - \hat{x}_{k-1}^+}{\mathtt{T}} &= A\hat{x}_{k-1}^+ + Bu_{k-1} + K(t_k)(y_k - C\hat{x}_{k-1}^+ \\ &\quad -\mathtt{T}CA\hat{x}_{k-1}^+ - \mathtt{T}CBu_{k-1})\end{aligned}$$

Hence, there exists

$$\begin{aligned}\delta\hat{x}(t_k) &= A\hat{x}(t_k) + Bu(t_k) + K(t_k)[y(t_k) - C\hat{x}(t_k)] \\ &\quad -\mathtt{T}K(t_k)C[A\hat{x}(t_k) + Bu(t_k)]\end{aligned}$$

Lastly, we have gotten the differential equations (7.7)-(7.9) which can be used to integrate the state estimate in DOKF. This completes the proof.

Remark 7.4. Note that (7.12) for $P(t_k)$ is called a delta operator Riccati differential equation, which can be used to compute the estimation-error covariance for the DOKF (7.7)-(7.9). Similar delta operator Riccati differential equation result has also been derived in [202]. However, the results in [202] did not provide detailed discrete-time Kalman filter for DOSs.

Remark 7.5. The DOKF in Theorem 7.3 can express both continuous-time and discrete-time cases. On one hand, we note from the characteristic of DOSs and the following equations

$$\begin{aligned}K_s &= \lim_{\mathtt{T}\to 0} P(t_k)C^T(CP(t_k)C^T\mathtt{T} + R)^{-1} = P(t)C_s^T R_s^{-1} \\ \dot{\hat{x}}(t) &= \lim_{\mathtt{T}\to 0} \delta\hat{x}(t_k) = A_s\hat{x}(t) + B_su(t) + K_s(y(t) - C_s\hat{x}(t)) \\ \dot{P}(t) &= \lim_{\mathtt{T}\to 0} \delta P(t_k) = -P(t)C_s^T R_s C_s P(t) + A_sP(t) + P(t)A_s^T + Q_s\end{aligned}$$

that when the sampling period $\mathtt{T}$ tends to 0 in system (7.3)-(7.4), the DOKF (7.7)-(7.9) in Theorem 7.3 reduces to the standard continuous-time Kalman filter in Lemma 7.1 for continuous system (7.1)-(7.2). On the other hand, a traditional discrete-time Kalman filter is obtained when sampling period $\mathtt{T} = 1$ from the proof of Theorem 7.3.

7.3.2 Convergence Analysis

In this section, we give an approach for setting up the convergence analysis of the DOKF (7.7)-(7.9) for system (7.3)-(7.4) which is equivalent to system

(7.10)-(7.11). By considering the state feedback control law $u_k = Kx_k$, system (7.10)-(7.11) is changed to the following closed-loop system:

$$x_{k+1} = F_K x_k + \varpi_k \tag{7.13}$$
$$y_k = Hx_k + \upsilon_k \tag{7.14}$$

where $\varpi_k = \Lambda\omega_k$ and $F_K = F + GK$, which implies that F_K is a bounded nonsingular matrix. According to system (7.13)-(7.14), the DOKF (7.7)-(7.9) can be written as the following form:

1) measurement update

$$K_{k+1} = P_{k+1}^- H^T (HP_{k+1}^- H^T + R)^{-1} \tag{7.15}$$
$$\hat{x}_{k+1}^+ = \hat{x}_{k+1}^- + K_{k+1}(y_{k+1} - H\hat{x}_{k+1}^-) \tag{7.16}$$
$$P_{k+1}^+ = (I - K_{k+1}H)P_{k+1}^- \tag{7.17}$$

2) time update

$$\hat{x}_{k+1}^- = F_K \hat{x}_k^+ \tag{7.18}$$
$$P_{k+1}^- = F_K P_k^+ F_K^T + Q \tag{7.19}$$

It has been shown form the proof of Theorem 7.3 that (7.7)-(7.9) are equivalent to (7.15)-(7.19) with $x(t_k) = x_k$. Hence, the objective in this part is to show $\hat{x}_k^+ \to x_k$ as $k \to \infty$.

Firstly, we will show that a Lyapunov functional set $\{V_k|_{k=1,2,\cdots}\}$ is a decreasing sequence.

Theorem 7.6. *We let*

$$\tilde{x}_{k+1}^+ = x_{k+1} - \hat{x}_{k+1}^+$$
$$\tilde{x}_{k+1}^- = x_{k+1} - \hat{x}_{k+1}^-$$

be the state error estimation and the state prediction error vectors, respectively. The Lyapunov functional set

$$\{V_k = (\tilde{x}_k^+)^T (P_k^+)^{-1} \tilde{x}_k^+ |_{k=1,2,\cdots}\}$$

is a decreasing sequence, if the following inequalities

$$-R_{k+1}^{-1} + R^{-1} H P_{k+1}^+ H^T R^{-1} \leq 0 \tag{7.20}$$

and

$$F_K^T (F_K P_k^+ F_K^T + Q)^{-1} F_K - (P_k^+)^{-1} \leq 0 \tag{7.21}$$

hold.

Proof. Choose a Lyapunov functional as follows:

$$V_{k+1}^{+} = (\tilde{x}_{k+1}^{+})^T (P_{k+1}^{+})^{-1} \tilde{x}_{k+1}^{+} \tag{7.22}$$

Setting $e_{k+1} = y_{k+1} - H\hat{x}_{k+1}^{-}$, we have

$$e_{k+1} \approx Hx_{k+1} - H\hat{x}_{k+1}^{-} = H\tilde{x}_{k+1}^{-}$$

and

$$\tilde{x}_{k+1}^{-} \approx F_K x_k - F_K \hat{x}_k^{-} = F_K \tilde{x}_k^{-}$$

By subtracting both sides of (7.16) from x_{k+1}, we obtain

$$\tilde{x}_{k+1}^{+} = \tilde{x}_{k+1}^{-} - P_{k+1}^{-} H^T (H P_{k+1}^{-} H^T + R)^{-1} e_{k+1}$$

Based on (7.15), there exist

$$\begin{aligned}
K_{k+1}(HP_{k+1}^{-}H^T + R) &= P_{k+1}^{-}H^T \\
K_{k+1}HP_{k+1}^{-}H^T + K_{k+1}R &= P_{k+1}^{-}H^T \\
K_{k+1}HP_{k+1}^{-}H^T - P_{k+1}^{-}H^T &= -K_{k+1}R \\
(K_{k+1}H - I)P_{k+1}^{-}H^T &= -K_{k+1}R
\end{aligned}$$

Considering (7.17), we have

$$\begin{aligned}
-P_{k+1}^{+}H^T &= -K_{k+1}R \\
P_{k+1}^{+}H^T R^{-1} &= K_{k+1}
\end{aligned}$$

Substituting K_{k+1} into (7.17), it is obtained

$$P_{k+1}^{+} = (I - P_{k+1}^{+} H^T R^{-1} H) P_{k+1}^{-}$$

Pre-multiplying $(P_{k+1}^{+})^{-1}$ and post-multiplying $(P_{k+1}^{-})^{-1}$ to the above equation, we get

$$(P_{k+1}^{+})^{-1} = (P_{k+1}^{-})^{-1} + H^T R^{-1} H \tag{7.23}$$

Following above equations, we have

$$\begin{aligned}
V_{k+1}^{+} &= (\tilde{x}_{k+1}^{-} - P_{k+1}^{+} H^T R^{-1} e_{k+1})^T (P_{k+1}^{+})^{-1} \\
&\quad \times (\tilde{x}_{k+1}^{-} - P_{k+1}^{+} H^T R^{-1} e_{k+1}) \\
&= (\tilde{x}_{k+1}^{-})^T (P_{k+1}^{+})^{-1} \tilde{x}_{k+1}^{-} - 2(\tilde{x}_{k+1}^{-})^T H^T R^{-1} e_{k+1} \\
&\quad + e_{k+1}^T R^{-1} H P_{k+1}^{+} H^T R^{-1} e_{k+1} \\
&= (\tilde{x}_{k+1}^{-})^T (P_{k+1}^{-})^{-1} \tilde{x}_{k+1}^{-} + e_{k+1}^T (-R^{-1} + R^{-1} H P_{k+1}^{+} H^T R^{-1}) e_{k+1}
\end{aligned}$$

In the following, V_{k+1}^- is defined as

$$\begin{aligned} V_{k+1}^- &= (\tilde{x}_{k+1}^-)^T (P_{k+1}^-)^{-1} \tilde{x}_{k+1}^- \\ &\approx (\tilde{x}_k^+)^T F_K^T (P_{k+1}^-)^{-1} F_K \tilde{x}_k^+ \\ &= (\tilde{x}_k^+)^T F_K^T (F_K P_k^+ F_K^T + Q)^{-1} F_K \tilde{x}_k^+ \end{aligned}$$

That is,

$$\begin{aligned} V_{k+1}^+ - V_k^+ &= e_{k+1}^T (-R^{-1} + R^{-1} H P_{k+1}^+ H^T R^{-1}) e_{k+1} \\ &+ (\tilde{x}_k^+)^T [F_K^T (F_K P_k^+ F_K^T + Q_k)^{-1} F_K - (P_k^+)^{-1}] \tilde{x}_k^+ \end{aligned}$$

A decreasing sequence $\{V_k|_{k=1,2,\cdots}\}$ means that

$$V_{k+1}^+ - V_k^+ = V_{k+1}^+ - V_{k+1}^- + V_{k+1}^- - V_k^+ \leq 0$$

A sufficient condition to ensure that is

$$\begin{aligned} -R^{-1} + R^{-1} H P_{k+1}^+ H^T R^{-1} &\leq 0 \\ F_K^T (F_K P_k^+ F_K^T + Q_k)^{-1} F_K - (P_k^+)^{-1} &\leq 0 \end{aligned}$$

This proof is finished.

Even if $\{V_k|_{k=1,2,\cdots}\}$ is a decreasing sequence, the convergence of the DOKF (7.7)-(7.9) is not ensured in Theorem 7.6.

Assume that system (7.13)-(7.14) satisfies the N-observability rank condition and is N-observable. We define the following manner:

$$\begin{aligned} Y_k = \begin{bmatrix} y_{k-N} \\ y_{k-N+1} \\ \vdots \\ y_k \end{bmatrix} &\approx \begin{bmatrix} H F_K^{k-N} x_k \\ H F_K^{k-N+1} x_k \\ \vdots \\ H x_k \end{bmatrix} = \begin{bmatrix} H F_K^{k-N} \\ H F_K^{k-N+1} \\ \vdots \\ H \end{bmatrix} x_k \\ &= \mathcal{O}_k(k-N, k) x_k \end{aligned} \tag{7.24}$$

Due to F_K is a bounded nonsingular matrix, there exist positive real numbers η_1 and η_2 such that for all $k \geq N$ and finite $N \geq 0$, the following Gramian matrix inequality holds

$$\eta_1 I_n \leq L_{k-N}(\mathcal{O}, \mathcal{R}) \leq \eta_2 I_n \tag{7.25}$$

with

$$\begin{aligned} \mathcal{R}_k(k-N, k) &= \mathrm{diag}\{R^{-1}, R^{-1}, \cdots, R^{-1}\} \\ L_{k-N}(\mathcal{O}, \mathcal{R}) &= \mathcal{O}_k^T(k-N, k) \mathcal{R}_k(k-N, k) \mathcal{O}_k(k-N, k) \end{aligned}$$

Then, the DOS (7.3)-(7.4) (and its associated the DOKF (7.7)-(7.9) for $\hat{x}_k^-$ and $\hat{x}_k^+$ sufficiently close to the true state x_k) is said to be reconstructible.

Lemma 7.7. *[21] If we assume that system (7.10)-(7.11) is reconstructible with (7.25), it is obtained that*

$$\lim_{k\to\infty} \lambda_{\min}\{(P_k^+)^{-1}\} = \infty \tag{7.26}$$

and

$$\sup \lim_{k\to\infty} \frac{\lambda_{\max}\{(P_k^+)^{-1}\}}{\lambda_{\min}\{(P_k^+)^{-1}\}} < \infty \tag{7.27}$$

It is easy to get Lemma 7.7 from the third lemma in [21] which includes the linear case such as system (7.10)-(7.11).

Theorem 7.8. *Suppose that (7.20), (7.21) and (7.25) hold, then the DOKF (7.7)-(7.9) used as an observer for the DOS (7.3)-(7.4), ensures that*

$$\lim_{k\to\infty} (x_k - \hat{x}_k^+) = 0$$

Proof. Because F_K is a bounded nonsingular matrix, it is easy to show that

$$(P_{k+1}^+)^{-1} = L_{k+1}(\mathcal{O}, \mathcal{R}) + \Psi(0, k)$$

with

$$\begin{aligned}\Psi(0, k) &= (F_K^{-k-1})^T P_0^{-1} F_K^{-k-1} \\ L_{k+1}(\mathcal{O}, \mathcal{R}) &= \mathcal{O}_k^T(1, k+1)\mathcal{R}_k(1, k+1)\mathcal{O}_k(1, k+1)\end{aligned}$$

On a horizon time kN, if the constructibility Gramian matrix

$$\mathcal{O}_k^T(1, kN)\mathcal{R}_k(1, kN)\mathcal{O}_k(1, kN)$$

is decomposed into k block matrices of horizon N, we obtain

$$\begin{aligned}(P_{kN}^+)^{-1} = \sum_{i=1}^{k} \big[&\mathcal{O}_k^T((i-1)N+1, iN) \\ &\times \mathcal{R}_k((i-1)N+1, iN)\mathcal{O}_k((i-1)N+1, iN)\big] + \Psi(0, kN-1)\end{aligned}$$

Since $(P_{kN}^+)^{-1} - \Psi(0, kN-1)$ is the sum of k constructibility Gramian matrix and under (7.25), it can be obtained that

$$0 < k\eta_1 \le \lambda\{(P_{kN}^+)^{-1} - \Psi(0, kN-1)\} \le k\eta_2 \tag{7.28}$$

where λ represents the eigenvalue symbol of $(P_{kN}^+)^{-1} - \Psi(0, kN-1)$. It is easy to see from (7.28) that $\lambda\{(P_{kN}^+)^{-1} - \Psi(0, kN-1)\}$ is bounded by $k\eta_1$ and $k\eta_2$. Therefore, we have that $k\eta_1 \to \infty$ and $k\eta_2 \to \infty$ as $k \to \infty$.

On the other hand, there exists

$$\frac{V_k^+}{tr\{(P_k^+)^{-1}\}} > \frac{\lambda_{\min}\{(P_k^+)^{-1}\}(\tilde{x}_k^+)^T\tilde{x}_k^+}{n\lambda_{\max}\{(P_k^+)^{-1}\}} > 0$$

According to (7.26) in Lemma 7.7, we obtain

$$\lim_{k\to\infty} tr\{(P_k^+)^{-1}\} = \infty.$$

Then, it is obtained that

$$\lim_{k\to\infty} \frac{\lambda_{\min}\{(P_k^+)^{-1}\}(\tilde{x}_k^+)^T\tilde{x}_k^+}{n\lambda_{\max}\{(P_k^+)^{-1}\}} = 0.$$

Thus, (7.27) yields to

$$\lim_{k\to\infty} \tilde{x}_k^+ = 0.$$

which means that $\lim_{k\to\infty}(x_k - \hat{x}_k^+) = 0$. This completes the proof.

Remark 7.9. It is well known that state feedback control method can not be used in general for the reason of that full states are hardly available such as in inverted pendulum model. Kalman filter as a state observer proves an effective technique to estimate the system states. In some practical systems [281], a continuous-time Kalman filter instead of the discrete-time case was used to accurately estimate system state. However, continuous-time Kalman filter can not be applied in a realistic situation for the application of digital computers. Therefore, we study how to apply DOKF for estimating system state of inverted pendulum model in the following experiment.

Remark 7.10. The main contribution of this chapter is the designed DOKF and its convergence analysis. Some results have been investigated on the problem of Kalman filter in delta domain such as in [171]. The convergence analysis of this kind of DOKFs is also an interesting idea to consider. The inverted pendulum control problem is challenging because it may go unstable for low sampling frequencies. In next section, some experimental results on inverted pendulum control problem will be given to show the effectiveness of the designed DOKF in this chapter.

7.4 Experimental Results

In order to show the effectiveness of the designed DOKF in Theorem 7.3 and convergence analysis in Theorems 7.6-7.8, we consider the control of an inverted pendulum model via a practical experiment.

Experiment: The difference between the real-time simulation and real-time experiments is that the plant model of the inverted pendulum model is replaced by D/A block Dac9, A/D block Adc9 and the real inverted-pendulum system. The real inverted-pendulum model considered here consists of the cart that is moving on a rail and the pole that is mounted on the cart via the pole shaft so that the pendulum can swing freely in the vertical plane. The inverted-pendulum model used in this experiment is shown in Fig. 7.1 The translation of the cart is enabled by the dc motor. The shaft of the

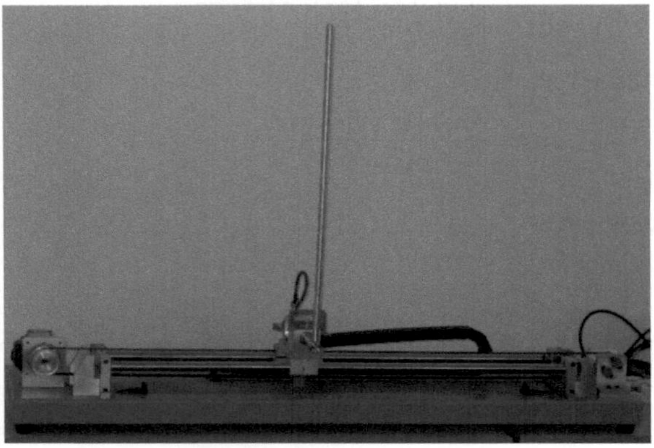

Fig. 7.1 Inverted pendulum construction.

motor is connected with the cart by a thin steel wire. Control algorithm and data acquisition and visualization are realized with MathWorks Matlab, Simulink, and Real-Time xPC target products on a personal computer with the 12-bit AD/DA converter. The two blocks Dac9 and Adc9 are the driver of the A/D and D/A channels in the embedded system and are designed in Matlab S-Function. Because of the limited pendulum-rail length (0.5 m), the pendulum-cart movement is also limited (0.25 m). The diagram of the real-time experiment is shown in Fig. 7.2.

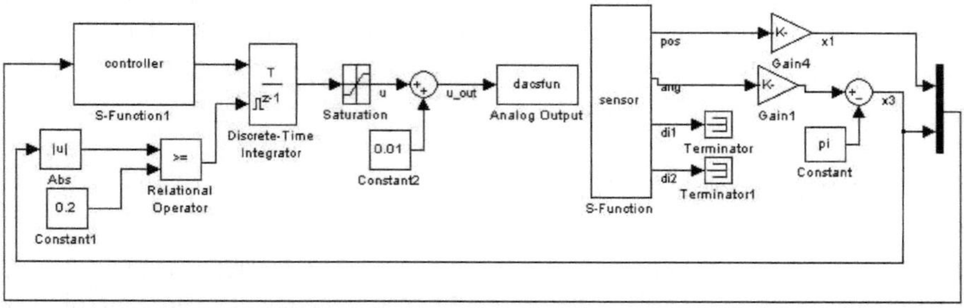

Fig. 7.2 Inverted pendulum diagram.

The physical structure of the inverted pendulum on a cart is shown in Fig. 7.3.

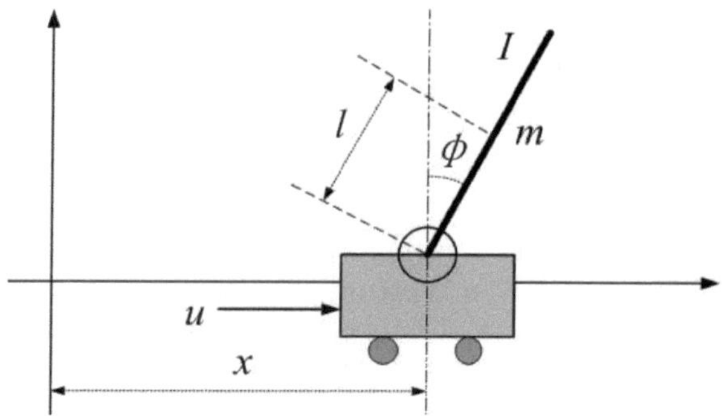

Fig. 7.3 Inverted pendulum Model.

In addition, the surface is friction free for model building simplicity. Symbols meaning in Fig. 7.3 are shown in the Table 7.1, where m, l, I and g are constants, ϕ, $\boldsymbol{x}$ and $\boldsymbol{u}$ are variables.We let state variables be $\boldsymbol{x}$, $\dot{\boldsymbol{x}}$, ϕ and $\dot{\phi}$. They correspond to the horizontal position, horizontal velocity of the cart, angle and angle velocity of the pole. Because the translation of the cart is enabled by a dc motor, acceleration variable has to be used as a control input variable, which means $\boldsymbol{u} = -\ddot{\boldsymbol{x}}$, where "$-$" is based on the turning direction of dc motor.

Table 7.1 parameters used in the inverted pendulum model

Symbol	Meaning
m	Mass of the pendulum rod
l	Length from the pendulum's axis to its centroid
I	Inertia of the pendulum rod
g	Gravitational constant
ϕ	Angle between the pendulum rod and the vertical
$\boldsymbol{x}$	Horizontal displacement of the cart
$\boldsymbol{u}$	Control input

The equation of motion can be obtained as follows:

$$\left(I + ml^2\right)\ddot{\phi} - mgl\phi = ml\ddot{\boldsymbol{x}}$$

For the swing link with uniform distribution quality, we have the following equation of inertia as

$$I = \frac{1}{3}ml^2$$

Based on above two equations, we have

$$\left(\frac{1}{3}ml^2 + ml^2\right)\ddot{\phi} - mgl\phi = ml\ddot{\boldsymbol{x}}$$

which is changed to

$$\ddot{\phi} = \frac{3g}{4l}\phi + \frac{3}{4l}\ddot{\boldsymbol{x}}$$

Let the state variables be $\boldsymbol{x} = x_1(t)$, $\dot{\boldsymbol{x}} = x_2(t)$, $\phi = x_3(t)$ and $\dot{\phi} = x_4(t)$, and input variable be $u(t) = \boldsymbol{u}$. Letting $g = 9.81\text{m/s}^2$ and $l = 0.25\text{m}$, the following continuous system is obtained to describe the inverted pendulum model:

$$\dot{x}(t) = A_s x(t) + B_s u(t) + \omega(t) \tag{7.29}$$

$$y(t) = C_s x(t) + \upsilon(t) \tag{7.30}$$

where

$$A_s = \begin{bmatrix} 0 & 1 & 0 & 0 \\ 0 & 0 & 0 & 0 \\ 0 & 0 & 0 & 1 \\ 0 & 0 & 29.4 & 0 \end{bmatrix}, \; B_s = \begin{bmatrix} 0 \\ -1 \\ 0 \\ 3 \end{bmatrix}, \; C_s = \begin{bmatrix} 1 & 0 & 0 & 0 \\ 0 & 0 & 1 & 0 \end{bmatrix}.$$

It can be seen from $|sI - A_s| = 0$ that there exist $s_1 = s_2 = 0$, $s_3 = -5.4222$ and $s_4 = 5.4222$. The linear system (7.29)-(7.30) of the inverted pendulum model is unstable. Furthermore, the controllability matrix is

$$\mathbf{Q}_s = \begin{bmatrix} B_s & A_s B_s & A_s^2 B_s & A_s^3 B_s \end{bmatrix} \tag{7.31}$$

and $rank(\mathbf{Q}_s) = 4$, so the linear model of the inverted pendulum is controllable.

In the following, the linear-quadratic regulator(LQR) is used for the calculation of the optimal gain matrix K such that the following state-feedback control law

$$u(t) = Kx(t)$$

minimizes the cost function

$$J_{LQG} = \int (x^T(t)\mathrm{Q}x(t) + u^T(t)\mathrm{R}u(t))dt.$$

For choosing the appropriate matrices as

$$\mathrm{Q} = \begin{bmatrix} 1000 & 0 & 0 & 0 \\ 0 & 0 & 0 & 0 \\ 0 & 0 & 200 & 0 \\ 0 & 0 & 0 & 0 \end{bmatrix}, \; \mathrm{R} = [1]$$

we can adjust the state-feedback LQR controller K as follows

$$K = \begin{bmatrix} -31.623 & -20.151 & -72.718 & -13.155 \end{bmatrix} \tag{7.32}$$

The stability of the inverted-pendulum model is guaranteed for the reason of that the LQR design for the proposed state-feedback controller is used.

The inverted-pendulum model is a continuous system with a rapid response. The following results will show that the inverted-pendulum system is only fit for small sampling periods. Firstly, by using the state-feedback controller (7.32), choosing the sampling period $\mathtt{T} = 0.2$s and the initial condition $u(t_0) = -4.4$, the control input of system (7.29)-(7.30) is given in Fig. 7.4.

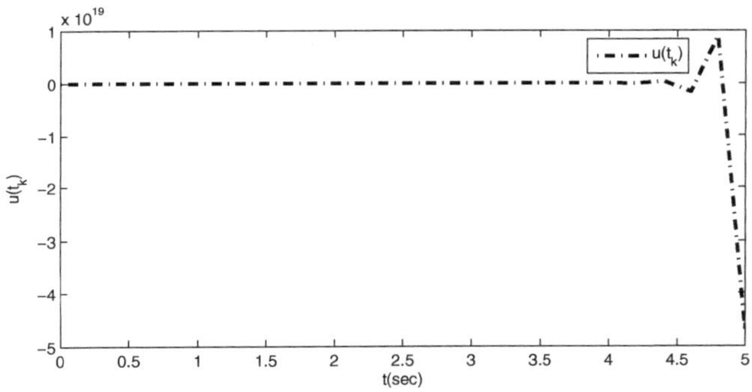

Fig. 7.4 Input of the inverted-pendulum model for sampling period is 0.2s.

The state vector of the closed-loop system (7.29)-(7.30) with initial conditions $x(t_0) = [0\ 0.2\ 0\ 0]$ is given in Fig. 7.5.

It is shown from Fig. 7.5 that the closed-loop system (7.29)-(7.30) with sampling period $\mathtt{T} = 0.2$s is divergent. Then, we choose a smaller sampling period $\mathtt{T} = 0.02$s. The control input of the closed-loop system (7.29)-(7.30) with an initial condition $u(t_0) = -4.4$ is given in Fig. 7.6.

The state vector of the closed-loop system (7.29)-(7.30) with initial conditions $x(t_0) = [0\ 0.2\ 0\ 0]$ is given in Fig. 7.7.

It is shown from Fig. 7.7 that the closed-loop system (7.29)-(7.30) with sampling period $\mathtt{T} = 0.02$s is convergent.

From above simulation results, we have that the sampling period $\mathtt{T} = 0.02$s is fit for the inverted-pendulum system. By sampling system (7.29)-(7.30), the following DOS is obtained

$$\delta x(t_k) = Ax(t_k) + Bu(t_k) + \omega(t_k) \tag{7.33}$$

$$y(t_k) = Cx(t_k) + \upsilon(t_k) \tag{7.34}$$

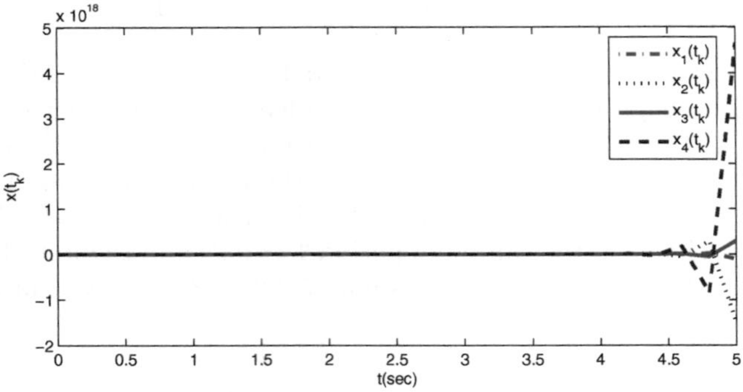

Fig. 7.5 States of the inverted-pendulum model for sampling period is 0.2s.

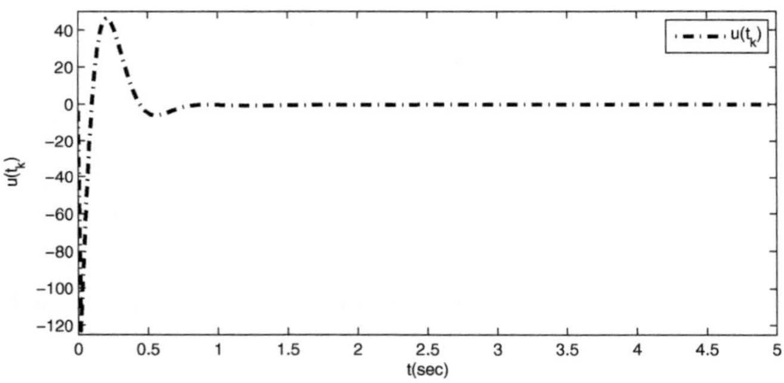

Fig. 7.6 Input of the inverted-pendulum model for sampling period is 0.02s.

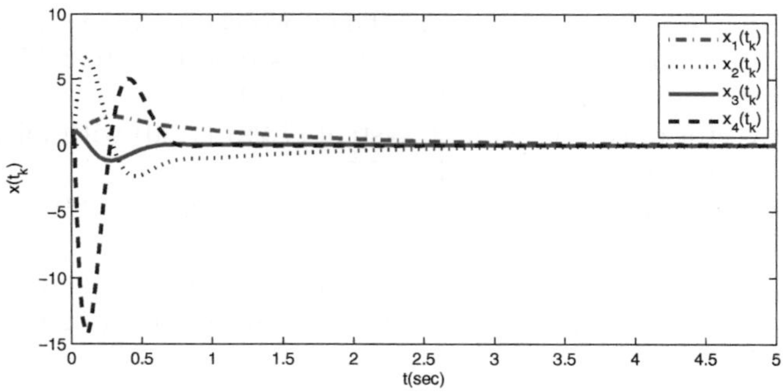

Fig. 7.7 States of the inverted-pendulum model for sampling period is 0.02s.

where

$$A = \begin{bmatrix} 0 & 1 & 0 & 0 \\ 0 & 0 & 0 & 0 \\ 0 & 0 & 0.2943 & 1.0020 \\ 0 & 0 & 29.4577 & 0.2943 \end{bmatrix}, \quad B = \begin{bmatrix} -0.0100 \\ -1.0000 \\ 0.0300 \\ 3.0059 \end{bmatrix}, \quad C = \begin{bmatrix} 1 & 0 & 0 & 0 \\ 0 & 0 & 1 & 0 \end{bmatrix}.$$

In the following, we give the following simulation results to show the effectiveness of the designed DOKF and its convergence analysis. Consider the initial condition of error covariance matrix $P(t_k)$ as follows:

$$P(t_0) = \begin{bmatrix} 1.5 & 2 & 1 & 2 \\ 2 & 1 & 0 & 3 \\ 1 & 0 & 2 & 1 \\ 2 & 3 & 1 & 1 \end{bmatrix}. \tag{7.35}$$

The estimated state $\hat{x}(t_k)$ in every sampling instant t_k can be gotten for system (7.33)-(7.34), with the initial condition $\hat{x}(0) = [0 \ \ -0.2 \ \ 0 \ \ 0]$. The estimated state vector is given in Fig. 7.8, in which $\hat{x}(t_k)$ consists of $xe_1(t_k)$, $xe_2(t_k)$, $xe_3(t_k)$ and $xe_4(t_k)$.

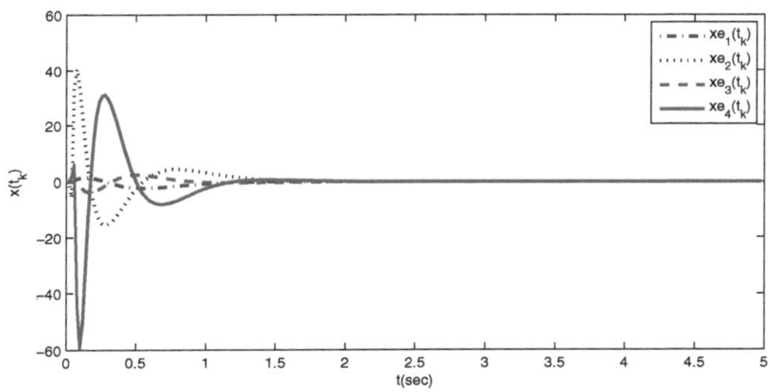

Fig. 7.8 Numerical simulation of estimated state vector $\hat{x}(t_k)$.

The simulation results of the biggest singular value of error covariance matrix $P(t_k)$ denoted by $\|P(t_k)\|$ is given in Fig. 7.9.

Fig. 7.10 shows the convergence rate of the state error norm $\|x(t_k)-\hat{x}(t_k)\|$.

Fig. 7.11 shows the simulation results of the control input for initial condition $u(0) = 4.4$.

For initial condition: $y(0) = [0 \ 0]^T$, the simulation results of the output for system (7.33)-(7.34) are given in Fig. 7.12, where $y(t_k)$ is denoted by $y_1(t_k)$ and $y_2(t_k)$.

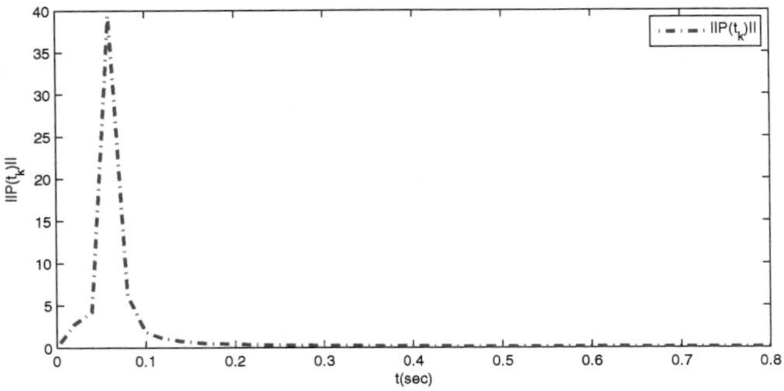

Fig. 7.9 Numerical simulation of rate of convergence of $\|P(t_k)\|$.

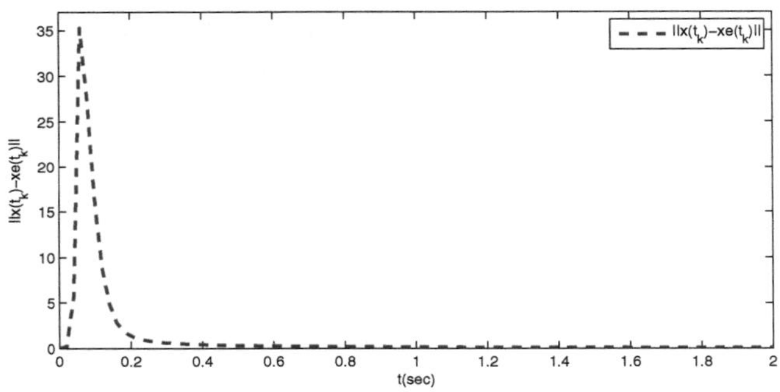

Fig. 7.10 Numerical simulation of rate of convergence of $\|x(t_k) - \hat{x}(t_k)\|$.

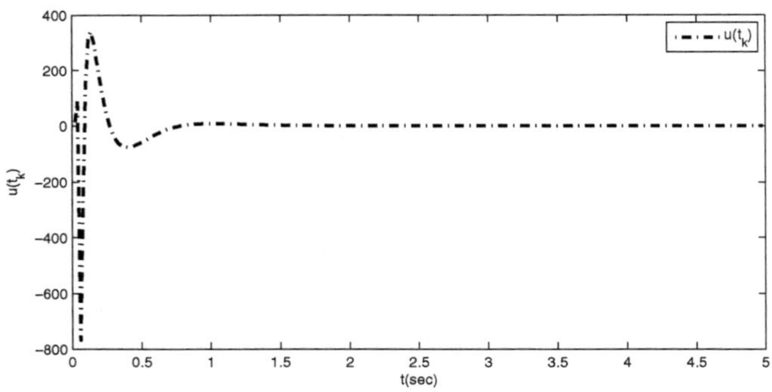

Fig. 7.11 Numerical simulation of control input $u(t_k)$.

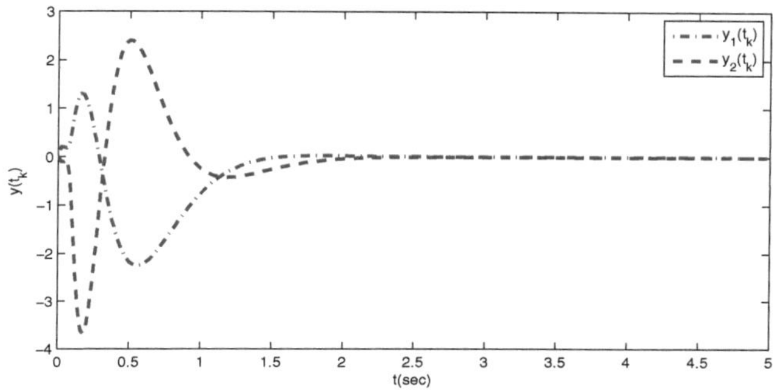

Fig. 7.12 Numerical simulation of output of $y(t_k)$.

It is clearly shown from the above simulation results that the estimated state vector is convergent to the real state vector in this inverted pendulum model. Hence, the DOKF obtained in Theorem 7.3 is convergent.

In the end, the real-time experiment results are given on the laboratory-scale setup in this section. The control signal of real inverted-pendulum system based on the real-time experiment results is shown and Fig. 7.13. The cart position of the real inverted-pendulum system is shown in Fig. 7.14. The pendulum angle of the real inverted-pendulum system is shown in Fig. 7.15.

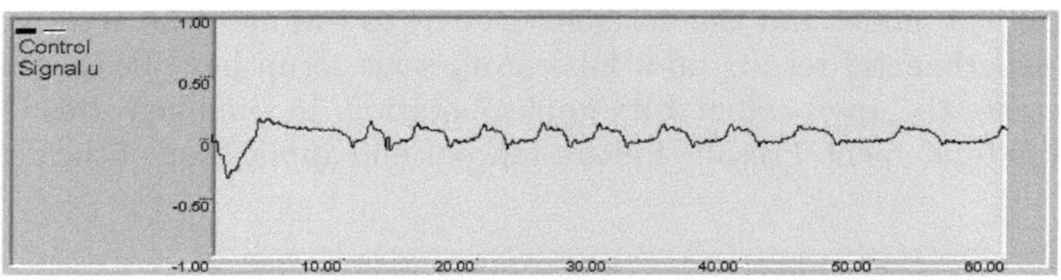

Fig. 7.13 Experimental result of the control signal.

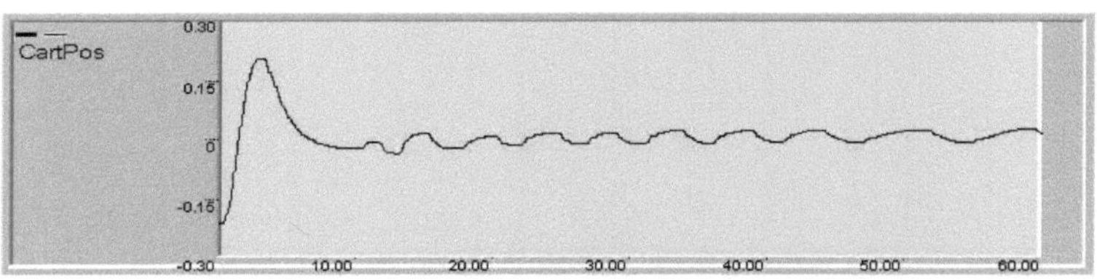

Fig. 7.14 Experimental result of the cart position.

The model used for control is very simple and neglects effects such as static friction. However, the inverted pendulum control problem is challenging because it may go unstable for sampling frequencies that are not sufficiently

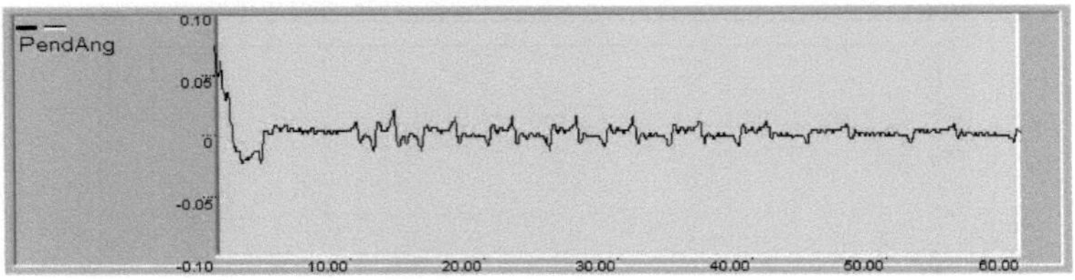

Fig. 7.15 Experimental result of the pendulum angle.

high. Therefore, the practical inverted pendulum model is presented as an interesting application of the DOKF in this chapter. And the experimental setup in this section has shown the developed techniques of the designed DOKF is practicable.

7.5 Summary

In this chapter, a general DOKF as a state observer has been introduced to estimate the full-state vectors of DOSs. The designed DOKF can express both continuous-time and discrete-time cases. By using Lyapunov approach, the convergence analysis of the designed DOKF has also been given. On the inverted pendulum experimental results, we have used a DOS to describe the practical model and the designed DOKF to estimate the state vectors. Some experimental results on a laboratory-scale setup have been presented to illustrate the approach and its implementation. In summary, this chapter has given fundamental results for the analysis and application of the designed DOKF.

Part IV
Finite Frequency Control of DOSs

Chapter 8
Strictly Positive Real Control for DOSs with a Low Frequency Range

8.1 Introduction

It is well known that each design specification is often given not for the entire frequency range but rather for a low frequency range of relevance, for example a closed-loop shaping control design typically requiring small sensitivity. Weighting method has been proven useful to deal with the finite frequency requirements [150]. However, the additional weights tend to increase the system complexity and the process of selecting appropriate weights is tedious and time-consuming. Another alternative approach has also been used to grid the frequency axis [286], but it suffers from the lack of a rigorous performance guarantee in the design process. Different from weighting function and frequency gridding methods, a generalized KYP lemma has been proposed in [119]. It can avoid computational burden and guarantee gain property performances simultaneously [115]. There are several results in the literature along this line, see for example [114, 116, 117]. Since the notion of positive realness plays a central role in systems theory [3], many researchers have considered positive real control for linear systems such as in [161, 162, 270], and references therein. The finite frequency positive realness approach for dynamical system design from a control perspective has been given in [118].

In frequency domain view, the poles are located in the stable boundary and the discrete systems will lose stability in finite word length computer using the traditional shift operator for fast sampling rate [170]. Delta operator method introduced in [82] is an effective approach for short sampling period. Delta operator requires smaller word length than shift operator [113]. The delta operator model can unify some previous related continuous and discrete systems into the framework of DOSs [194] and it also has the advantage of better numerical properties at high sampling rates [196]. Therefore, the study of DOSs has received much attention during the past years, including robust control [215], filter design [285], sliding mode control [254], and so on. However, to the best of our known there have been few papers on the finite frequency positive real control problem in a low frequency range in delta domain, which motivates us to make an effort in this chapter.

H. Yang et al.: Analysis and Synthesis of Delta Operator Systems, LNCIS 430, pp. 121–133.
springerlink.com © Springer-Verlag Berlin Heidelberg 2012

8.2 Problem Statement

In order to show the relations of linear system characteristics in s-domain, z-domain and δ-domain, we give a linear continuous system as follows

$$\dot{x}(t) = A_s x(t) + B_s u(t) \tag{8.1}$$
$$y(t) = C_s x(t) + D_s u(t) \tag{8.2}$$

where $x(t) \in R^n$ is the plant state, $u(t) \in R^m$ is the control input, $y(t) \in R^l$ is the measured output, A_s, B_s, C_s and D_s are constant matrices. Furthermore, transfer function matrix $G(s)$ of system (8.1)-(8.2) is $G(s) = C_s(sI - A_s)^{-1}B_s + D_s$. If $u(t)$ is processed via a zero-order-hold and letting $\mathtt{T}$ be a sample period, by using traditional shift operator approach system (8.1)-(8.2) is sampled to the following discrete system in as follows

$$x(k+1) = A_z x(k) + B_z u(k) \tag{8.3}$$
$$y(k) = C_z x(k) + D_z u(k) \tag{8.4}$$

The system parameters are given as follows:

$$A_z = e^{A_s \mathtt{T}}, \ B_z = \int_0^{\mathtt{T}} e^{A_s(\mathtt{T}-v)} B_s dv, \ C_z = C_s, \ D_z = D_s$$

A problem with this representation is that regardless of the underlying the linear continuous system (8.1)-(8.2), there exist

$$\lim_{\mathtt{T}\to 0} A_z = \lim_{\mathtt{T}\to 0} e^{A_s \mathtt{T}} = I, \ \lim_{\mathtt{T}\to 0} B_z = \lim_{\mathtt{T}\to 0} \int_0^{\mathtt{T}} e^{A_s(\mathtt{T}-v)} B_s dv = 0.$$

The z-domain transfer function matrix $G(z)$ of the linear discrete system (8.3)-(8.4) is given by $G(z) = C_z(zI - A_z)^{-1}B_z + D_z$ with $\lim_{\mathtt{T}\to 0} G(z) = 0$. As a result, for fast sampling the standard forward-shift representation of a discrete-time system becomes extremely sensitive to roundoff errors.

Before presenting the delta operator design equations, we will define some notations. Note that $\rho = (z-1)/\mathtt{T}$ given in [82], the variable ρ is the delta operator variable and is analogous to the Laplace variable s in continuous-time system transfer function matrix $G(s)$ and the forward-shift transform variable z in discrete-time system transfer function matrix $G(z)$. We will consider the following linear DOS in this chapter

$$\delta x(t_k) = Ax(t_k) + Bu(t_k) \tag{8.5}$$
$$y(t_k) = Cx(t_k) + Du(t_k) \tag{8.6}$$

with

$$A = (e^{A_s \mathrm{T}} - I)/\mathrm{T}, \; B = (\int_0^{\mathrm{T}} e^{A_s(\mathrm{T}-v)} B_s dv)/\mathrm{T}, \; C = C_s, \; D = D_s$$

Hence, it can be obtained that

$$\lim_{\mathrm{T}\to 0} A = \lim_{\mathrm{T}\to 0} \frac{e^{A_s \mathrm{T}} - I}{\mathrm{T}} = A_s \lim_{\mathrm{T}\to 0} e^{A_s \mathrm{T}} = A_s$$

$$\lim_{\mathrm{T}\to 0} B = \lim_{\mathrm{T}\to 0} \frac{\int_0^{\mathrm{T}} e^{A_s(\mathrm{T}-v)} B_s dv}{\mathrm{T}} = B_s e^0 = B_s$$

The delta domain transfer function matrix $G(\rho)$ for the linear DOS (8.5)-(8.6) is shown as $G(\rho) = C(\rho I - A)^{-1}B + D$, from which we have $\lim_{\mathrm{T}\to 0} G(\rho) = G(s)$.

It is shown from Fig. 8.1 that a frequency range discrete-time setting is an arc of the circle with center $-\frac{1}{\mathrm{T}}$ and radius $\frac{1}{\mathrm{T}}$ in δ-plant, which can be mapped to a line segment on the imaginary axis containing the origin in z-plane and an arc of the unit circle in s-plane.

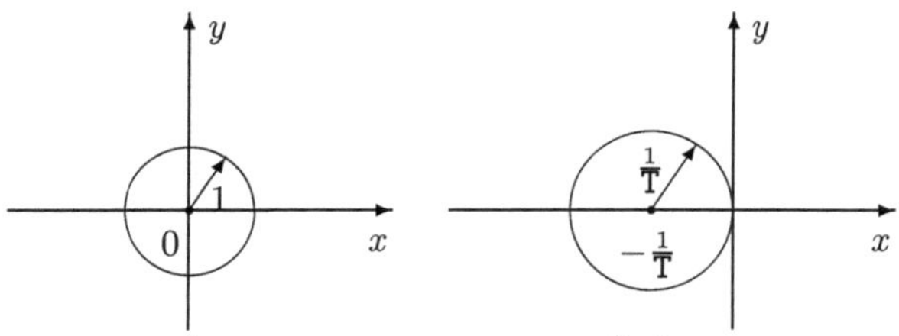

Fig. 8.1 The mapping of discrete shift and delta planes.

One of the key developments in this chapter is a unified characterization of low frequency range for the linear DOS (8.5)-(8.6). Therefore, we consider the frequency variable $\rho = (e^{j\theta} - 1)/\mathrm{T}$ with $|\theta| \le \vartheta$, which means that the low frequency transfer function matrix $G(\rho)$ in delta domain is rewritten as

$$G(\rho) = G\left(\frac{e^{j\theta} - 1}{\mathrm{T}}\right) = C\left(\frac{e^{j\theta} - 1}{\mathrm{T}} I - A\right)^{-1} B + D, \; \forall \; |\theta| \le \vartheta$$

which will be used to develop our main results.

Definition 8.1. A stable square transfer function $G(\rho)$ with $\rho = (e^{j\theta} - 1)/\mathrm{T}$ is said to be finite frequency strictly positive real with bandwidth ϑ if

i) All the elements of $G(\rho)$ are analytic in $|\rho| > 1/\mathrm{T}$.

ii) $G^*(\rho) + G(\rho) > 0$, for all $|\theta| \le \vartheta$ and $|\rho| > 1/\mathrm{T}$.

Remark 8.2. Above results imply that the delta representation can converge to the continuous-time representation as the sampling rate increases. In frequency domain view, the poles are located in the stable boundary and the discrete systems will lose stability in finite word length computer using the traditional shift operator for fast sampling rate. Hence, we will investigate the kind of DOSs to replace traditional discrete system in this chapter.

The following lemma which captures certain properties of dynamical system expressed in terms of rational transfer function will be used in this chapter to obtain a special case for DOSs.

Lemma 8.3. *[114] Let complex matrices A, B, $\Theta = \Theta^*$ and $(\Phi, \Psi) \in \Omega$ be given where*

$$\Omega := \left\{ (\Phi, \Psi) \middle| \Phi = M^* \begin{bmatrix} 0 & \alpha \\ \alpha & 0 \end{bmatrix} M,\ \Psi = M^* \begin{bmatrix} -1 & \beta \\ \beta & 1 \end{bmatrix} M \right\} \tag{8.7}$$

where $\alpha, \beta \in R$, $M \in C^{2\times 2}$ and $\alpha \det(M) \neq 0$. Define

$$\Lambda := \{\lambda \in C \,|\sigma(\lambda, \Phi) = 0, \sigma(\lambda, \Psi) \geq 0\} \tag{8.8}$$

Suppose A has no eigenvalues λ such that $\sigma(\lambda, \Phi) = 0$. Then the following statements are equivalent.

(i) The following frequency domain condition

$$\begin{bmatrix} (\lambda I - A)^{-1}B \\ I \end{bmatrix}^* \Pi \begin{bmatrix} (\lambda I - A)^{-1}B \\ I \end{bmatrix} < 0, \tag{8.9}$$

holds for all $\lambda \in \Lambda$.

(ii)There exist Hermitian matrices P and $Q > 0$ such that

$$\begin{bmatrix} A & B \\ I & 0 \end{bmatrix}^T L(P,Q) \begin{bmatrix} A & B \\ I & 0 \end{bmatrix} + \Theta < 0 \tag{8.10}$$

where $L(P,Q) = \Phi \otimes P + \Psi \otimes Q$.

8.3 Main Results

The key problem of using Lemma 8.3 in delta domain is choosing appropriate matrices Φ and Ψ such that $(\Phi, \Psi) \in \Omega$. It is a difficult step for the reason of that matrices Φ and Ψ are not unique due to freedom in the choice of α and β in (8.7). Note that the set Λ in delta domain is the intersection of the following two sets

$$\Lambda_\phi := \left\{ \frac{e^{j\theta} - 1}{\mathrm{T}} \middle| \theta \in R, \sigma\left(\frac{e^{j\theta} - 1}{\mathrm{T}}, \Phi\right) = 0 \right\} \tag{8.11}$$

$$\Lambda_\psi := \left\{ \frac{e^{j\theta}-1}{\mathrm{T}} \middle| \theta \in R, \sigma\left(\frac{e^{j\theta}-1}{\mathrm{T}}, \Psi\right) \geq 0 \right\} \tag{8.12}$$

We can choose Φ and Ψ by using (8.11) and (8.12), respectively. By using the term $L(P,Q) = \Phi \otimes P + \Psi \otimes Q$ in statement (ii) of Lemma 8.3, it can be obtained that Φ and Ψ satisfy (8.8) which is written in delta domain as follows

$$\Lambda := \left\{ \frac{e^{j\theta}-1}{\mathrm{T}} \middle| \theta \in R, \sigma(\lambda, \Phi) = 0, \sigma(\lambda, \Psi) \geq 0 \right\} \tag{8.13}$$

By the same, the set Ω in delta domain is also the intersection of Ω_ϕ and Ω_ψ, where

$$\Omega_\phi := \left\{ \Phi \middle| \Phi = M^* \begin{bmatrix} 0 & \alpha \\ \alpha & 0 \end{bmatrix} M, \ \alpha \det(M) \neq 0 \right\} \tag{8.14}$$

$$\Omega_\psi := \left\{ \Psi \middle| \Psi = M^* \begin{bmatrix} -1 & \beta \\ \beta & 1 \end{bmatrix} M \right\} \tag{8.15}$$

where $\alpha, \beta \in R$ and $M \in C^{2\times 2}$. Then, Φ and Ψ are parametrized by $M = NK$ and $N \in \mathcal{N}_\gamma$, in which $\mathcal{N}_\gamma$ is defined as

$$\mathcal{N}_\gamma := \left\{ J^* F J e^{j\theta} \middle| F \in R^{2\times 2}, \theta \in R, \det(F) = \gamma \right\}$$

where $\gamma := 1/\alpha$ and $J = \mathrm{diag}\{1, j\}$, and K is an arbitrary matrix satisfying

$$\Phi = K^* \begin{bmatrix} 0 & 1 \\ 1 & 0 \end{bmatrix} K \tag{8.16}$$

The following lemma gives an easily verifiable necessary and sufficient condition for $(\Phi, \Psi) \in \Omega$.

Lemma 8.4. *[114] Let $\Phi, \Psi \in \mathbf{H}$ be given and consider the set Ω in (8.7), Then, $\Phi, \Psi \in \Omega$ holds if and only if $\det(\Phi) < 0$, $\det(\Phi_o) < 0$ hold, where $\Phi_o := \mathrm{Re}((JK)^{-*}\Psi(JK)^{-1})$ with K being an arbitrary matrix such that (8.16).*

Theorem 8.5. *Let matrices $A \in C^{n\times n}$ and $B \in C^{n\times m}$ and a symmetric matrix $\Theta \in \mathbf{H}_{n+m}$ be given. Suppose that Ω in Lemma 8.3 represents curves on the complex plane. The following strictly positive real statements for the DOS (8.5)-(8.6) are equivalent:*

(i) The finite frequency inequality

$$\begin{bmatrix} (\rho I - A)^{-1}B \\ I \end{bmatrix}^* \Theta \begin{bmatrix} (\rho I - A)^{-1}B \\ I \end{bmatrix} < 0, \ \theta \in \Lambda \tag{8.17}$$

holds for all $\rho = (e^{j\theta}-1)/T$ such that $\theta \in \Lambda$ in Lemma 8.3.

(ii)There exist matrices $P, Q \in \mathbf{H}_n$ satisfying $Q > 0$, and

$$\begin{bmatrix} A & B \\ I & 0 \end{bmatrix}^T \Xi \begin{bmatrix} A & B \\ I & 0 \end{bmatrix} + \Theta < 0 \tag{8.18}$$

where

$$\Xi = \begin{bmatrix} T^2 P & TP + TQ \\ TP + TQ & (2 - 2\cos\vartheta_l)Q \end{bmatrix}, \quad \Theta = -\begin{bmatrix} 0 & C^* \\ C & D + D^* \end{bmatrix}$$

Proof. Considering the linear DOS (8.5)-(8.6) and it's frequency variable ρ, matrix Θ in (8.17) can be rewritten as follows

$$\Theta = \begin{bmatrix} C & D \\ 0 & I \end{bmatrix}^* \Pi \begin{bmatrix} C & D \\ 0 & I \end{bmatrix}$$

Then, (8.17) can be rewritten as

$$\begin{bmatrix} G(\rho) \\ I \end{bmatrix}^* \Pi \begin{bmatrix} G(\rho) \\ I \end{bmatrix} < 0 \tag{8.19}$$

The strictly positive real condition can be obtained by setting

$$\Pi := \Pi_\alpha = \begin{bmatrix} 0 & -I \\ -I & 0 \end{bmatrix}$$

Substituting Π_α into (8.19), it can be given that $G^*(\rho) + G(\rho) > 0$.

By an appropriate choice of Φ, the delta domain frequency variable $\rho = (e^{j\theta} - 1)/\mathrm{T}$ in the set Λ in Lemma 8.3 is characterized by

$$\sigma(\rho, \Phi) = 0, \ \Phi := \begin{bmatrix} \mathrm{T}^2 & \mathrm{T} \\ \mathrm{T} & 0 \end{bmatrix}$$

Note that $\det(\Phi) = \mathrm{T}^2 \cdot 0 - \mathrm{T} \cdot \mathrm{T} = -\mathrm{T}^2 < 0$. Furthermore, the matrix K satisfying (8.16) is chosen as

$$K = \begin{bmatrix} \mathrm{T} & 0 \\ \frac{\mathrm{T}}{2} & 1 \end{bmatrix}, \ JK = \begin{bmatrix} \mathrm{T} & 0 \\ \frac{\mathrm{T}}{2} j & j \end{bmatrix} \tag{8.20}$$

Then, we have

$$(JK)^{-1} = \begin{bmatrix} \frac{1}{\mathrm{T}} & 0 \\ -\frac{1}{2} & -j \end{bmatrix}, \ (JK)^{-*} = \begin{bmatrix} \frac{1}{\mathrm{T}} & -\frac{1}{2} \\ 0 & j \end{bmatrix}$$

By an appropriate choice of Ψ in Lemma 8.3, the set Λ is specialized to define low frequency variable range. We consider the low frequency condition with

$$\Lambda_l = \left\{ \frac{e^{j\theta} - 1}{\mathrm{T}} \middle| \theta \in R, |\theta| \leq \vartheta_l \right\} \tag{8.21}$$

where $\vartheta_l \in R$ being a given positive scalar. Noting that $|\theta| \leq \vartheta_l$ if and only if $\rho = (e^{j\theta} - 1)/\mathrm{T}$ satisfies $\rho + \bar{\rho} = \frac{2\cos\theta - 2}{\mathrm{T}} \geq \frac{2\cos\vartheta_l - 2}{\mathrm{T}}$, which means that $\cos\theta \geq \cos\vartheta_l$ holds. If we let $\gamma_l := 2\cos\vartheta_l - 2$ and choose

$$\Psi = \begin{bmatrix} 0 & \mathrm{T} \\ \mathrm{T} & -\gamma_l \end{bmatrix}$$

there exists $\sigma(\rho, \Psi) = 2(\cos\theta - \cos\vartheta_l) \geq 0$.

In this case, we have that the set Λ_l in (8.21) is equal to Λ in Lemma 8.3. By using the matrix K in (8.20), we have

$$(JK)^{-*}\Psi(JK)^{-1} = \begin{bmatrix} -\frac{1}{2}(1 + \cos\vartheta_l) & -j\cos\vartheta_l \\ j\cos\vartheta_l & -\gamma_l \end{bmatrix},$$

from which it can be obtained

$$\Psi_o = \mathrm{Re}[(JK)^{-*}\Psi(JK)^{-1}] = \begin{bmatrix} -\frac{1}{2}(1 + \cos\vartheta_l) & 0 \\ 0 & -\gamma_l \end{bmatrix}$$

Therefore, there exists $\det(\Psi_o) = \cos^2\vartheta_l - 1 < 0$, if and only if $0 < \vartheta_l < \pi$ holds. Due to Lemma 8.4, we conclude that $(\Phi, \Psi) \in \Omega$ satisfies Lemma 8.3 and the set Λ_l in (8.21) is equal to Λ in Lemma 8.3. The corresponding state space condition in delta domain is given by (8.10) with

$$L(P, Q) = \Phi \otimes P + \Psi \otimes Q = \begin{bmatrix} \mathrm{T}^2 P & \mathrm{T}P + \mathrm{T}Q \\ \mathrm{T}P + \mathrm{T}Q & (2 - 2\cos\vartheta_l)Q \end{bmatrix}$$

From Definition 8.1 we get that the positive real condition in Theorem 8.5 is obtained.

The following theorem provides a physical interpretation of being low frequency strictly positive in terms of the passivity property with respect to a certain class of input signals.

Theorem 8.6. *Consider the real symmetric matrix Θ in Theorem 8.5 and the DOS (8.5)-(8.6), the following strictly positive real statements are equivalent:*

(i) The statement (i) and (ii) in Theorem 8.5 hold.
(ii)The property

$$\sum_{k=-\infty}^{\infty} \begin{bmatrix} x(t_k) \\ u(t_k) \end{bmatrix}^T \Theta \begin{bmatrix} x(t_k) \\ u(t_k) \end{bmatrix} < 0$$

holds for all square summable input $u(t_k)$ such that

$$-\mathrm{T}\sum_{k=-\infty}^{\infty} x(t_k)\delta^T x(t_k) \leq (1-\cos\vartheta)\sum_{k=-\infty}^{\infty} x(t_k)x^T(t_k) \qquad (8.22)$$

Proof. For notational simplicity, we firstly denote

$$H(\rho) = \begin{bmatrix} (\rho I - A)^{-1}B \\ I \end{bmatrix}^* \Theta \begin{bmatrix} (\rho I - A)^{-1}B \\ I \end{bmatrix},$$

$$\psi(t_k) = \begin{bmatrix} x(t_k) \\ u(t_k) \end{bmatrix}^T \Theta \begin{bmatrix} x(t_k) \\ u(t_k) \end{bmatrix}$$

Suppose (i) holds. Multiplying the inequality in (ii) of Theorem 8.5 by $[x^T(t_k)\ u^T(t_k)]$ from the left and by its transpose from the right, we have

$$\begin{bmatrix} \delta x(t_k) \\ x(t_k) \end{bmatrix}^T \begin{bmatrix} \mathrm{T}^2 P & \mathrm{T}P + \mathrm{T}Q \\ \mathrm{T}P + \mathrm{T}Q & 2(1-\cos\vartheta)Q \end{bmatrix} \begin{bmatrix} \delta x(t_k) \\ x(t_k) \end{bmatrix} + \psi(t_k) < 0$$

which can be rewritten as $\mathrm{T}\delta(x^T(t_k)Px(t_k)) + 2\mathrm{T}x^T(t_k)Q\delta x(t_k) + 2x^T(t_k)(1-\cos\vartheta)x(t_k) + \psi(t_k) < 0$. Taking the summation from $k=-\infty$ to ∞, using the stability property, and completing the square, it is obtained

$$\sum_{k=-\infty}^{\infty} \left(tr[Q(2-2\cos\vartheta)x(t_k)x^T(t_k) + 2\mathrm{T}x(t_k)\delta^T x(t_k)] + \psi(t_k)\right) < 0$$

Based on $Q \geq 0$ and inequality (8.22), we have statement (ii) holds.

In order to show (ii)⇒(i), we suppose (i) is not true. Then, there exists $\theta_\star \in R$ such that $|\theta_\star| \leq \vartheta$ and the maximum eigenvalue of $H(\rho)$ is strictly positive. Since A, B and Θ are all real, such that $\theta_\star$ can always be chosen to be nonnegative. Thus, we have $\lambda_{\max} H(\rho) > 0,\ 0 \leq \theta_\star \leq \vartheta$. Now due to stability of the system, $\lambda_{\max} H(\rho)$ is a continuous function of θ and hence a small perturbation of $\theta_\star$ does not alter the sign of $\lambda_{\max} H(\rho)$. Therefore, we can assume $0 \leq \theta_\star \leq \vartheta$ without loss of generality. It can be obtained that inequality (8.22) is rewritten as $\sum_{k=-\infty}^{\infty}\left[(2-2\cos\vartheta)x(t_k)x^T(t_k) + 2\mathrm{T}x(t_k)\delta^T x(t_k)\right] \geq 0$. Based on the relation of z-transform and δ-transform, we have $H(e^{j\theta}) = H(\rho)$ and

$$\sum_{k=-\infty}^{\infty} \left[2x(t_k)x^T(t_{k+1}) - 2\cos\vartheta x(t_k)x^T(t_k)\right] \geq 0 \qquad (8.23)$$

Let $u_\star$ be the unit eigenvector of $H(\rho)$ corresponding to the maximum eigenvalue, and define the input signal $u := \eta$ by its z-transform $\hat{\eta}$ as follows:

$$\hat{\eta}(e^{j\theta}) := \begin{cases} u_\star \ (\text{when } |\theta - \theta_\star| \leq \varepsilon) \\ \bar{u}_\star \ (\text{when } |\theta + \theta_\star| \leq \varepsilon) \\ 0 \qquad (\text{otherwise}) \end{cases}$$

where $\bar{u}_\star$ is the complex complex conjugate of $u_\star$. Now, letting $\Phi(z)$ be the transfer function from u to x, it can be obtained that

$$\begin{aligned}&\sum_{k=-\infty}^{\infty}\left[2x(t_k)x^T(t_{k+1})-2\cos\vartheta x(t_k)x^T(t_k)\right]\\&=\frac{1}{\pi}\int_{\pi}^{\pi}(\cos\theta-\cos\vartheta)\Phi(e^{j\theta})\hat{\eta}(e^{j\theta})\hat{\eta}(e^{j\theta})^*\Phi(e^{j\theta})^*d\theta\end{aligned} \tag{8.24}$$

which is positive semi-definite for small $\varepsilon > 0$ such that $\theta_\star + \varepsilon < \vartheta$. On the other hand, the left hand of inequality (8.22) is equal to

$$\sum_{k=-\infty}^{\infty}\begin{bmatrix}x(t_k)\\u(t_k)\end{bmatrix}^T\Theta\begin{bmatrix}x(t_k)\\u(t_k)\end{bmatrix}=\frac{1}{\pi}\int_{\theta_\star-\varepsilon}^{\theta_\star+\varepsilon}u_\star^*H(e^{j\theta})u_\star d\theta$$

Since $u_\star^*H(e^{j\theta})u_\star$ is a real-valued, continuous function of θ and we have $u_\star^*H(e^{j\theta_\star})u_\star > 0$. Hence, there exists a sufficiently small $\varepsilon > 0$ such that $u_\star^*H(e^{j\theta})u_\star > 0$ for all $\theta \in [\theta_\star-\varepsilon, \theta_\star+\varepsilon]$. For such ε, the above integral takes a positive value, implying that statement (ii) is false. Therefore, if statement (i) is false, statement (ii) must be false.

In the following, we will solve a finite frequency strictly positive real state feedback controller design problem by using the proposed results. The controller is given as: $u(t_k) = Kx(t_k)$. The closed-loop DOS is expressed as follows:

$$\delta x(t_k) = (A+BK)x(t_k) \tag{8.25}$$

Theorem 8.7. *The closed-loop DOS (8.25) is asymptotically stable and it's transfer function satisfies finite frequency strictly positive realness, if there exist symmetric positive definite matrices $Q > 0$ and $X > 0$, real matrices P and Y, such that the following two LMIs hold:*

$$\begin{bmatrix}(T-2)X & AX+BY\\ * & AX+BY+XA^T+Y^TB^T\end{bmatrix}<0 \tag{8.26}$$

$$\begin{bmatrix}\Omega(P,Q) & T^2A^TPB+T(P+Q)B-C^T\\ * & T^2B^TPB-D-D^T\end{bmatrix}<0 \tag{8.27}$$

where

$$\Omega(P,Q)=T^2A^TPA+T(P+Q)A+TA^T(P+Q)+2(1-\cos\vartheta)Q.$$

Then, the strictly positive real controller is given by $K = YX^{-1}$.

Proof. It is easy to show that LMI (8.27) can be obtained from LMI (8.18) in Theorem 8.5. In the next, we construct a simple Lyapunov functional in delta domain as: $V(x(t_k)) = x^T(t_k)Rx(t_k)$. Taking delta operator manipulations of $V(x(t_k))$ along the trajectory of the closed-loop system (8.5), we obtain:

$$\delta V(x(t_k)) = \delta^T x(t_k)Rx(t_k) + x^T(t_k)R\delta x(t_k) + \mathtt{T}\delta^T x(t_k)R\delta x(t_k).$$

For the positive definite real matrix R, one has that $-2\delta^T x(t_k)R\delta x(t_k) + 2\delta^T x(t_k)R(A+BK)x(t_k) = 0$. Combining above results, we have $\delta V(x(t_k)) \leq \xi^T(t_k)\Xi_1\xi(t_k) < 0$, where $\xi^T(t_k) = \left[\delta^T x(t_k)\ x^T(t_k)\right]$ and

$$\Xi_1 = \begin{bmatrix} (\mathtt{T}-2)R & RA+RBK \\ * & RA+RBK+A^TR+K^TB^TR \end{bmatrix}.$$

Pre-multiplying and post-multiplying $\Xi_1 < 0$ by $diag\{R^{-1},\ R^{-1}\}$, respectively, letting $R^{-1} = X$ and $KR^{-1} = Y$, $\Xi_1 < 0$ is equivalent to LMI (8.26).

We will prove that $x(t_k) \to 0$ as $k \to \infty$. Due to LMI (8.26) hold, there exists

$$\delta V_0(x(t_k)) = \frac{1}{\mathtt{T}}\left[x^T(t_k+\mathtt{T})Rx(t_k+\mathtt{T}) - x^T(t_k)Rx(t_k)\right] < 0$$

It can be seen that for $x(t_k) \neq 0$ the following inequality holds

$$x^T(t_k+\mathtt{T}_k)Rx(t_k+\mathtt{T}_k) - x^T(t_k)Rx(t_k) < 0$$

which implies that the Lyapunov functional $x^T(t_k)Rx(t_k)$ in delta domain is a strictly decreasing function. Thus, we obtain $x^T(t_k)Rx(t_k) \to 0$, as $k \to \infty$. Since R is a positive definite constant matrix, we have $x(t_k) \to 0$, as $k \to \infty$. Therefore, the feasible strictly positive real state feedback control law obtained in this theorem asymptotically stabilizes the closed-loop DOS (8.6). This completes the proof.

8.4 Numerical Simulation

Example 8.8. In this example, the SISO mathematical model of one-link PWM -controlled joint driven by pneumatic muscles can be taken for a 3th-order SISO system, please refer to Fig. 8.2.

In this control system, a degree of freedom (DOF) of arthrosis is controlled by two PMA which are products of FESTO, MAS-20N-AA-MCGK. The two PMA is connected by a pulley using a flexible nonelastic synchronous belt. The angle sensor is set coaxially with pulley. Data by angle sensor is input into computer by DAQ cards with sampling frequency 100HZ. The control

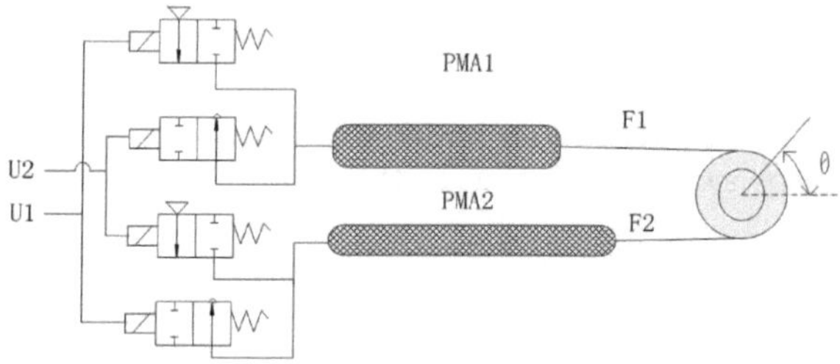

Fig. 8.2 The one-link PWM -controlled joint model.

signal is put out into PWM - high-speed on-off solenoid valve (FESTO, MHE2 -MS1H-3/2G-QS-4-K) by DAQ cards. The weight of the joint is 2.45kg, and maximal turning angle is about $\pm 50°$. A continuous linear system can be obtained by the PMA model in [230] and joint dynamics equation in [268] as

$$\begin{aligned}\dot{x}_1(t) &= x_2(t),\\ \dot{x}_2(t) &= x_3(t),\\ \dot{x}_3(t) &= -a_1x_1(t) - a_2x_2(t) - a_3x_3(t) + b_0u(t),\end{aligned}$$

where $x_1(t) = \theta(t)$, $x_2(t) = \dot{\theta}(t)$ and $x_3(t) = \ddot{\theta}(t)$ are state variables, $u(t)$ is a input variables, $a_1 = 922.5$, $a_2 = 185.5$, $a_3 = 17$ and $b_0 = 1325$. By using the delta operator approach and letting the sampling period be $\mathtt{T} = 0.01$, the following DOS is obtained for sampling the above continuous system as

$$\delta x(t_k) = Ax(t_k) + Bu(t_k)$$

where

$$A = \begin{bmatrix} -0.0147 & 0.9970 & 0.0047 \\ -4.3551 & -0.8905 & 0.9167 \\ -845.6961 & -174.4111 & -16.4751 \end{bmatrix}, \ B = \begin{bmatrix} 0 \\ 6.3 \\ 1214.7 \end{bmatrix}.$$

Let $C = [1\ 1\ 1]$, $D = [1]$ and $\theta \leq 0.5$ in this system. By using the Matlab LMI Control Toolbox to solve LMIs (8.26) and (8.27) in Theorem 8.7, we obtain the solutions as follows:

$$P = \begin{bmatrix} 18.0367 & 1.9075 & 0.2827 \\ 1.9075 & 0.7416 & 0.0511 \\ 0.2827 & 0.0511 & 0.0106 \end{bmatrix}, \ Q = \begin{bmatrix} 110.2266 & 13.8451 & 1.3926 \\ 13.8451 & 1.7969 & 0.1743 \\ 1.3926 & 0.1743 & 0.0179 \end{bmatrix},$$

$$X = \begin{bmatrix} 747.9657 & -177.5039 & -119.7912 \\ -177.5039 & 307.8429 & -178.5564 \\ -119.7912 & -178.5564 & 392.1043 \end{bmatrix},$$

$$Y = \begin{bmatrix} 493.8085 & -81.8910 & -104.0333 \end{bmatrix}.$$

Furthermore, the low frequency strictly positive real controller in the model is given as $u(t_k) = \left[0.6959\ 0.1422\ 0.0120\right] x(t_k)$. For initial condition: $x_1(0) = x_2(0) = x_3(0) = 1$, the simulation results of the state response of the closed-loop delta operator switched system in this example is given in Fig. 8.3.

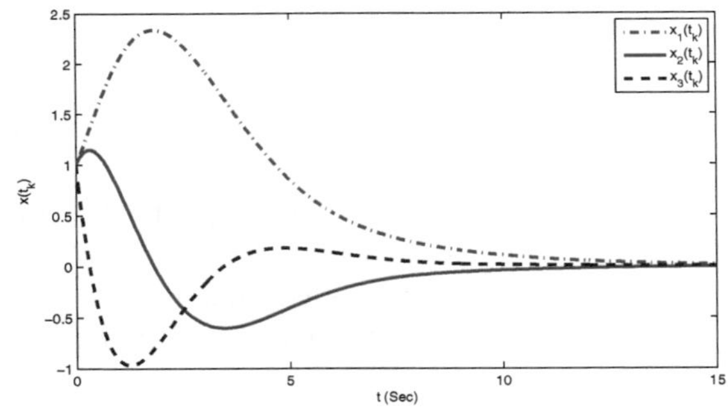

Fig. 8.3 The state response of $x(t_k)$.

Furthermore, Fig. 8.4 shows the simulation results of the control input for initial condition $u(0) = 0.9$.

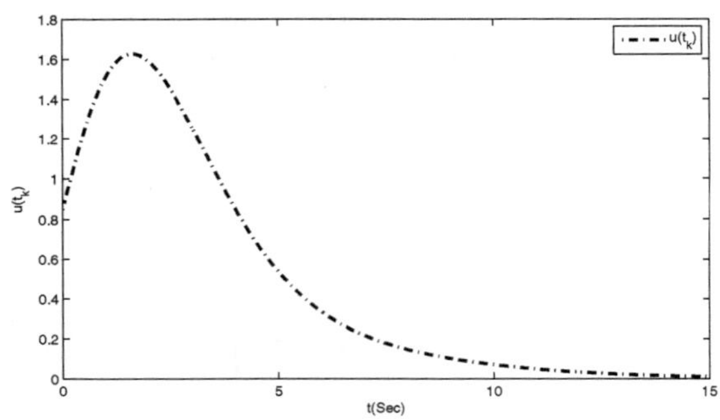

Fig. 8.4 The control input $u(t_k)$.

From these simulation results, it has been clearly shown that the designed positive real state feedback digital controller ensures the asymptotic stability of the closed-loop DOS in this example.

8.5 Summary

In this chapter, the KYP lemma on a low frequency range has been generalized in delta domain to treat some strictly positive real control problems. The objective of the strictly positive real control problem is to design a controller such that the transfer function is strictly positive real and the resulting closed-loop system is stable. Sufficient conditions for the low frequency strictly positive real controller of the closed-loop DOSs have been presented in terms of solutions to a set of LMIs. A numerical example has been given to illustrate the effectiveness and potential for the developed techniques.

Chapter 9
Generalized Finite Frequency KYP Lemma and Applications to Fault Detection

9.1 Introduction

In the last two decades, fault detection has attracted considerable attention from many researchers due to the increasing demand for reliability and safety in industrial processes, for example [135, 188, 288]. Some results on fault detection in frequency domain have been reported in literatures, such as [1, 56, 302]. The effects of modeling disturbances and faults usually occupy in different frequency ranges. Therefore, designing an filter to detect the fault by satisfying rigorous performances in finite frequency domain is a key task in this topic. Though weighting functions [150] and frequency gridding [286] can deal with the finite frequency requirements. However, both of the two methods are hard to avoid computational burden and guarantee gain property performances simultaneously. With the aid of the generalized KYP Lemma, new LMI conditions were given for the finite frequency performance index [115, 119]. There are several results in the literature along this line, please refer to [116, 117, 118].

H_∞ norm has been used to deal with disturbances which may result in significant changes in the residual and lead to false alarms [197]. H_- norm has been defined as the smallest nonzero singular value of transfer function matrix from fault to residual [93]. Fault detection filters considering both H_- and H_∞ performance indexes have been investigated in [121, 239]. There has been a little results on fault finite frequency ranges detection filters with H_∞ and H_- performance indices for discrete systems [236] and continuous system [238], respectively. Delta operator is introduced to deal with fast sampling in [82, 170]. A number of results on DOSs have been reported in literatures, including nonlinear modelling [4], filter design [285], adaptive control [284], and so on. The H_∞ control problems in delta domain have been presented in [42, 83, 171, 218]. Up to now, there have been few papers on the generalized KYP lemma in delta domain to treat conditions on restricted frequency

H. Yang et al.: Analysis and Synthesis of Delta Operator Systems, LNCIS 430, pp. 135–155.
springerlink.com © Springer-Verlag Berlin Heidelberg 2012

ranges and the results developed for the purpose of fault detection filter design, which motivates us to make the effort in this chapter.

9.2 Problem Statement

In this paper, the main results obtained in the next section will be developed for the purpose of fault detection filter design in delta domain.

Firstly, we consider the following linear DOS

$$\delta x(t_k) = Ax(t_k) + Bu(t_k) + B_d d(t_k) + B_f f(t_k) \tag{9.1}$$

$$y(t_k) = Cx(t_k) + Du(t_k) + D_d d(t_k) + D_f f(t_k) \tag{9.2}$$

where $x(t_k) \in R^n$ is the state variable, $u(t_k) \in R^{n_u}$ is the control input, $y(t_k) \in R^{n_y}$ is the measured output, $d(t_k) \in R^{n_d}$ represents the disturbance vector, $f(t_k) \in R^{n_f}$ denotes the fault input vector. A, B_d, B_f, C, D_d and D_f are known constant real matrices with appropriate dimensions. Furthermore, we let (A, C) be observable in this paper.

In order to detect the system fault, the following fault detection filter will be designed as

$$\delta \hat{x}(t_k) = A\hat{x}(t_k) + Bu(t_k) + L(y(t_k) - \hat{y}(t_k)) \tag{9.3}$$

$$\hat{y}(t_k) = C\hat{x}(t_k) + Du(t_k) \tag{9.4}$$

$$r(t_k) = y(t_k) - \hat{y}(t_k) \tag{9.5}$$

where $r(t_k) \in R^{n_y}$ is the residual signal which carries information on the time and location of the occurrence of the faults, $\hat{x}(t_k) \in R^n$ and $\hat{y}(t_k) \in R^{n_y}$ denote the state and output estimation vectors, respectively. The design parameter is the filter gain matrix L.

By defining the state estimate error $\bar{x}(t_k) = x(t_k) - \hat{x}(t_k)$, we have the following state error system:

$$\delta \bar{x}(t_k) = (A - LC)\bar{x}(t_k) + (B_d - LD_d)d(t_k) + (B_f - LD_f)f(t_k) \tag{9.6}$$

$$r(t_k) = C\bar{x}(t_k) + D_d d(t_k) + D_f f(t_k) \tag{9.7}$$

The objective of fault detection is to design a suitable observer (9.3)-(9.4) and to determine the residual evaluation function and an appropriate threshold to realize the following three conditions:

AaThe estimation error approaches asymptotically to zero.
AbThe effects of disturbances to the residual $r(t_k)$ is reduced.
AcThe effects of faults to the residual $r(t_k)$ is increased.

Letting $\rho = (e^{j\theta} - 1)/\mathtt{T}$, we consider the following two middle frequency ranges for disturbances and faults in this paper.

$$\Omega_d = \left\{ \frac{e^{j\theta} - 1}{\text{T}} \middle| \ \theta \in R, \ \vartheta_{d1} \leq \theta \leq \vartheta_{d2} \right\}$$
$$\Omega_f = \left\{ \frac{e^{j\theta} - 1}{\text{T}} \middle| \ \theta \in R, \ \vartheta_{f1} \leq \theta \leq \vartheta_{f2} \right\}$$

where ϑ_{d1}, ϑ_{d2}, ϑ_{f1} and ϑ_{f2} are given positive scalars which reflect the individual frequency ranges of disturbances and faults. The dynamics of the residual generator is formatted as

$$r(\rho) = r_d(\rho) + r_f(\rho) = G_{rd}(\rho)d(\rho) + G_{rf}(\rho)f(\rho)$$

where

$$G_{rd}(\rho) = C[\rho I - (A - LC)]^{-1}(B_d - LD_d) + D_d, \ \forall \ \theta \in \Omega_d$$
$$G_{rf}(\rho) = C[\rho I - (A - LC)]^{-1}(B_f - LD_f) + D_f, \ \forall \ \theta \in \Omega_f$$

Integrating the performance indexes for discrete systems in [236] and continuous systems in [238], the following H_∞ and H_- performance indexes in delta domain are introduced in this paper. The performance H_∞ is used to denote the effect of residual generator on disturbances as well as unknown signals:

$$H_\infty := \|G_{rd}(\rho)\|_\infty = \sup_\theta \lambda_{\min}(G_{rd}((e^{j\theta} - 1)/\text{T})) < \gamma, \ \forall \ \theta \in \Omega_d \quad (9.8)$$

where γ denotes the worst-case criterion for the effect of disturbances on the residual $r(t_k)$, the smaller γ is, the more robust the generator becomes and the more unknown input is restrained.

While H_- index is used as measurement of the worst case fault sensitively of residual generator on fault:

$$H_- := \|G_{rf}(\rho)\|_- = \inf_\theta \lambda_{\max}(G_{rf}((e^{j\theta} - 1/\text{T}))) > \beta, \ \forall \ \theta \in \Omega_f \quad (9.9)$$

where β is a measurement of the fault sensitivity in the worst case from fault $f(t_k)$ to residual $r(t_k)$. The larger β is, the more sensitive the generator becomes and the more detectable fault information residual generator can capture.

The conditions A(a-c) can be rewritten as follows

BaSystem (9.6) is asymptotically stable.
BbH_∞ performance index (9.8) holds for system (9.6)-(9.7).
BcH_- performance index (9.9) holds for system (9.6)-(9.7).

Remark 9.1. In this paper, we consider middle frequency ranges both in disturbances and faults for fault detection filter design. The reason of that is given in the following.

Note that $\vartheta_1 \leq \theta \leq \vartheta_2$ is equivalent to $|\theta_l| \leq \vartheta_l$, where

$$\theta_l = \theta - \vartheta_c,\ \vartheta_l = (\vartheta_2 - \vartheta_1)/2,\ \vartheta_c = (\vartheta_1 + \vartheta_2)/2$$

Then, there exists

$$|\theta_h| \geq \vartheta_h \Longleftrightarrow \vartheta_h \leq |\theta_h| \leq \pi \Longleftrightarrow \vartheta_h \leq \theta_h \leq 2\pi - \theta_h$$

As a result, both low and high frequency ranges can be changed to middle frequency ranges.

9.3 Frequency Rang Characterization

In this section, we consider the following linear DOS

$$\delta x(t_k) = Ax(t_k) + Bu(t_k) \tag{9.10}$$
$$y(t_k) = Cx(t_k) + Du(t_k) \tag{9.11}$$

where A, B, C and D are the matrix coefficients of the system. One of the key developments in this section is a unified characterization of finite frequency ranges for the linear DOS (9.10)-(9.11). By using $\rho = (e^{j\theta} - 1)/\mathrm{T}$, the delta domain transfer function matrix $G(\rho)$ is written as

$$G(\rho) = C(\rho I - A)^{-1} B + D,\ \forall\, \theta \in R.$$

The following lemma which captures certain properties of dynamical system expressed in terms of rational transfer function will be used in this paper to obtain a special case for DOSs.

Lemma 9.2. *[114] Let complex matrices A, B, $\Theta = \Theta^*$ and $(\Phi, \Psi) \in \Omega$ be given where*

$$\Omega := \left\{ (\Phi, \Psi) \middle| \Phi = M^* \begin{bmatrix} 0 & \alpha \\ \alpha & 0 \end{bmatrix} M,\ \Psi = M^* \begin{bmatrix} -1 & \beta \\ \beta & 1 \end{bmatrix} M \right\} \tag{9.12}$$

where $\alpha, \beta \in R$, $M \in C^{2\times 2}$ and $\alpha \det(M) \neq 0$. Define

$$\Lambda := \{\lambda \in C \,|\sigma(\lambda, \Phi) = 0, \sigma(\lambda, \Psi) \geq 0\}$$

Suppose A has no eigenvalue λ such that $\sigma(\lambda, \Phi) = 0$. Then the following statements are equivalent.

(i) The following frequency domain condition

$$\begin{bmatrix} (\lambda I - A)^{-1} B \\ I \end{bmatrix}^* \Theta \begin{bmatrix} (\lambda I - A)^{-1} B \\ I \end{bmatrix} < 0, \tag{9.13}$$

holds for all $\lambda \in \Lambda$.

(ii)There exist Hermitian matrices P and $Q > 0$ such that

$$\begin{bmatrix} A & B \\ I & 0 \end{bmatrix}^T L(P,Q) \begin{bmatrix} A & B \\ I & 0 \end{bmatrix} + \Theta < 0 \tag{9.14}$$

where $L(P,Q) = \Phi \otimes P + \Psi \otimes Q$.

The key problem of using Lemma 9.2 in delta domain is how to choose two appropriate matrices Φ and Ψ such that $(\Phi, \Psi) \in \Omega$. It is a difficult step for the reason of that matrices Φ and Ψ are not unique due to freedom in the choice of α and β in (9.12). Then, Φ and Ψ are parametrized by $M = NK$ with $N \in \mathcal{N}_\gamma$, where $\mathcal{N}_\gamma$ is defined as

$$\mathcal{N}_\gamma := \left\{ J^* F J e^{j\theta} \,\middle|\, F \in R^{2\times 2}, \theta \in R, \det(F) = \gamma \right\}$$

where $\gamma := 1/\alpha$ and $J = diag\{1, j\}$, and K is an arbitrary matrix satisfying

$$\Phi = K^* \begin{bmatrix} 0 & 1 \\ 1 & 0 \end{bmatrix} K \tag{9.15}$$

The following lemma gives an easily verifiable necessary and sufficient condition for $(\Phi, \Psi) \in \Omega$.

Lemma 9.3. *[114] Let $\Phi, \Psi \in \mathbf{H}$ be given and consider the set Ω in (9.12). Then $\Phi, \Psi \in \Omega$ holds if and only if*

$$\det(\Phi) < 0, \ \det(\Phi_o) < 0$$

hold, where

$$\Phi_o := \mathrm{Re}((JK)^{-*} \Psi (JK)^{-1})$$

with K being an arbitrary matrix such that (9.15) holds.

9.3.1 Frequency Restrictions

In this section, we specialize a general result of Lemma 9.2 to obtain a generalized KYP lemma for the linear DOS (9.10)-(9.11) in different frequency ranges.

Theorem 9.4. *Let matrices $A \in C^{n\times n}$ and $B \in C^{n\times m}$ and a symmetric matrix $\Theta \in \mathbf{H}_{n+m}$ be given. Suppose that Ω in Lemma 1 represents curves on the complex plant. The following statements are equivalent:*

(i) The finite frequency inequality

$$\begin{bmatrix} (\rho I - A)^{-1}B \\ I \end{bmatrix}^* \Theta \begin{bmatrix} (\rho I - A)^{-1}B \\ I \end{bmatrix} < 0, \ \theta \in \Lambda \tag{9.16}$$

holds for all $\rho = (e^{j\theta} - 1)/T$ *such that* $\theta \in \Lambda$ *in Lemma 9.2.*
(ii)There exist matrices $P, Q \in \mathbf{H}_n$ *satisfying* $Q > 0$, *and*

$$\begin{bmatrix} A & B \\ I & 0 \end{bmatrix}^T \Xi \begin{bmatrix} A & B \\ I & 0 \end{bmatrix} + \Theta < 0 \tag{9.17}$$

where Ξ *is given in Table 9.1 which is given as follows*

Table 9.1 The matrix Ξ in Theorem 9.4

	LF ($\lvert\theta\rvert \le \vartheta_l$)	MF ($\vartheta_1 \le \theta \le \vartheta_2$)	HF($\lvert\theta\rvert \ge \vartheta_h$)
Ξ	$\begin{bmatrix} \mathrm{T}^2 P & \mathrm{T}P + \mathrm{T}Q \\ \mathrm{T}P + \mathrm{T}Q & 2\kappa_l Q \end{bmatrix}$	$\begin{bmatrix} \mathrm{T}^2 P & \mathrm{T}P + \mathrm{T}e^{j\vartheta_c}Q \\ \mathrm{T}P + \mathrm{T}e^{-j\vartheta_c}Q & 2\kappa_m Q \end{bmatrix}$	$\begin{bmatrix} \mathrm{T}^2 P & \mathrm{T}P - \mathrm{T}Q \\ \mathrm{T}P - \mathrm{T}Q & 2\kappa_h Q \end{bmatrix}$

with $\vartheta_c := (\vartheta_1 + \vartheta_2)/2$ *and* $\vartheta_c := (\vartheta_2 - \vartheta_1)/2$, $\kappa_l = 1 - \cos\vartheta_l$, $\kappa_m = \cos\vartheta_c - \cos\vartheta_o$ *and* $\kappa_h = \cos\vartheta_h - 1$, *LF, HF and MF stand for low, high and middle frequency ranges, respectively.*

Proof. By an appropriate choice of Φ, the frequency variable $\rho = (e^{j\theta} - 1)/\mathrm{T}$ in the set Λ in delta domain is characterized by

$$\sigma(\rho, \Phi) = 0, \ \Phi := \begin{bmatrix} \mathrm{T}^2 & \mathrm{T} \\ \mathrm{T} & 0 \end{bmatrix}$$

Note that

$$\det(\Phi) = \mathrm{T}^2 \cdot 0 - \mathrm{T} \cdot \mathrm{T} = -\mathrm{T}^2 < 0$$

By considering the linear DOS (9.10)-(9.11), the matrix K in (9.15) is chosen as

$$K = \begin{bmatrix} \mathrm{T} & 0 \\ \frac{\mathrm{T}}{2} & 1 \end{bmatrix}, \ JK = \begin{bmatrix} \mathrm{T} & 0 \\ \frac{\mathrm{T}}{2}j & j \end{bmatrix} \tag{9.18}$$

Then, we have

$$(JK)^{-1} = \begin{bmatrix} \frac{1}{\mathrm{T}} & 0 \\ -\frac{1}{2} & -j \end{bmatrix}, \ (JK)^{-*} = \begin{bmatrix} \frac{1}{\mathrm{T}} & -\frac{1}{2} \\ 0 & j \end{bmatrix}$$

By an appropriate choice of Ψ in Lemma 9.2, the set Λ is specialized to define some different frequency variable ranges.

Firstly, we consider the low frequency condition with

$$\Lambda_l = \left\{ \frac{e^{j\theta}-1}{\mathrm{T}} \middle| \theta \in R, |\theta| \le \vartheta_l \right\} \tag{9.19}$$

where ϑ_l is a given positive scalar. Noting that $|\theta| \le \vartheta_l$ if and only if $\rho = (e^{j\theta}-1)/\mathrm{T}$ satisfies

$$\rho + \bar{\rho} = \frac{2\cos\theta - 2}{\mathrm{T}} \ge \frac{2\cos\vartheta_l - 2}{\mathrm{T}}$$

which means that $\cos\theta \ge \cos\vartheta_l$ holds. If we let $\gamma_l := 2\cos\vartheta_l - 2$ and choose

$$\Psi = \begin{bmatrix} 0 & \mathrm{T} \\ \mathrm{T} & -\gamma_l \end{bmatrix}$$

there exists

$$\sigma(\rho, \Psi) = 2(\cos\theta - \cos\vartheta_l) \ge 0$$

In this case, we have that the set Λ_l in (9.19) is equal to Λ in Lemma 9.2. By using the matrix K in (9.18), we have

$$(JK)^{-*}\Psi(JK)^{-1} = \begin{bmatrix} -\frac{1}{2}(1+\cos\vartheta_l) & -j\cos\vartheta_l \\ j\cos\vartheta_l & -\gamma_l \end{bmatrix}$$

from which it can be obtained

$$\Psi_o = \mathrm{Re}[(JK)^{-*}\Psi(JK)^{-1}] = \begin{bmatrix} -\frac{1}{2}(1+\cos\vartheta_l) & 0 \\ 0 & -\gamma_l \end{bmatrix}$$

Therefore, there exists

$$\det(\Psi_o) = \cos^2\vartheta_l - 1 < 0$$

if and only if $0 < \vartheta_l < \pi$ holds in (9.19). Due to Lemma 9.3, we conclude that $(\Phi, \Psi) \in \Omega$ satisfies Lemma 9.2 and the set Λ_l in (9.19) is equal to Λ in Lemma 9.2. The corresponding state space condition in delta domain is given by (9.14) with

$$L(P,Q) = \Phi \otimes P + \Psi \otimes Q = \begin{bmatrix} \mathrm{T}^2 P & \mathrm{T}P + \mathrm{T}Q \\ \mathrm{T}P + \mathrm{T}Q & (2 - 2\cos\vartheta_l)Q \end{bmatrix}$$

Secondly, the high frequency condition with

$$\Lambda_h = \left\{ \frac{e^{j\theta}-1}{\mathrm{T}} \middle| \theta \in R, \vartheta_h \le |\theta| \le \pi \right\} \tag{9.20}$$

can be treated similarly. We have that $\vartheta_h \le |\theta| \le \pi$ holds if and only if $\rho = (e^{j\theta}-1)/\mathrm{T}$ satisfies

$$\rho + \bar{\rho} = \frac{2\cos\theta - 2}{\mathrm{T}} \le \frac{2\cos\vartheta_h - 2}{\mathrm{T}}$$

which means that $\cos\vartheta_h \geq \cos\theta$ holds. If define $\gamma_h := 2\cos\vartheta_h - 2$ and choose

$$\Psi = \begin{bmatrix} 0 & -\mathrm{T} \\ -\mathrm{T} & \gamma_h \end{bmatrix}$$

there exists

$$\sigma(\rho,\Psi) = -e^{j\theta} - e^{-j\theta} + 2 + \gamma_h = 2(\cos\vartheta_h - \cos\theta) \geq 0$$

In this case, we have that the set Λ_h in (9.20) is equal to Λ in Lemma 9.2. Furthermore, we have that

$$(JK)^{-*}\Psi(JK)^{-1} = \begin{bmatrix} \frac{1}{2}(1+\cos\vartheta_h) & j\cos\vartheta_h \\ -j\cos\vartheta_h & \gamma_h \end{bmatrix}$$

Hence, it can be obtained

$$\Psi_o = \mathrm{Re}[(JK)^{-*}\Psi(JK)^{-1}] = \begin{bmatrix} \frac{1}{2}(1+\cos\vartheta_h) & 0 \\ 0 & \gamma_h \end{bmatrix}$$

There exists

$$\det(\Psi_o) = \cos^2\vartheta_h - 1 < 0$$

if and only if $0 < \vartheta_h < \pi$ holds in (9.20). From Lemma 9.3, we obtain that $(\Phi,\Psi) \in \Omega$ satisfying Lemma 9.2 and the set Λ_h in (9.20) is equal to Λ in Lemma 9.2. The corresponding state space condition is given by (9.14) with $L(P,Q)$ in delta domain as

$$L(P,Q) = \Phi \otimes P + \Psi \otimes Q = \begin{bmatrix} \mathrm{T}^2P & \mathrm{T}P - \mathrm{T}Q \\ \mathrm{T}P - \mathrm{T}Q & (2\cos\vartheta_h - 2)Q \end{bmatrix}$$

Lastly, we consider the middle frequency condition

$$\Lambda_m = \left\{ \frac{e^{j\theta}-1}{\mathrm{T}} \,\middle|\, \theta \in R, \vartheta_1 \leq \theta \leq \vartheta_2 \right\} \tag{9.21}$$

Note that the frequency interval condition can be written as

$$|\theta - \vartheta_c| \leq \vartheta_o \quad \text{or} \quad \cos(\theta - \vartheta_c) \geq \cos\vartheta_o \tag{9.22}$$

where $0 \leq \vartheta_o \leq \pi$ and

$$\vartheta_c := \frac{\vartheta_1 + \vartheta_2}{2}, \quad \vartheta_o := \frac{\vartheta_2 - \vartheta_1}{2}$$

Then, we let $\gamma_m := 2\cos\vartheta_o - 2\cos\vartheta_c$ and choose

$$\Psi = \begin{bmatrix} 0 & \mathrm{T}e^{j\vartheta_c} \\ \mathrm{T}e^{-j\vartheta_c} & -\gamma_m \end{bmatrix}$$

there exists

$$\sigma(\rho, \Psi) = 2\cos(\theta - \vartheta_c) - 2\cos\vartheta_o$$

Based on $\cos(\theta - \vartheta_c) \leq \cos\vartheta_o$, we have that $\sigma(\rho, \Psi) \geq 0$. In this case, we have that the set Λ_m in (9.21) is equal to Λ in Lemma 9.2. Next, using the matrix K in (9.18), we can obtain that

$$(JK)^{-*}\Psi(JK)^{-1} = \begin{bmatrix} \frac{1}{4}\gamma_m - \cos\vartheta_c & \sin\vartheta_c - j\cos\vartheta_o \\ \sin\vartheta_c + j\cos\vartheta_o & -\gamma_m \end{bmatrix}$$

and

$$\Psi_o = \mathrm{Re}[(JK)^{-*}\Psi(JK)^{-1}] = \begin{bmatrix} \frac{1}{4}\gamma_m - \cos\vartheta_c & \sin\vartheta_c \\ \sin\vartheta_c & -\gamma_m \end{bmatrix}$$

There exists

$$\det(\Psi_o) = -1 + \cos^2\vartheta_o$$

Due to $\cos^2\vartheta_o \leq 1$, if and only if $0 < \vartheta_o < \pi$. Hence, we have $\det(\Psi_o) < 0$.

From Lemma 9.3, we conclude that $(\Phi, \Psi) \in \Omega$ satisfying Lemma 9.2 and the set Λ_m in (9.21) is equal to Λ in Lemma 9.2. The corresponding state space condition can be obtained from (9.14) with

$$L(P,Q) = \Phi \otimes P + \Psi \otimes Q = \begin{bmatrix} \mathrm{T}^2 P & \mathrm{T}P + \mathrm{T}e^{j\vartheta_c}Q \\ \mathrm{T}P + \mathrm{T}e^{-j\vartheta_c}Q & (2\cos\vartheta_c - 2\cos\vartheta_o)Q \end{bmatrix}$$

In summary, for the delta operator discrete-time setting, we have

$$\Phi := \begin{bmatrix} \mathrm{T}^2 & \mathrm{T} \\ \mathrm{T} & 0 \end{bmatrix}, \; \Lambda = \left\{ \frac{e^{j\theta} - 1}{\mathrm{T}} \middle| \theta \in \Upsilon \right\}$$

where θ is a subset of real numbers specified by Ψ in Table 9.2.

Table 9.2 Comparison of the obtained results in different frequency domain ranges

	LF	MF	HF
Υ	$\lvert\theta\rvert \leq \vartheta_l$	$\vartheta_1 \leq \theta \leq \vartheta_2$	$\lvert\theta\rvert \geq \vartheta_h$
Ψ	$\begin{bmatrix} 0 & \mathrm{T} \\ \mathrm{T} & -2(\cos\vartheta_l - 1) \end{bmatrix}$	$\begin{bmatrix} 0 & \mathrm{T}e^{j\vartheta_c} \\ \mathrm{T}e^{-j\vartheta_c} & -2(\cos\vartheta_o - \cos\vartheta_c) \end{bmatrix}$	$\begin{bmatrix} 0 & -\mathrm{T} \\ -\mathrm{T} & 2(\cos\vartheta_h - 1) \end{bmatrix}$

where $\vartheta_c := (\vartheta_1 + \vartheta_2)/2$ and $\vartheta_c := (\vartheta_2 - \vartheta_1)/2$, and LF, HF and MF stand for low, high and middle frequency ranges, respectively.

Remark 9.5. For the Table 9.1, the homologous conclusions in s-domain and z-domain were shown in [115]. It has been shown that the use of the delta operator not only leads to better numerical conditioning of numerical algorithms but also ensures that the discrete-time results converge to the continuous results as $\mathtt{T} \to 0$ [82]. This latter feature is useful in unifying discrete-time and continuous-time control theory and software. Thus we have obtained interesting and valuable results in Table 9.1 of this subsection. So do as in Table 9.2.

9.3.2 Gain Properties

In this subsection, we consider some gain conditions which will be used in fault detection filter design. For the linear DOS (9.10)-(9.11), matrix Θ in (9.17) can be chosen as follows

$$\Theta = \begin{bmatrix} C & D \\ 0 & I \end{bmatrix}^* \Pi \begin{bmatrix} C & D \\ 0 & I \end{bmatrix}$$

Then, (9.16) can be rewritten as

$$\begin{bmatrix} G(\rho) \\ I \end{bmatrix}^* \Pi \begin{bmatrix} G(\rho) \\ I \end{bmatrix} < 0 \tag{9.23}$$

Firstly, the small gain condition (H_∞ performance index) is obtained by choosing

$$\Pi := \Pi_\gamma = \begin{bmatrix} I & 0 \\ 0 & -\gamma^2 I \end{bmatrix}$$

Substituting Π_γ into (9.23), we have

$$\lambda_{\max}(G(\rho)) < \gamma.$$

It means that the small gain condition (H_∞ performance index) is satisfied.

Then, we consider the high gain condition (H_- performance index) by letting

$$\Pi := \Pi_\beta = \begin{bmatrix} -I & 0 \\ 0 & \beta^2 I \end{bmatrix}$$

Substituting Π_β into (9.23), there exists

$$\lambda_{\max}(G(\rho)) > \beta.$$

which means that the high gain condition (H_- performance index) is satisfied.

In summary, the above gain properties can be summed up for matrix Π giving in Table 9.3, where SG and HG stand for small gain condition and high gain condition, respectively.

Table 9.3 The matrix Π in (9.23)

	SG	HG
Π	$\begin{bmatrix} I & 0 \\ 0 & -\gamma^2 I \end{bmatrix}$	$\begin{bmatrix} -I & 0 \\ 0 & \beta^2 I \end{bmatrix}$

Remark 9.6. By choosing different Π in (9.23), we have different gain properties. In general, any gain properties expressed in terms of a quadratic form can be characterized as there is no restriction on the choice of Π.

9.4 Application to Fault Detection

The following lemmas will be used to prove our main results.

Lemma 9.7. *[285] The property of delta operator: for any time-varying function $x(t)$ and $y(t)$*

$$\delta(x(t)y(t)) = \delta(x(t))y(t) + x(t)\delta(y(t)) + T\delta(x(t))\delta(y(t)),$$

where T is a sampling period.

Lemma 9.8. *[62] (Projection Lemma) Let Γ, Θ and Λ be given. There exists a matrix F satisfying*

$$\Gamma F \Lambda + (\Gamma F \Lambda)^T + \Theta < 0$$

if and only if the following two conditions hold

$$\Gamma^{\perp} \Theta \Gamma^{\perp^T} < 0, \quad \Lambda^{T^{\perp}} \Theta \Lambda^{T^{\perp^T}} < 0$$

Lemma 9.9. *[238] Considering the same transfer function matrix $G(\rho) = C(\rho I - A)^{-1}B + D$, giving R with appropriate dimensions, the frequency inequality (9.16) in Theorem 9.4 is satisfied if there exist matrix variables X, Y and V of appropriate dimensions, satisfying $\mathcal{X} \in X(\mathcal{C}, R)$ and*

$$T \begin{bmatrix} \Xi & 0 \\ 0 & \Pi \end{bmatrix} T^T < \begin{bmatrix} -\mathcal{X} \\ \mathcal{A}\mathcal{X} + \mathcal{B}YR \end{bmatrix} + \begin{bmatrix} -\mathcal{X} \\ \mathcal{A}\mathcal{X} + \mathcal{B}YR \end{bmatrix}^T$$

where

$$X(\mathcal{C}, R) = \{\mathcal{C}^{\dagger} X R + (I - \mathcal{C}^{\dagger}\mathcal{C})V \mid \det(X) \neq 0\}$$
$$\mathcal{A} = \begin{bmatrix} A^T & C^T \\ B^T & D^T \end{bmatrix}, \ \mathcal{B} = \begin{bmatrix} -C^T \\ -D^T \end{bmatrix}, \ \mathcal{C} = [I \ 0]$$

And T is the permutation matrix defined as

$$\begin{bmatrix} M_1 \; M_2 \; M_3 \; M_4 \end{bmatrix} T = \begin{bmatrix} M_1 \; M_3 \; M_2 \; M_4 \end{bmatrix}$$

9.4.1 Fault Detection Filter

Theorem 9.10. *The error DOS (9.6) is asymptotically stable. It also satisfies the H_∞ norm bound γ and H_- norm bound β in finite middle frequency range sets Ω_d and Ω_f, respectively. If there exist some symmetric positive definite matrices $P > 0$, $Q_d > 0$ and $Q_f > 0$, and some real matrices P_d, P_f, X, Y, V_{id} and V_{if} $(i = 1, 2, \cdots, 6)$, such that the following LMIs hold:*

$$\begin{bmatrix} \mathcal{T}P - qX - qX^T & P + pX + qX^T A - qY^T C \\ * & -pX^T A + pY^T C - pA^T X + pC^T Y \end{bmatrix} < 0 \tag{9.24}$$

$$\begin{bmatrix} \mathcal{T}^2 P_d & V_{1d}^T & \Phi(1,3) & -V_{1d}^T D_d \\ * & I + V_{2d}^T + V_{2d} & V_{3d} - V_{2d}^T C & V_{4d} - V_{2d}^T D_d \\ * & * & \Phi(3,3) & \Phi(3,4) \\ * & * & * & -\gamma^2 I - D_d^T V_{4d} - V_{4d}^T D_d \end{bmatrix} < 0 \tag{9.25}$$

$$\begin{bmatrix} \mathcal{T}^2 P_f + X^T + X & V_{1f}^T & \Psi(1,3) & \Psi(1,4) \\ * & -I + V_{2f}^T + V_{2f} & V_{3f} - V_{2f}^T C & V_{4f} - V_{2f}^T D_f \\ * & * & \Psi(3,3) & \Psi(3,4) \\ * & * & * & \Psi(4,4) \end{bmatrix} < 0 \tag{9.26}$$

where p and q are arbitrary fixed scalars satisfying $p(\mathcal{T}p + 2q) < 0$, and with

$$\begin{aligned}
\Phi(1,3) &= X - V_{1d}^T C + \mathcal{T}P_d + \mathcal{T}e^{j\vartheta_{dc}} Q_d, \\
\Phi(3,4) &= -C^T V_{4d} - X^T B_d - V_{3d}^T D_d + Y^T D_d \\
\Phi(3,3) &= -A^T X - X^T A - C^T V_{3d} - V_{3d}^T C + C^T Y + Y^T C \\
&\quad -2(\cos\vartheta_{do} - \cos\vartheta_{dc}) Q_d, \\
\Psi(1,4) &= -X B_f - X^T B_f + Y^T D_f - V_{1f}^T D_f \\
\Psi(1,3) &= X - V_{1f}^T C - X^T A + Y^T C + \mathcal{T}P_f + \mathcal{T}e^{j\vartheta_{fc}} Q_f \\
\Psi(3,3) &= -A^T X - X^T A - C^T V_{3f} - V_{3f}^T C + C^T Y + Y^T C \\
&\quad -2(\cos\vartheta_{fo} - \cos\vartheta_{fc}) Q_f, \\
\Psi(3,4) &= A^T X B_f - C^T V_{4f} - C^T Y B_f - X^T B_f - V_{3f}^T D_f + Y^T D_f \\
\Psi(4,4) &= \beta^2 I + B_f^T X B_f + B_f^T X^T B_f - D_f^T Y B_f - B_f^T Y^T D_f \\
&\quad -D_f^T V_{4f} - V_{4f}^T D_f.
\end{aligned}$$

Then, we have $Y = L^T X$.

Proof. (i) Asymptotic stability condition. Define the following Lyapunov function in delta domain:

$$V(x(t_k)) = x^T(t_k)Px(t_k)$$

By using Lemma 9.8 and taking the delta operator manipulations of $V(x(t_k))$ along the trajectory of system (9.6), it can be obtained that

$$\begin{aligned}\delta V(x(t_k)) &= \delta^T x(t_k)Px(t_k) + x^T(t_k)P\delta x(t_k) + \mathrm{T}\delta^T x(t_k)P\delta x(t_k)\\ &= x^T(t_k)(A-LC)^T Px(t_k) + x^T(t_k)P(A-LC)x(t_k)\\ &+\mathrm{T}\delta^T x(t_k)P\delta x(t_k).\end{aligned}$$

Based on Definition 1.2, there exists

$$(A-LC)^T P + P(A-LC) + \mathrm{T}(A-LC)^T P(A-LC) < 0 \qquad (9.27)$$

such that $\delta V(x(k)) < 0$. It is clearly shown from (9.27) that

$$\begin{bmatrix} A-LC \\ I \end{bmatrix}^T \begin{bmatrix} \mathrm{T}P & P \\ P & 0 \end{bmatrix} \begin{bmatrix} A-LC \\ I \end{bmatrix} < 0 \qquad (9.28)$$

Since the nullspace of $\begin{bmatrix} -qI & pI \end{bmatrix}^T$ is $\begin{bmatrix} pI & qI \end{bmatrix}$, we get

$$\begin{bmatrix} pI & qI \end{bmatrix} \begin{bmatrix} \mathrm{T}P & P \\ P & 0 \end{bmatrix} \begin{bmatrix} pI \\ qI \end{bmatrix} = \mathrm{T}p^2P + 2pqP < 0$$

Note that $\begin{bmatrix} (A-LC)^T & I \end{bmatrix}$ is the nullspace of $\begin{bmatrix} -I & A-LC \end{bmatrix}^T$. By using Lemma 9.8, the inequality (9.28) is equivalent to

$$\begin{bmatrix} \mathrm{T}P & P \\ P & 0 \end{bmatrix} < \begin{bmatrix} -I \\ (A-LC)^T \end{bmatrix} X \begin{bmatrix} -qI \\ pI \end{bmatrix}^T + \begin{bmatrix} -qI \\ pI \end{bmatrix} X^T \begin{bmatrix} -I \\ (A-LC)^T \end{bmatrix}^T \qquad (9.29)$$

The inequality (9.29) implies LMI (9.24) holds. Therefore, we have that the error delta operator dynamic system (9.6) with $d(k) = 0$ and $f(k) = 0$ is asymptotically stable.

(ii) Disturbance attenuation condition. Based on Condition Bb, the disturbance to residual transfer function matrix $G_{rd}(\frac{e^{j\theta}-1}{\mathrm{T}})$ satisfies

$$\lambda_{\min}\left(G_{rd}\left(\frac{e^{j\theta}-1}{\mathrm{T}}\right)\right) < \gamma, \ \forall\, \theta \in \Omega_d.$$

It can be obtain that the disturbance attenuation condition can be rewritten as follows

$$\begin{bmatrix} G_{rd}(\frac{e^{j\theta}-1}{\mathrm{T}}) \\ I \end{bmatrix}^T \Pi_d \begin{bmatrix} G_{rd}(\frac{e^{j\theta}-1}{\mathrm{T}}) \\ I \end{bmatrix} < 0, \ \vartheta_{d1} \le \theta \le \vartheta_{d2} \qquad (9.30)$$

By using Theorem 9.4, we have the frequency condition (9.30) is equivalent to

$$\begin{bmatrix} \mathcal{A} & \mathcal{B}_d \\ I & 0 \end{bmatrix}^T \Xi_d \begin{bmatrix} \mathcal{A} & \mathcal{B}_d \\ I & 0 \end{bmatrix} + \begin{bmatrix} C & D_d \\ 0 & I \end{bmatrix}^T \Pi_d \begin{bmatrix} C & D_d \\ 0 & I \end{bmatrix} < 0 \tag{9.31}$$

where $\vartheta_{dc} = (\vartheta_{d1} + \vartheta_{d2})/2$, $\vartheta_{do} = (\vartheta_{d2} - \vartheta_{d1})/2$ and

$$\Xi_d = \begin{bmatrix} \mathrm{T}^2 P_d & \mathrm{T} P_d + \mathrm{T} e^{j\vartheta_{dc}} Q_d \\ \mathrm{T} P_d + \mathrm{T} e^{-j\vartheta_{dc}} Q_d & -2(\cos\vartheta_{do} - \cos\vartheta_{dc}) Q_d \end{bmatrix}, \ \Pi_d = \begin{bmatrix} I & 0 \\ 0 & -\gamma^2 I \end{bmatrix}$$

in which P_d and Q_d are two Hermitian matrices with $Q_d > 0$.

By using Lemma 9.9, we have that (9.31) is rewritten as the following inequalities

$$T \begin{bmatrix} \Xi_d & 0 \\ 0 & \Pi_d \end{bmatrix} T^T < \begin{bmatrix} -\mathcal{X} \\ \mathcal{A}\mathcal{X} + \mathcal{B}_d Y E \end{bmatrix} + \begin{bmatrix} -\mathcal{X} \\ \mathcal{A}\mathcal{X} + \mathcal{B}_d Y E \end{bmatrix}^T \tag{9.32}$$

with

$$\mathcal{X}(\mathcal{C}, E_d) = \{\mathcal{C}^\dagger X E_d + (I - \mathcal{C}^\dagger \mathcal{C})\mathcal{V}_d \mid \det(X) \neq 0\}$$
$$\mathcal{A} = \begin{bmatrix} A^T & C^T \\ B_d^T & D_d^T \end{bmatrix}, \ \mathcal{B}_d = \begin{bmatrix} -C^T \\ -D_d^T \end{bmatrix}, \ \mathcal{C} = \begin{bmatrix} I & 0 \end{bmatrix}$$

Next we denote

$$\mathcal{X} = \begin{bmatrix} I \\ 0 \end{bmatrix} X E_d + \begin{bmatrix} 0 & 0 \\ 0 & I \end{bmatrix} \mathcal{V}_d, \quad \mathcal{V}_d = \begin{bmatrix} \mathcal{V}_d^{\imath} \\ \mathcal{V}_d^{\jmath} \end{bmatrix}$$
$$\mathcal{V}_d^{\jmath} = \begin{bmatrix} V_{1d} & V_{2d} & V_{3d} & V_{4d} \end{bmatrix}, \ E_d = \begin{bmatrix} 0 & 0 & I & 0 \end{bmatrix}, \ Y = L^T X.$$

After some matrix manipulations, we obtain that (9.25) is equivalent to (9.32). That is, LMI condition (9.25) is satisfied, if inequality condition (9.32) holds. Therefore, we have that the error delta operator dynamic system (9.6) with $f(k) = 0$ satisfies the H_∞ norm bound γ for the finite middle frequency range in Ω_d if LMI condition (9.25) holds.

(iii) Lastly, we give the fault sensitivity condition by using the matrix $\Pi_f = \Pi_\beta$ and letting

$$E_f = \begin{bmatrix} I & 0 & I & -B_f \end{bmatrix}.$$

Following the same lines to that for the disturbance attenuation condition, it is immediate.

9.4.2 Filter Algorithm

According to the method in [30], the combined constraints A(1-3) such as in the way of

$$\frac{\|G_{rf}\|_-}{\|G_{rf}\|_\infty} > \frac{\beta}{\gamma} > \alpha$$

can, of course, be chosen and this combination also makes sense. However, this is not recommended. The reasons are as follows. Firstly, the Constraints B(a-c) provide directly quantitative measures for sensitivity of a fault detection observer. The value γ is very useful for threshold selection in detection decision-making. The ratio β/γ indicates how good a designed fault detection filter is and therefore can be used for evaluation of fault detection filters. Secondly, the present problem formulation enables the direct time-domain solution of the fault detection filter problem by using the LMI approach. Finally, as will be shown, the fault detection is equivalent to a constrained H_∞ estimation problem, the latter can be further reformulated as standard problem of constrained optimization.

Thus, we give the following algorithm:

Algorithm 9.1. *Given a scalar β, the search problem of the lowest possible value of γ and make the error delta operator dynamic system (9.6) be asymptotically stable and formulated as the following convex optimization problem:*

$$\begin{gathered} \min\ \gamma \\ s.t.\ (9.24),\ (9.25)\ and\ (9.26) \end{gathered} \tag{9.33}$$

which can be effectively solved by the existing Matlab LMI toolbox [63].

In the end, the fault detection filter gain matrix L can be given as $L = X^{-T}Y^T$. That is, we have finished the design of fault detection filter (9.3)-(9.4) for the DOS (9.1)-(9.2).

9.4.3 Threshold Design

Based on the above analysis, the fault detection problem for the DOS (9.1)-(9.2) can be viewed as finding the appropriate fault detection filter to make the system be asymptotically stable, minimize the effects of disturbances and enhance the effects of faults.

In order to detect the faults, the widely adopted approach is to choose an appropriate threshold J_{th} and determine the evaluation function $J_r(n)$. That is, after the detection filter is designed, a threshold J_{th} and a residual evaluation function $J_r(n)$ can be selected

$$J_r(n) = \sqrt{\frac{\mathtt{T}}{n}\sum_{k=k_0}^{k_0+n} r^T(t_k)r(t_k)} \tag{9.34}$$

$$J_{th} = \sup_{f=0} J_r(n) \tag{9.35}$$

where k_0 denotes the initial evaluation time instant, n denotes the evaluation time steps. Based on this, the occurrence of faults can be detected by the following logic rule

$$\begin{cases} J_r(n) \le J_{th} & \text{no alarm} \\ J_r(n) > J_{th} & \text{alarm} \end{cases} \tag{9.36}$$

Usually, a threshold function is chosen according to the test. It has been point in page 28 of [30] that there are many ways of defining evaluation functions and determining thresholds. We choose the threshold as a constant positive value in the paper. As shown in (9.36), if the threshold is exceeded by the residual evaluation function, we can hypothesize that a fault is likely. This finished the threshold design.

9.5 Numerical Examples

In the following, we provide a numerical example to demonstrate the effectiveness of the proposed methods in this paper.

Example 9.11. we consider a longitudinal dynamics of the VTOL aircraft described in [174]. The states are the horizontal velocity (knot), vertical velocity (knot), pitch rate (degree/s) and pitch angle (degree), respectively. The inputs are the collective pitch control and the longitudinal cyclic pitch control. By using the delta operator approach and letting the sampling period be $\mathtt{T} = 0.01$, the following DOS is obtained as

$$\delta x(t_k) = Ax(t_k) + Bu(t_k) + B_d d(t_k) + B_f f(t_k) \tag{9.37}$$

$$y(t_k) = Cx(t_k) + D_d d(t_k) + D_f f(t_k) \tag{9.38}$$

The system parameters are given as follows:

$$A = \begin{bmatrix} -9.0056 & -0.6622 & 0.2243 & 4.5150 \\ 48.8375 & 2.5039 & 5.2740 & -23.1504 \\ 22.8476 & 2.3091 & -3.9177 & -17.1989 \\ 0.3584 & 0.0362 & 0.9399 & -0.2680 \end{bmatrix}, \quad B_f = \begin{bmatrix} 0.3242 & 0.2460 \\ 3.5687 & -7.3930 \\ -4.9013 & 4.0082 \\ -0.0765 & 0.0625 \end{bmatrix},$$

$$B = B_d = \begin{bmatrix} 0.0036 & -0.0103 \\ 0.0805 & 1.0387 \\ 0.9399 & 0.0362 \\ 0.0144 & 0.0004 \end{bmatrix}, \quad C = \begin{bmatrix} 1 & 0 & 0 & 0 \\ 0 & 1 & 0 & 0 \\ 0 & 0 & 1 & 0 \\ 0 & 1 & 1 & 1 \end{bmatrix}, \quad D_d = D_f = \begin{bmatrix} 0 & 0.2 \\ 0 & 0.1 \\ 0.3 & 0 \\ 0 & 0 \end{bmatrix}.$$

Assume that the frequency range of disturbances is $-0.1 \le \theta_d \le 0.1$ and the frequency range of faults is $0.1 \le \theta_f \le 0.2$. Letting $p = -1$, $q = 1$ and $\beta = 0.4$, using Algorithm 9.1, we can solving the optimization problem (9.33). Then, the fault detection observer gain matrix L is obtained as follows

$$L = \begin{bmatrix} -3.8772 & -0.1365 & 1.3652 & 6.6254 \\ -13.1802 & -4.6767 & 7.0575 & -43.5879 \\ 24.1047 & 4.8022 & -12.5520 & -13.1925 \\ -0.2136 & 0.8867 & 1.5964 & -1.1116 \end{bmatrix}.$$

To analyze the effects of faults and disturbances on residual of the detection filter, we consider the following struck faults

$$f(k) = \begin{cases} \begin{bmatrix} 0.05 \ 0.04 \end{bmatrix}^T & 1 \ge t_k \ge 3 \\ \begin{bmatrix} 0 \ 0 \end{bmatrix}^T & \text{elsewhere} \end{cases}$$

Let the disturbance be

$$d(t_k) = \begin{bmatrix} 0.02\cos(k) \ 0.02\sin(k) \end{bmatrix}^T$$

Considering the fact that a real state vector in DOS can be replaced by an estimated state vector using the fault detection observer obtained in Theorem 9.10. We firstly give the simulation results of the state estimate response of the DOS in this example for initial condition: $\hat{x}_1(0) = \hat{x}_2(0) = \hat{x}_3(0) = \hat{x}_4(0) = 0$, shown in Fig. 9.1, where $\hat{x}_1(k)$, $\hat{x}_2(k)$, $\hat{x}_3(k)$, $\hat{x}_4(k)$ are denoted by xo_1(k), xo_2(k), xo_3(k) and xo_4(k), respectively.

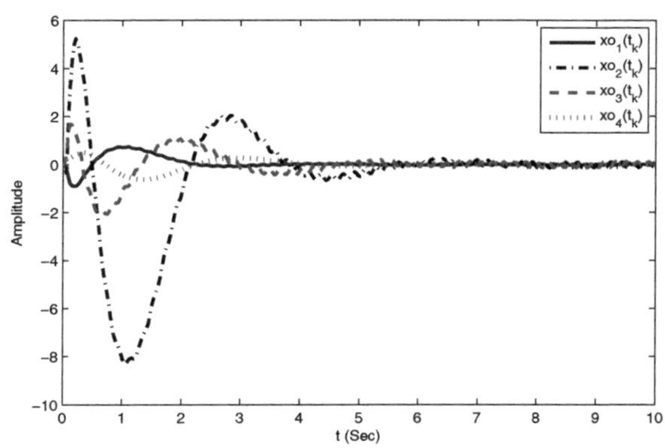

Fig. 9.1 The state estimate response of $\hat{x}(t_k)$.

With zero initial condition, the state response are given in Fig. 9.2.

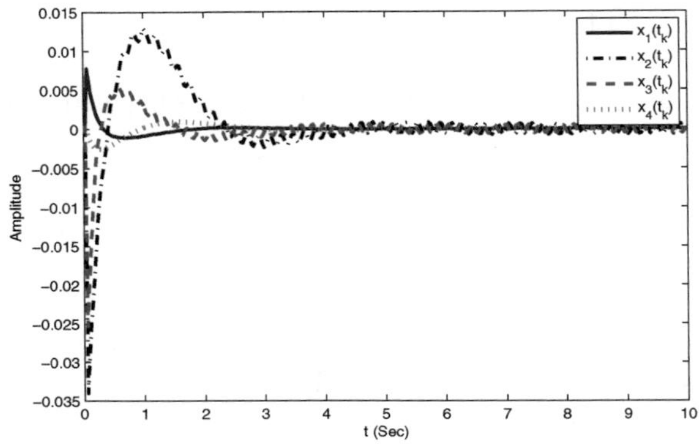

Fig. 9.2 The state response of $x(t_k)$.

Then, the residual outputs are shown in Fig. 9.3 with initial condition $r(0) = 0$, from which we can see that the faults are well discriminated from disturbances.

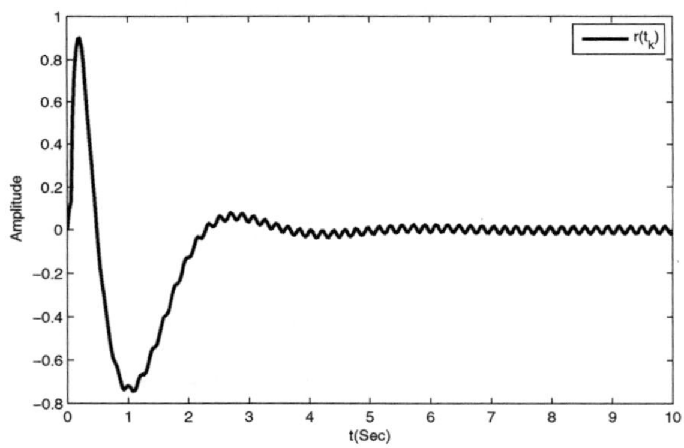

Fig. 9.3 The residual output $r(t_k)$.

To detect the fault, we choose the residual evaluation function as stated in (9.34), and the residual evaluation output is shown in Fig. 9.4, where $J_r(n)$ and J_{th} are denoted by Jrn and Jth, respectively.

In this figure, the dashed line denotes the threshold which is 15.5. We can conclude that the fault can be detected as $t_k = 1.2$ by using the logic rule (9.36).

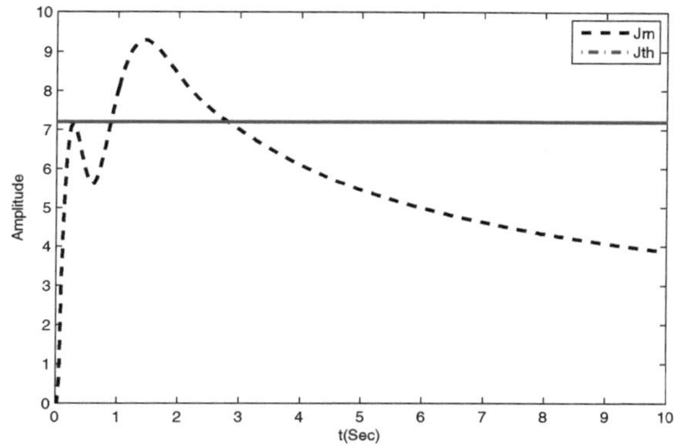

Fig. 9.4 Residual evaluation $J_r(n)$.

Furthermore, the frequency behaviors of H_∞ and H_- performance indexes for maximal singular value of $G_{rd}(\rho)$ and minimal singular value of $G_{rf}(\rho)$ are given to show the potential of the finite frequency approach in delta domain in Fig. 9.5 and Fig. 9.6, respectively.

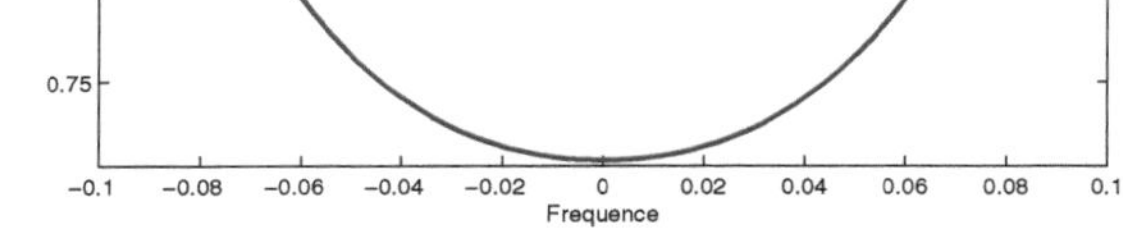

Fig. 9.5 Frequency behaviors of maximal singular value of $G_{rd}(\rho)$.

It has to be point out that we have chosen $\beta = 0.4$ for using Algorithm 9.1 in this example, thus we adjust the scale of vertical axis in complex plane such that the minimal singular value of $\|G_{rf}(\rho)\|_-$ is 0.4 shown in Fig. 9.6. Though there will exist some deviations when adjust the scale of vertical axis, the shape of the curve is unchanged. It can be seen from Fig. 9.6 that the the minimal value of $\|G_{rf}(\rho)\|_-$ is being in the interval $[0.1,\ 0.2]$. The same manipulate is also done for Fig. 9.5, which shows that the maximal value of $\|G_{rf}(\rho)\|_\infty$ is 0.95 being in the interval $[-0.1,\ 0.1]$.

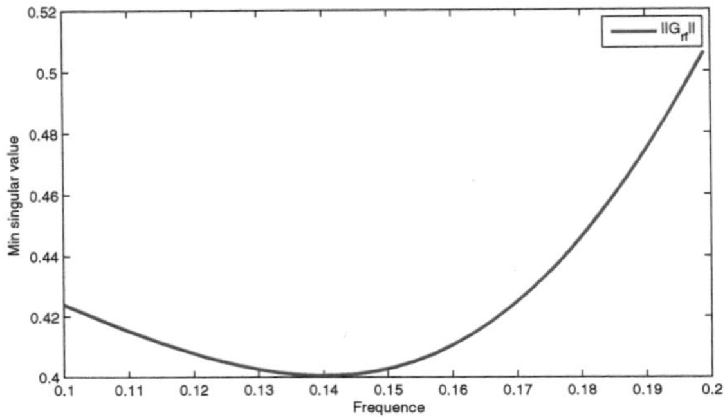

Fig. 9.6 Frequency behaviors of minimal singular value of $G_{rf}(\rho)$.

Remark 9.12. The emphasis of this paper is the theoretical analysis on the fault detection method in finite frequency for DOSs. The effects of modeling disturbances and faults may occupy in different frequency ranges and the numerical advantage of the delta operator approach is clear for short sampling period. Hence, the combination of fault detection and delta operator approach in finite frequency is an interesting idea to consider. Furthermore, the fault isolation with H_∞ performance has been considered for a class of nonlinear systems in [2]. However, the fault isolation in this paper is based on the modeling disturbances and faults occupy in different frequency ranges. The drawbacks of the suggested scheme in this paper appears if the modeling disturbances and faults occupy in the same frequency ranges. This paper has also some highlight advantages with respect to different approaches available in the related literature. Some comparison results are given in Table 9.4.

Table 9.4 Comparisons of the obtained results with others

References	Contributions
[56, 302]	Fault detection for full frequency.
[93, 121]	H_-/H_∞ fault detection.
[4, 285]	Modelling or filtering in delta domain.
[236, 238]	H_-/H_∞ fault detection for finite frequency.
[2]	H_∞ fault detection and isolation on an aircraft.
This paper	H_-/H_∞ fault detection for finite frequency in delta domain.

9.6 Summary

This chapter has generalizeed the KYP lemma in delta domain to treat some conditions on different restricted frequency ranges. The frequency ranges are characterized by two quadratic forms. This characterization encompasses low/middle/high frequency conditions for delta operator systems. By utilizing the obtained delta domain generalized KYP lemma, a fault detection filter design problem has been dealt with for DOSs in finite frequency domain. The fault detection filter design problem has been formulated as a two-objective optimization algorithm with H_- and H_∞ performance indexes. A numerical example has been given to illustrate the effectiveness and potential for the developed techniques.

Part V
Networked Control Using DOSs

Chapter 10
Stability of Markovian Jump Systems over Networks

10.1 Introduction

Delta operator based implementations have gained interest due to their excellent finite word length performance under fast sampling [82, 170]. Delta operator models have been studied extensively, for example [137, 142, 176, 177, 194]. Some control problems have also been obtained in recent years, such as [33, 123, 215, 222, 307]. Recently, a robust SMC method has been considered for a class of time-delay DOSs [254]. Furthermore, Markovian jump systems have attracted a lot of researchers from control and operations research communities, for example [95, 208, 211, 245, 247], The stability and stabilization problems for both continuous-time and discrete-time Markovian jump systems were investigated in [298]. Some results on Markovian jump systems with time delays have also been given in [299, 300].

In the last decades, the general theory for NCSs have been widely investigated, for example [66, 73, 169, 308], and the references therein. A control performance using the H_∞ gain was measured for NCSs [205]. The stabilization problems of NCSs with random delays described by discrete-time systems and continuous-time systems were investigated in [301] and [108], respectively. The stability analysis of discrete-time NCSs subject to packet loss was concerned in [97]. The kind of Markovian jump systems has been widely used in NCSs. A stabilization problem for NCSs with Markovian characterization has been investigated in [157]. The kind of mode-dependent time delays, which means the combination of switching mechanism and time delays, is more in line with the actual situation than others for networked control models [271]. To the best of our knowledge, there have been few papers on stability problems of Markovian jump systems with mode-dependent time delays via delta operator approach, and less papers using the kind of Markovian jump DOSs describe a NCS. Both of the two sides motivate us to make an effort in this chapter.

H. Yang et al.: Analysis and Synthesis of Delta Operator Systems, LNCIS 430, pp. 159–174.
springerlink.com
© Springer-Verlag Berlin Heidelberg 2012

The main contributions of this chapter are summarized as below:

i A Markov process transition rate matrix and some new results on the Markov chain in delta domain are given in this chapter.
ii We describe NCSs by a class of Markovian jump DOSs with mode-dependent time delays.
iii A stochastic asymptotically stable criterion is given for Markovian jump DOSs with mode-dependent time delays.

10.2 Problem Statement

The networked control model is shown as in Fig. 10.1, in which we assume that there only exists networks in feedback channel.

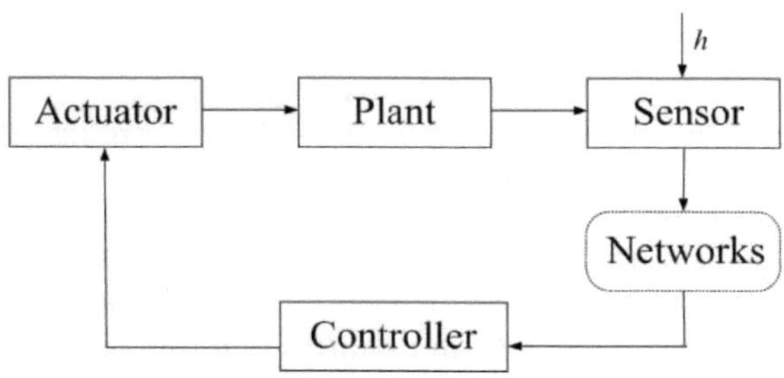

Fig. 10.1 Schematic diagram of NCS.

In the NCS, the plant is continuous and controlled by a digital controller, the signals of the plant are synchronously measured with the sensor at a sampling rate $1/h$. The digital controller uses the information of the plant transmitted through the networks to generate a digital control action. In this scenario, the controller and the sensor are working in time-driven mode, while the actuator is working in event-driven mode. The Markovian jump continuous system plant is described as

$$\dot{x}(t) = A_s(\eta_t)x(t) + B_s(\eta_t)u(t) \tag{10.1}$$

where $x(t) \in R^p$ and $u(t) \in R^q$ are the state vector and control input vector in s-domain. $A_s(\eta_t)$ and $B_s(\eta_t)$ are known constant matrices for each fixed $\eta_t = i$. Here, (η_t) is a discrete-state Markov process with values in a finite set $S = \{1,\ 2,\ \cdots, s\}$. The transition probability from mode i at time t to the mode j at time $t + \Delta t,\ i,\ j \in S$, is

$$Pr(\eta_{t+\Delta t} = j|\ \eta_t = i) = \begin{cases} \vartheta_{ij}\Delta t + o(\Delta t), & i \neq j, \\ 1 + \vartheta_{ii}\Delta t + o(\Delta t), & i = j, \end{cases} \tag{10.2}$$

where $\Delta t > 0$, $\lim_{\Delta t \to 0} o(\Delta t)/\Delta t = 0$. The continuous Markov process transition rate matrix Λ_s is defined by

$$\Lambda_s = \begin{bmatrix} \vartheta_{11} & \vartheta_{12} & \cdots & \vartheta_{1s} \\ \vartheta_{21} & \vartheta_{22} & \cdots & \vartheta_{2s} \\ \vdots & \vdots & \ddots & \vdots \\ \vartheta_{s1} & \vartheta_{s2} & \cdots & \vartheta_{ss} \end{bmatrix}, \quad \text{with} \quad \vartheta_{ij} \geq 0, \quad \vartheta_{ii} = -\sum_{j=1, j \neq i}^{s} \vartheta_{ij}. \tag{10.3}$$

Letting $\Pi_s(t)$ be the transition probability matrix of continuous Markov chain, there exists

$$\Lambda_s = \lim_{\Delta t \to 0} \frac{\Pi_s(\Delta t) - \Pi_s(0)}{\Delta t} = \Pi_s'(0) \quad \text{and} \quad \Pi_s(\Delta t) = e^{\Lambda_s \Delta t}. \tag{10.4}$$

Firstly, utilizing traditional shift operator to sample the Markovian jump continuous system (10.1), when sampling period h is 1, the following Markovian jump discrete system can be obtained

$$x(k+1) = A_z(\gamma_k)x(k) + B_z(\gamma_k)u(k), \tag{10.5}$$

where $x(k) \in R^p$ and $u(k) \in R^q$ are the state vector and control input vector in z-domain. $A_z(\gamma_k) = e^{A_s(\eta_t)h}$ and $B_z(\gamma_k) = \int_0^h e^{A_s(\eta_t)(h-s)} B_s(\eta_t) ds$ are known constant matrices. Moreover, when sampling is fast, that is $h \to 0$, there exist

$$\lim_{h \to 0} A_z(\gamma_k) = \lim_{h \to 0} e^{A_s(\eta_t)h} = I, \tag{10.6}$$

$$\lim_{h \to 0} B_z(\gamma_k) = \lim_{h \to 0} \int_0^h e^{A_s(\eta_t)(h-s)} B_s(\eta_t) ds = \mathbf{0}. \tag{10.7}$$

For the Markovian jump discrete system (10.5), the system mode $\gamma_k = k$ is a time homogeneous Markov chain taking values in a finite set $S = \{1,\ 2,\ \cdots, s\}$ with stationary transition probabilities from mode i at time k to mode j at time $k+1$

$$Pr(\gamma_{k+1} = j |\ \gamma_k = i) = \pi_{ij} \tag{10.8}$$

The Markov process transition probability matrix Π_z is defined by

$$\Pi_z = \begin{bmatrix} \pi_{11} & \pi_{12} & \cdots & \pi_{1s} \\ \pi_{21} & \pi_{22} & \cdots & \pi_{2s} \\ \vdots & \vdots & \ddots & \vdots \\ \pi_{s1} & \pi_{s2} & \cdots & \pi_{ss} \end{bmatrix}, \quad \text{with} \quad \pi_{ij} \geq 0, \quad \sum_{j=1}^{s} \pi_{ij} = 1. \tag{10.9}$$

where $\Pi_z = \Pi_s(1) = e^{\Lambda_s}$.

Secondly, delta operator approach is used to sample the Markovian jump continuous system (10.1). For the reason of that only discrete system model is valuable for NCSs, the delta operator in this chapter is defined as

$$\delta x(kh) = \frac{x(kh+h) - x(kh)}{h} \tag{10.10}$$

It can be seen from (10.10) that the delta operator method can unify continuous and discrete systems into the DOSs framework. By using the delta operator approach, the sampling system in δ-domain of the Markovian jump continuous system (10.1) is given as follows

$$\delta x(kh) = A(\eta_k)x(kh) + B(\eta_k)u(kh), \tag{10.11}$$

where $x(kh) \in R^p$ and $u(kh) \in R^q$ are the state vector and control input vector in δ-domain. $A(\eta_k) = (A_z(\gamma_k) - I)/h$ and $B(\eta_k) = B_z(\gamma_k)/h$ are known constant matrices. When $h \to 0$, there exist

$$\lim_{h\to 0} A(\eta_k) = \lim_{h\to 0}(e^{A_s(\eta_t)h} - I)/h = A_s(\eta_t), \tag{10.12}$$

$$\lim_{h\to 0} B(\eta_k) = \lim_{h\to 0} \int_0^h e^{A_s(\eta_t)(h-s)} B_s(\eta_t)ds/h = B_s(\eta_t). \tag{10.13}$$

where (η_k) is a discrete-state Markov process with values in a finite set $S = \{1,\ 2,\ \cdots, s\}$. The Markov process transition rate matrix Λ_δ is defined by

$$\Lambda_\delta = \frac{\Pi_s(h) - \Pi_s(0)}{h - 0} = \frac{e^{\Lambda_s h} - I}{h} = \begin{bmatrix} \lambda_{11} & \lambda_{12} & \cdots & \lambda_{1s} \\ \lambda_{21} & \lambda_{22} & \cdots & \lambda_{2s} \\ \vdots & \vdots & \ddots & \vdots \\ \lambda_{s1} & \lambda_{s2} & \cdots & \lambda_{ss} \end{bmatrix}. \tag{10.14}$$

Remark 10.1. The Markov process transition rate matrix Λ_s is a special case of Λ_δ, since there exists

$$\lim_{h\to 0} \Lambda_\delta = \lim_{h\to 0} \frac{e^{\Lambda_s h} - I}{h} = \lim_{h\to 0} \Lambda_s e^{\Lambda_s h} = \Lambda_s. \tag{10.15}$$

Furthermore, the transition rate matrix Λ_δ also covers the transition probability matrix Π_z which is relative to Π_s for $\Pi_z = \Pi_s(1)$. Therefore, it can be said that the continuous-time Markov process transition rate matrix Λ_s and the Markov process transition probability matrix Π_z are the special case of the Markov process transition rate matrix Λ_δ in δ-domain when sampling period $h \to 0$ and $h = 1$, respectively.

It is well known that network-induced time delays and data packets dropout are two important issues to be dealt with. In this chapter, for convenience of analysis, it is assumed that random communication delay occurs only in feedback channel. Together with each mode in the Markov process, the corresponding delay is assumed to be time-varying but upper bounded a known constant. As is shown in Fig. 10.1, system state $x(t)$ and system mode η_k are transmitted over networks. Considering the effect of random communication delay $\tau(\eta_k)$, the following mode-dependent state-feedback control law is described by

$$u(kh) = u(i_k h) = K(\eta_k)x(kh - \tau(\eta_k)), \tag{10.16}$$

where i_k, $(k = 1,\ 2,\ 3,\cdots)$ are some integers and $\{i_1,\ i_2,\ i_3,\cdots\} \subset \{1,\ 2,\ 3,\cdots\}$. Since $i_k h = kh - (kh - i_k h)$, define $\tau(\eta_k) = kh - i_k h$, which denotes the time-varying delay in the control signal. And $d_m \le \tau(\eta_k) \le d_M$ with $d_m = n_m h$ and $d_M = n_M h$, n_m and n_M are two known positive and finite integers, from which we have $n_m \le n(\eta_k) \le n_M$. Obviously, there exists

$$\begin{aligned} &\tau_k \le \tau(\eta_k) \le (i_{k+1} - i_k)h + \tau_{k+1}, \\ &kh \in [i_k h + \tau_k,\ i_{k+1}h + \tau_{k+1}). \end{aligned}$$

where time-delay τ_k denotes the time from the instant $i_k h$ when sensor nodes sample sensor data from a plant to the instant when actuator transfers data to the plant. It can be obtained that

$$\bigcup_{k=1}^{\infty} [i_k h + \tau_k,\ i_{k+1}h + \tau_{k+1}) = [t_0,\ \infty),\ t_0 \ge 0.$$

Remark 10.2. $\{i_1,\ i_2,\ i_3,\ \cdots\}$ is a subset of $\{1,\ 2,\ 3,\ \cdots\}$, and it denotes the sequence of time points of successful data transmissions from the sensor to the actuator. Moreover, it is not required that $i_{k+1} > i_k$, $\{i_1,\ i_2,\ i_3,\ \cdots\} = \{1,\ 2,\ 3,\ \cdots\}$ means that no packet dropout occurs in the transmission. If $i_{k+1} = i_k + 1$, it implies that $h + \tau_{k+1} > \tau_k$.

These assumptions on network-induced time delays and data packets dropout are reasonable and they have also been employed in [292]. By introducing parameter uncertainties and defining $A_d(\eta_k) = B(\eta_k)K(\eta_k)$ where $K(\eta_k)$ is a given Markovian jump controller, the Markovian jump DOS (10.11) is rewritten as

$$\begin{aligned} \delta x(kh) &= (A(\eta_k) + \Delta A(\eta_k, kh))x(kh) + (A_d(\eta_k) \\ &\quad + \Delta A_d(\eta_k, kh))x(kh - \tau(\eta_k)), && (10.17) \\ x(kh) &= \phi(kh),\ kh \in [-d_M,\ 0]. && (10.18) \end{aligned}$$

where $x(kh) \in R^p$ is the state variable; $u(kh) \in R^q$ is the input variable; $\phi(kh)$ is a vector-value initial function. For notational simplicity, in the sequel, for $\eta_k = i \in S$, we denote $A(\eta_k)$ by A_i, $\Delta A(\eta_k, kh)$ by $\Delta A_i(kh)$, $A_i(kh) = A_i + \Delta A_i(kh)$, and so on. The uncertainty parameters $\Delta A_i(kh)$ and $\Delta A_{di}(kh)$ are called linear fractional uncertainty parameters, which are time-varying matrices. They are given as follows:

$$\begin{aligned} \left[\Delta A_i(kh)\ \Delta A_{di}(kh)\right] &= H_i \hat{F}_i(kh) \left[E_{1i}\ E_{2i}\right], && (10.19) \\ \hat{F}_i(kh) &= F_i(kh) \left[I - G_i F_i(kh)\right]^{-1}, && (10.20) \end{aligned}$$

where H_i, E_{1i}, E_{2i} and G_i are known constant real matrices, $F_i(kh)$ is a class of unknown time-varying matrices satisfying $F_i^T(kh)F_i(kh) \leq I$, $\forall k \geq 0$. It is assumed that the matrix $[I - G_iF_i(kh)]^{-1}$ is invertible for any $F_i(kh)$ and $I - G_i^TG_i > 0$.

The object of this chapter is to consider the stability for the Markovian jump DOS (10.17) with mode-dependent time delays.

Remark 10.3. The terms (10.19) and (10.20) describe a wider class of parameter uncertainties than norm-bounded parameter uncertainties. It is easy to see that the linear fractional parameter uncertainties can be reduced to norm-bounded parameter uncertainties when $G_i = 0$.

Definition 10.4. The Markovian jump DOS (10.17) with mode-dependent time delays is stochastic robustly asymptotically stable, if the following conditions hold for a delta domain Lyapunov function $V(x(kh))$:

i $V(x(kh)) \geq 0$, with equality if and only if $x(kh) = 0$;
ii $\sigma V(x(kh)) < 0$,

where $\sigma(\cdot) = [\mathbf{E}[V(x(kh+h))] - V(x(kh))]/h$ is a special delta operator for a stochastic process.

Remark 10.5. The numerical advantage of the delta operator approach has been clearly shown in [196]. The case of sampling period being zero in (10.10) is omitted since only discrete system model is valuable for NCSs. It has been shown that the operator $\delta(\cdot)$ in (10.10) is different from the above delta operator definition $\sigma(\cdot)$ because of the stochastic Markov process. Consequently, an extend delta operator definition is used in this chapter.

Before ending this section, the following lemmas are recalled to prove our main results in this chapter.

Lemma 10.6. *[124] For any constant positive semi-definite symmetric matrix W, two positive integers r and r_0 satisfying $r \geq r_0 \geq 1$, the following inequality holds*

$$\left(\sum_{i=r_0}^{r} x(i)\right)^T W \left(\sum_{i=r_0}^{r} x(i)\right) \leq (r - r_0 + 1) \sum_{i=r_0}^{r} x^T(i)Wx(i).$$

Lemma 10.7. *[269] Given constant matrices D, E and a symmetric constant matrix Υ of appropriate dimensions, the following inequality holds:*

$$\Upsilon + D\hat{F}(kh)E + E^T\hat{F}^T(kh)D^T \leq 0,$$

with $\hat{F}(kh)$ as in (10.20), if and only if there exists a scalar $\varepsilon > 0$ such that

$$\Upsilon + \begin{bmatrix} \varepsilon^{-1}E^T & \varepsilon D \end{bmatrix} \begin{bmatrix} I & -G \\ -G^T & I \end{bmatrix}^{-1} \begin{bmatrix} \varepsilon^{-1}E \\ \varepsilon D^T \end{bmatrix} \leq 0.$$

10.3 Main Results

In this section, we will develop a new LMI approach to solve the robust stability problem for the uncertain Markovian jump DOS (10.17) with mode-dependent time delays. Our main result is given as follows.

Theorem 10.8. *The uncertain Markovian jump DOS with mode-dependent time delays described by (10.17) is stochastic robustly asymptotically stable, if there exist symmetric positive definite matrices $P_i > 0$, $Q > 0$ and $R > 0$, some real matrices L_i, as well as positive scalars α_i, such that the following LMIs hold:*

$$\Sigma = \begin{bmatrix} \Gamma_{11} & \Gamma_{12} & L_i A_{di} & 0 & L_i H_i \\ * & \Gamma_{22} & \frac{1}{d_M} R + P_i A_{di} & \alpha_i E_{1i}^T & P_i H_i \\ * & * & -Q - \frac{1}{d_M} R & \alpha_i E_{2i}^T & 0 \\ * & * & * & -\alpha_i I & \alpha_i G_i \\ * & * & * & * & -\alpha_i I \end{bmatrix} < 0, \ \forall\, i \in \mathcal{S} \tag{10.21}$$

with

$$\Gamma_{11} = h^2 \sum_{j=1}^{s} \lambda_{ij} P_j + h P_i + d_M R - L_i - L_i^T,$$

$$\Gamma_{22} = \sum_{j=1}^{s} \lambda_{ij} P_j + P_i A_i + \frac{d_M - d_m}{h} Q - d_M^{-1} R,$$

$$\Gamma_{12} = h \sum_{j=1}^{s} \lambda_{ij} P_j + L_i A_i.$$

Proof. Construct Lyapunov-Krasovskii functional in delta domain as follows:

$$\begin{aligned} V(x, kh, \eta_k) = {} & V_1(x, kh, \eta_k) + V_2(x, kh, \eta_k) + V_3(x, kh, \eta_k) \\ & + V_4(x, kh, \eta_k) \end{aligned} \tag{10.22}$$

with

$$V_1(x, kh, \eta_k) = x^T(kh) P(\eta_k) x(kh),$$

$$V_2(x, kh, \eta_k) = h \sum_{j=1}^{n(\eta_k)} x^T(kh - jh) Q x(kh - jh),$$

$$V_3(x, kh, \eta_k) = h \sum_{j=n_m}^{n_M - 1} \sum_{l=1}^{j} x^T(kh - lh) Q x(kh - lh),$$

$$V_4(x, kh, \eta_k) = \sum_{j=1}^{n(\eta_k)} \sum_{l=1}^{j} e^T(kh - lh) R e(kh - lh)$$

where $e(j) = x(j) - x(j+h)$, so there exist $\delta x(j) = -e(j)/h$ and $e(kh-ih) = x(kh-ih) - x(kh-(i-1)h)$. From Definition 10.4, the delta operator of the stochastic process $\{x(kh), \eta_k, kh \geq 0\}$, acting on $V(x, kh, \eta_k)$ at the point $\{x, kh, \eta_k = i\}$, can be expressed as follows:

$$\sigma(V(x, kh, \eta_k)) = \frac{\mathbf{E}[V(x, kh+h, \eta_{k+1})] - V(x, kh, \eta_k)}{h}. \tag{10.23}$$

Let the mode at time kh be i, that is $\eta_k = i$. Recall that at time $kh+h$, the system may jump to any mode $\eta_{k+1} = j$. By letting $\mathbf{E}(P_j) = h\sum_{j=1}^{s} \lambda_{ij} P_j + P_i$ and taking the stochastic delta operator manipulations of $V_1(x, kh, \eta_k)$ along the trajectory of system (10.17), we can obtain:

$$\begin{aligned}
\sigma V_1(x, kh, \eta_k) &= \frac{1}{h}\{\mathbf{E}[x^T(kh+h)P(\eta_{k+1})x(kh+h)] \\
&\qquad -x^T(kh)P(\eta(kh))x(kh)\} \\
&= \sum_{j=1}^{s} \lambda_{ij} x^T(kh+h)P_j x(kh+h) + \frac{1}{h}[x^T(kh+h)P_i x(kh+h) \\
&\quad -x^T(kh)P_i x(kh)] \\
&= h^2 \sum_{j=1}^{s} \lambda_{ij} \delta^T(x(kh))P_j \delta(x(kh)) + h\sum_{j=1}^{s} \lambda_{ij}\delta^T(x(kh))P_j x(kh) \\
&\quad +h\sum_{j=1}^{s} \lambda_{ij} x^T(kh)P_j \delta(x(kh)) \\
&\quad +\sum_{j=1}^{s} \lambda_{ij} x^T(kh)P_j x(kh) + h\delta^T(x(kh))P_i\delta(x(kh)) \\
&\quad +\delta^T(x(kh))P_i x(kh) + x^T(kh)P_i\delta(x(kh)).
\end{aligned} \tag{10.24}$$

where

$$x^T(kh)P_i\delta(x(kh)) = x^T(kh)P_i[A_i(kh)x(kh) + A_{di}(kh)x(kh-\tau_i)].$$

Remark 10.9. If we let $h \to 0$, the result of stochastic delta operator for $V_1(x, kh, \eta_k)$ will be equal to the ones of weak infinitesimal operator which was employed in [211]. Therefore, it can be said that the weak infinitesimal operator is an especial case of the stochastic delta operator defined in (10.23). We have gotten a less conservative result by using delta operator approach.

Taking the stochastic delta operator manipulations of $V_2(x, kh, \eta_k)$ and $V_3(x, kh, \eta_k)$, it can be obtained that

$$\sigma V_2(x, kh, \eta_k) = \frac{1}{h}\left\{\mathbf{E}\left[h\sum_{j=1}^{n_i} x^T(kh-(j-1)h)Qx(kh-(j-1)h)\right]\right.$$

$$
\begin{aligned}
&\qquad - h\sum_{j=1}^{n_i} x^T(kh-jh)Qx(kh-jh)\Bigg\} \\
&\le \frac{1}{h}\Bigg\{h\sum_{j=1}^{n_M} x^T(kh-(j-1)h)Qx(kh-(j-1)h) \\
&\qquad -h\sum_{j=1}^{n_i} x^T(kh-jh)Qx(kh-jh)\Bigg\} \\
&\le x^T(kh)Qx(kh) - x^T(kh-\tau_i)Qx(kh-\tau_i) \\
&\qquad + \sum_{j=n_i}^{n_M-1} x^T(kh-jh)Qx(kh-jh).
\end{aligned} \tag{10.25}
$$

$$
\begin{aligned}
\sigma V_3(x,kh,\eta_k) &= \sum_{j=n_m}^{n_M-1}\left(\sum_{l=1}^{j} x^T(kh-(l-1)h)Qx(kh-(l-1)h)\right. \\
&\qquad \left. -\sum_{l=1}^{j} x^T(kh-lh)Qx(kh-lh)\right) \\
&= -\sum_{j=n_m}^{n_M-1} x^T(kh-jh)Qx(kh-jh) \\
&\qquad + \frac{d_M-d_m}{h}x^T(kh)Qx(kh).
\end{aligned} \tag{10.26}
$$

From $n_i \ge n_m$, it can be obtained that

$$
\sum_{j=n_i}^{n_M-1} x^T(kh-jh)Qx(kh-jh) - \sum_{j=n_m}^{n_M-1} x^T(kh-jh)Qx(kh-jh) \le 0. \tag{10.27}
$$

Based on Lemma 10.6 and $n_i \le n_M$, take the stochastic delta operator manipulations of $V_4(x,kh,\eta_k)$ as follows

$$
\begin{aligned}
\sigma V_4(x,kh,\eta_k) &= \frac{1}{h}\left[\sum_{j=1}^{n_i}\sum_{l=1}^{j} e^T(kh-(l-1)h)Re(kh-(l-1)h)\right. \\
&\qquad \left. -\sum_{j=1}^{n_i}\sum_{l=1}^{j} e^T(kh-lh)Re(kh-lh)\right] \\
&\le \frac{1}{h}\left[\sum_{j=1}^{n_M} e^T(kh)Re(kh) - \sum_{j=1}^{n_i} e^T(kh-jh)Re(kh-jh)\right]
\end{aligned}
$$

$$\begin{aligned}
&\le \frac{n_M}{h} e^T(kh)Re(kh) - \frac{1}{nh}\left[\sum_{j=1}^{n_i} e(kh-jh)\right]^T R\left[\sum_{j=1}^{n_i} e(kh-jh)\right] \\
&= -d_M^{-1}[x(kh-\tau_i)-x(kh)]^T R\,[x(kh-\tau_i)-x(kh)] \\
&\quad +d_M\delta^T(x(kh))R\delta(x(kh))
\end{aligned} \tag{10.28}$$

For any real matrix L_i and system (10.17), we have

$$\begin{aligned}
0 &= 2\delta^T(x(kh))L_i\left[-\delta x(kh) + A_i(kh)x(kh) + A_{di}(kh)x(kh-\tau_i)\right] \\
&= -2\delta^T(x(kh))L_i\delta(x(kh)) + 2\delta^T(x(kh))L_iA_i(kh)x(kh) \\
&\quad +2\delta^T(x(kh))L_iA_{di}(kh)x(kh-\tau_i).
\end{aligned} \tag{10.29}$$

From Definition 10.4, there exists $\sigma V(x, kh, \eta_k) < 0$. Combining (10.24)-(10.29) results in

$$\sigma V(x, kh, \eta_k) \le \psi^T(kh)\Sigma_1\psi(kh) < 0, \tag{10.30}$$

where

$$\psi^T(kh) = \left[\,\delta^T(x(kh))\ x^T(kh)\ x^T(kh-\tau_i)\,\right],$$

and

$$\Sigma_1 = \begin{bmatrix} \Sigma_1(1,1) & \Sigma_1(1,2) & L_iA_{di}(kh) \\ * & \Sigma_1(2,2) & \frac{1}{d_M}R + P_iA_{di}(kh) \\ * & * & -Q - \frac{1}{d_M}R \end{bmatrix}, \tag{10.31}$$

with

$$\Sigma_1(1,1) = h^2\sum_{j=1}^{s}\lambda_{ij}P_j + hP_i + d_MR - L_i - L_i^T,$$

$$\Sigma_1(1,2) = h\sum_{j=1}^{s}\lambda_{ij}P_j + L_iA_i(kh),$$

$$\Sigma_1(2,2) = \sum_{j=1}^{s}\lambda_{ij}P_j + P_iA_i(kh) + \frac{d_M - d_m}{h}Q - d_M^{-1}R.$$

From (10.19) and (10.30), it can be gotten that

$$\Sigma_1 = \Upsilon + \xi_i\hat{F}_i(kh)\zeta_i + \zeta_i^T\hat{F}_i^T(kh)\xi_i^T < 0, \tag{10.32}$$

with

$$\Upsilon = \begin{bmatrix} \Upsilon(1,1) & \Upsilon(1,2) & L_iA_{di} \\ * & \Upsilon(2,2) & \frac{1}{d_M}R + P_iA_{di} \\ * & * & -Q - \frac{1}{d_M}R \end{bmatrix},$$

$$\xi_i = \begin{bmatrix} L_i H_i \\ P_i H_i \\ 0 \end{bmatrix}, \qquad \zeta_i^T = \begin{bmatrix} 0 \\ E_{1i}^T \\ E_{2i}^T \end{bmatrix},$$

with

$$\Upsilon(1,1) = h^2 \sum_{j=1}^{s} \lambda_{ij} P_j + h P_i + d_M R - L_i - L_i^T,$$

$$\Upsilon(1,2) = h \sum_{j=1}^{s} \lambda_{ij} P_j + L_i A_i,$$

$$\Upsilon(2,2) = \sum_{j=1}^{s} \lambda_{ij} P_j + P_i A_i + \frac{d_M - d_m}{h} Q - d_M^{-1} R.$$

By Lemma 10.7, there exists a scalar $\varepsilon_i > 0$ such that inequality (10.32) is equal to the following inequality

$$\Upsilon + \begin{bmatrix} \varepsilon_i \zeta_i^T & \varepsilon_i^{-1} \xi_i \end{bmatrix} \begin{bmatrix} I & -G_i \\ -G_i^T & I \end{bmatrix}^{-1} \begin{bmatrix} \varepsilon_i \, \zeta_i \\ \varepsilon_i^{-1} \xi_i^T \end{bmatrix} < 0. \tag{10.33}$$

Using Schur's complement, $\Sigma_1 < 0$ can be changed to $\Sigma_2 < 0$, where

$$\Sigma_2 = \begin{bmatrix} \Sigma_2(1,1) & \Sigma_2(1,2) & L_i A_{di} & 0 & \varepsilon_i^{-1} L_i H_i \\ * & \Sigma_2(2,2) & \frac{1}{d_M} R + P_i A_{di} & \varepsilon_i E_{1i}^T & \varepsilon_i^{-1} P_i H_i \\ * & * & -Q - \frac{1}{d_M} R & \varepsilon_i E_{2i}^T & 0 \\ * & * & * & -I & G_i \\ * & * & * & * & -I \end{bmatrix}, \tag{10.34}$$

with

$$\Sigma_2(1,1) = h^2 \sum_{j=1}^{s} \lambda_{ij} P_j + h P_i + d_M R - L_i - L_i^T,$$

$$\Sigma_2(1,2) = h \sum_{j=1}^{s} \lambda_{ij} P_j + L_i A_i,$$

$$\Sigma_2(2,2) = \sum_{j=1}^{s} \lambda_{ij} P_j + P_i A_i + \frac{d_M - d_m}{h} Q - d_M^{-1} R.$$

Pre-multiplying and post-multiplying $\Sigma_2 < 0$ by $\text{diag}\{I,\ I,\ I,\ \varepsilon_i I,\ \varepsilon_i I\}$, respectively, letting $\varepsilon_i^2 = \alpha_i$, the inequality $\Sigma_2 < 0$ is equivalent to $\Sigma < 0$. From Definition 10.4, the Markovian jump DOS (10.17) with mode-dependent time delays is stochastic robustly asymptotically stable. This completes the proof.

Based on Theorem 10.8, we deal with the stability of system (10.17) with $\Delta A_i(kh) = 0$ and $\Delta A_{di}(kh) = 0$, i.e.

$$\delta x(kh) = A(\eta_t)x(kh) + A_d(\eta_t)x(kh - \tau(\eta_t)), \tag{10.35}$$
$$x(kh) = \phi(kh), \qquad kh \in [-d_M,\ 0]. \tag{10.36}$$

where $\phi(kh)$ is a vector-value initial function. The following corollary is given as:

Corollary 10.10. *The Markovian jump DOS described by (10.35) is asymptotically stable, if there exist symmetric positive definite matrices P_i, Q and R, some real matrices L_i, such that the following LMIs hold:*

$$\begin{bmatrix} \Phi_{11} & h\sum_{j=1}^{s}\lambda_{ij}P_j + L_iA_i & L_iA_{di} \\ * & \Phi_{22} & \frac{1}{d_M}R + P_iA_{di} \\ * & * & -Q - \frac{1}{d_M}R \end{bmatrix} < 0, \tag{10.37}$$

with

$$\Phi_{11} = h^2\sum_{j=1}^{s}\lambda_{ij}P_j + hP_i + d_MR - L_i - L_i^T,$$
$$\Phi_{22} = \sum_{j=1}^{s}\lambda_{ij}P_j + P_iA_i + \frac{d_M - d_m}{h}Q - d_M^{-1}R.$$

If the system mode set $S = \{1\}$, the Markovian jump DOS (10.35) can be simplified to a general DOS which is given as below:

$$\delta x(kh) = (A + \Delta A(kh))x(kh) + (A_d + \Delta A_d(kh))x(kh - \tau_k), \tag{10.38}$$
$$x(kh) = \phi(kh), \qquad kh \in [-d_M,\ 0]. \tag{10.39}$$

where τ_k is a time-varying delay and $\phi(kh)$ is a vector-value initial function.

Corollary 10.11. *The DOS with time-varying delays described by (10.38) is robustly asymptotically stable, if there exist symmetric positive definite matrices P, Q and R, a real matrix L, as well as a positive scalar α, such that the following LMI holds:*

$$\begin{bmatrix} \Psi_{11} & hP + LA & LA_d & 0 & LH \\ * & \Psi_{22} & \frac{1}{d_M}R + PA_d & \alpha E_1^T & PH \\ * & * & -Q - \frac{1}{d_M}R & \alpha E_2^T & 0 \\ * & * & * & -\alpha I & \alpha G \\ * & * & * & * & -\alpha I \end{bmatrix} < 0, \tag{10.40}$$

with

$$\Psi_{11} = h^2P + hP + d_MR - L - L^T,\ \Psi_{22} = P + PA + \frac{d_M - d_m}{h}Q - d_M^{-1}R.$$

Some remarks are given as follows:

Remark 10.12. It can be seen that the DOS with time-varying delays described by (10.38) is the same as the one considered in [196], and Corollary 10.11 is similar with the Theorem 1 in [196]. However, the DOS (10.38) with time-varying delays and Corollary 10.11 are only the simplified results in this chapter. Hence, it can be said that we get more extensive results in this chapter than [196].

Remark 10.13. To the best of our knowledge, there have been few chapters on stability problem of Markovian jump DOSs with mode-dependent time delays. Therefore, it is hard to compare the conservatism of this chapter with others. We only discuss the Lyapunov functional used in this chapter. If we only use $V_1(x, kh, \eta_k)$ and $V_2(x, kh, \eta_k)$ in this chapter, the results would also be feasible. The add terms $V_3(x, kh, \eta_k)$ and $V_4(x, kh, \eta_k)$ will decrease the conservatism of stability criterion in this chapter.

Remark 10.14. We want to make a bridge between the discrete-time Markov chain and the continuous-time Markov chain by using the Markov chain in delta domain. That is, the proposed Markovian jump method in this chapter can unify some previous related continuous and discrete Markovian jump systems into Markovian jump DOS framework. It is obvious that some other issues on the Markov chain in delta domain can be easily obtained to further considerations.

Remark 10.15. It is well known that delta operator method has the advantage of better numerical properties at high sampling rates and Markovian jump models have a special effect of dealing with NCSs. A kind of Markovian jump DOSs with mode-dependent time delays is used to describe the NCSs in this chapter. Some comparison results are given in Table 10.1.

Table 10.1 Comparisons of the obtained results with others

References	Contributions
[208]	Markov chain for discrete systems
[211]	Markov chain for continuous systems
[298]	Markov chain for NCSs
This chapter	Markov chain in delta domain for NCSs

10.4 Numerical Example

In this section, we give an example to demonstrate the results obtained in previous sections.

Example 10.16. The continuous Markovian jump system as NCS plant is described as

$$\dot{x}(t) = A_{s1}x(t) + B_{s1}u(t), \quad \dot{x}(t) = A_{s2}x(t) + B_{s2}u(t).$$

with

$$A_{s1} = \begin{bmatrix} -0.75 & -0.75 \\ 1.50 & -1.50 \end{bmatrix}, \ B_{s1} = \begin{bmatrix} 1.2 \\ -1 \end{bmatrix},$$
$$A_{s2} = \begin{bmatrix} -0.15 & -0.49 \\ 1.50 & -2.10 \end{bmatrix}, \ B_{s2} = \begin{bmatrix} 0.9 \\ -1 \end{bmatrix}.$$

The continuous transition rate matrix Λ_s is given by

$$\Lambda_s = \begin{bmatrix} -\vartheta_1 & \vartheta_1 \\ \vartheta_2 & -\vartheta_2 \end{bmatrix}, \text{ where } \vartheta_1 = \frac{4}{5}, \ \vartheta_2 = \frac{2}{3}.$$

Utilizing shift operator and choosing $h = 0.01$, a discrete Markovian jump system of the sampling result of the above continuous Markovian jump system in z-domain is obtained as follows:

$$x(k+1) = A_{z1}x(k) + B_{z1}u(k), \quad x(k+1) = A_{z2}x(k) + B_{z2}u(k)$$

with

$$A_{z1} = \begin{bmatrix} 0.9227 & -0.0669 \\ 0.1338 & 0.8557 \end{bmatrix}, \ B_{z1} = \begin{bmatrix} 0.1189 \\ -0.0843 \end{bmatrix},$$
$$A_{z2} = \begin{bmatrix} 0.9817 & -0.0438 \\ 0.1341 & 0.8074 \end{bmatrix}, \ B_{z2} = \begin{bmatrix} 0.0915 \\ -0.0838 \end{bmatrix}.$$

The Markov process transition probability matrix Π_z is defined by

$$\Pi_z = \begin{bmatrix} \pi_1 & \pi_2 \\ \pi_3 & \pi_4 \end{bmatrix} = \begin{bmatrix} 0.5804 & 0.4196 \\ 0.3497 & 0.6503 \end{bmatrix} \text{ where } \pi_1 + \pi_2 = 1, \ \pi_3 + \pi_4 = 1.$$

By using delta operator and choosing $h = 0.1$, a Markovian jump DOS of the sampling result of the above continuous Markovian jump system in δ-domain is given as follows:

$$\delta x(kh) = A_1x(kh) + B_1u(kh), \quad \delta x(kh) = A_2x(kh) + B_2u(kh)$$

with

$$A_1 = \begin{bmatrix} -0.7734 & -0.6691 \\ 1.3382 & -1.4425 \end{bmatrix}, \ B_1 = \begin{bmatrix} 1.1888 \\ -0.8434 \end{bmatrix},$$
$$A_2 = \begin{bmatrix} -0.1828 & -0.4380 \\ 1.3409 & -1.9260 \end{bmatrix}, \ B_2 = \begin{bmatrix} 0.9150 \\ -0.8382 \end{bmatrix}.$$

Then, the transition rate matrix in δ-domain is

$$\Lambda_\delta = \begin{bmatrix} -\lambda_1 & \lambda_1 \\ \lambda_2 & -\lambda_2 \end{bmatrix}, \text{ where } \lambda_1 = 0.7942,\ \lambda_2 = 0.6618.$$

In the following, we consider the linear system in this example overs a communications network as in Fig. 10.1. That is, the Markovian jump DOS (10.17) mode-dependent time delays is used to describe the NCS in this example. In order to investigate the stability analysis of system (10.17), the networked controller is chosen as $K_1 = \begin{bmatrix} -0.11 & -0.25 \end{bmatrix}$ and $K_2 = \begin{bmatrix} -0.02 & -1.31 \end{bmatrix}$. Then, for system (10.17) we obtain the following parameters

$$A_{d1} = B_1 K_1 = \begin{bmatrix} -0.1308 & -0.2972 \\ 0.0928 & 0.2109 \end{bmatrix}, \ A_{d2} = B_2 K_2 = \begin{bmatrix} -0.0183 & -1.1986 \\ 0.0168 & 1.0981 \end{bmatrix}.$$

Next, choosing $d_m = 0.1$ and $d_M = 1.1$, letting $\eta = [1\ 1]^T$ and $\Omega = [\eta\ \eta]$, the following uncertain parameters for the Markovian jump DOS (10.17) with $i = 1, 2$ are given as

$$G_1 = 0.03 \cdot \Omega,\ H_1 = 0.12 \cdot \Omega,\ E_{11} = 0.01 \cdot \Omega,\ E_{21} = 0.02 \cdot \Omega,$$
$$G_2 = 0.02 \cdot \Omega,\ H_2 = 0.23 \cdot \Omega,\ E_{12} = 0.03 \cdot \Omega,\ E_{22} = 0.01 \cdot \Omega.$$

Fig. 10.2 is one of the possible realizations of Markovian jumping mode for system (10.17) considered in this example.

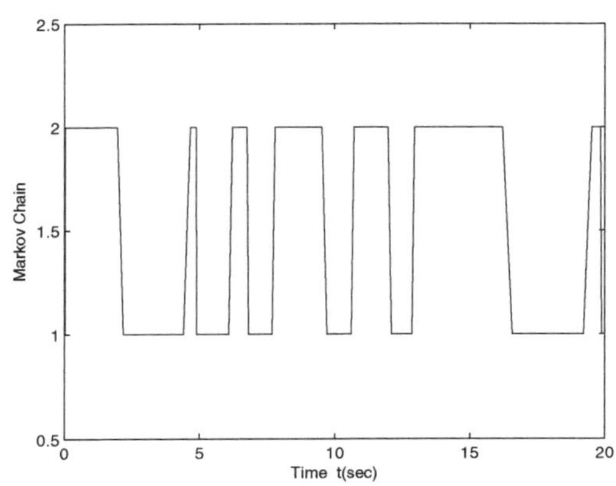

Fig. 10.2 Random Markovian jump mode.

Lastly, by using Theorem 10.8 and LMI control toolbox in Matlab, we get that system (10.17) with mode-dependent time delays is robustly asymptotically stable and the solutions of LMI (10.21) is given as follows:

$$P_1 = \begin{bmatrix} 16.8863 & 0.5481 \\ 0.5481 & 35.3498 \end{bmatrix}, \ L_1 = \begin{bmatrix} 1.7866 & -0.8478 \\ 1.3609 & 5.0778 \end{bmatrix}, \ Q = \begin{bmatrix} 0.0049 & 0.0013 \\ 0.0013 & 0.0146 \end{bmatrix},$$
$$P_2 = \begin{bmatrix} 16.3312 & 0.5321 \\ 0.5321 & 32.3973 \end{bmatrix}, \ L_2 = \begin{bmatrix} 1.7424 & -0.9034 \\ 1.4192 & 4.7565 \end{bmatrix}, \ R = \begin{bmatrix} 1.2269 & 1.2258 \\ 1.2258 & 3.3056 \end{bmatrix},$$
$$\alpha_1 = 239.2297, \ \alpha_2 = 252.1907.$$

Under the mode sequence shown in Fig. 10.2 and initial condition $x_1(0) = x_2(0) = 0.5$, the corresponding state trajectories of system (10.17) with is shown in Fig. 10.3.

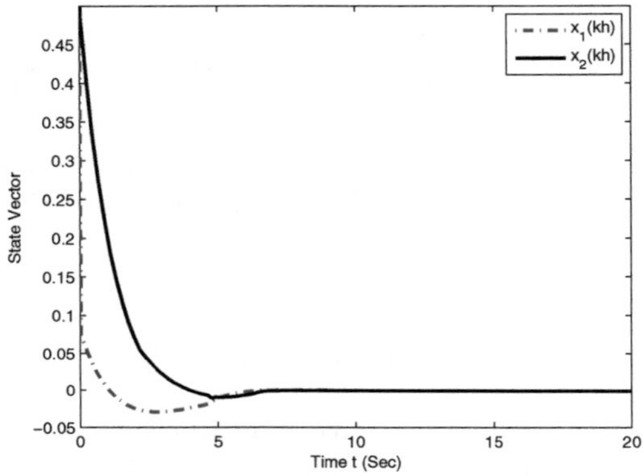

Fig. 10.3 State of Markovian jump DOS.

It is shown that the closed-loop Markovian jump DOS (10.17) is stochastically stable. Furthermore, the numerical simulation of getting the Markov process transition rate matrix Λ_δ in δ-domain has also been given in this section. The jump relation (10.14) is also clearly shown in this example.

10.5 Summary

In this chapter, based on the Lyapunov-Krasovskii functional method in delta domain and the LMI technique, the problem of robust stability for Markovian jump DOSs with mode-dependent time delays has been studied. This class jump DOSs come from NCSs which contain sensor-to-controller random network-induced delay and arbitrary packet loss. Moreover, we have used a class of linear fractional parameter uncertainties, which means that a less conservative result would be obtained than by using norm-bounded parameter uncertainties. Numerical example has been given to illustrate the feasibility and effectiveness of the developed technique.

Chapter 11
Stabilization of NCSs with Nonuniform Sampling Periods

11.1 Introduction

NCSs is defined as a feedback control system that the control loops are closed through a real-time network [40, 60]. The general theory for NCSs has received much attention in recent years, see for examples [73, 97, 108, 167, 244], and the references therein. A switched linear model has also been proposed to describe NCS in [304]. Predictive NCSs have also been investigated in [155, 156]. A sampling period scheduling of NCSs with multiple control loops has been addressed in [187]. The values of packets dropout and network-induced delays in NCSs are derived from the values of the input delay by the viewpoint of a logic ZOH [272]. The logic ZOH is assumed to be both time-driven and event-driven and has the logical capability of comparing the time steps of the arrived control input packets and choosing the latest one to control the process.

In a NCS, all signals are sampled by a sampler [94], and the sampling rate for each signal may be varying from sample to sample according to actual situations. The kind of sampling periods was called as time-varying sampling periods or nonuniform sampling periods. The nonuniform sampled-data control has been extensively studied in the past years, for example [65, 168, 201, 220, 287]. H_∞ controllers for NCSs have been designed by using active-varying sampling periods in [241]. In [77], the problem of robust H_∞ control was investigated for sampled-data systems with probabilistic sampling, but the control plant could not be discretized so that the result is hard to be applied in NCS frameworks. Felicitously, the sampling-period in DOSs is an explicit parameter which can be chosen as an appropriate value based on networks load conditions [82, 170]. Many results on DOSs have been obtained in [53, 176, 177, 215, 307]. Using delta operator instead of traditional shift operator for sampling NCSs is an interesting idea, which motivates us to make an effort in this chapter.

H. Yang et al.: Analysis and Synthesis of Delta Operator Systems, LNCIS 430, pp. 175–202.
springerlink.com © Springer-Verlag Berlin Heidelberg 2012

The main contributions of this chapter are summarized as below:

i Based on nonuniform sampling periods, a novel delayed switched DOS is introduced to solve the stabilization problem of NCSs.
ii According to networks load conditions, an appropriate sampling period interval can be chosen since sampling period is an explicit parameter in a DOS.
iii A new general asymptotically stable theorem is given for the switched DOSs with time-varying delays by Lyapunov functional method in delta domain.

11.2 Problem Statement

In this chapter, we consider a NCS model with packets dropout and network-induced delays both in forward and backward channels, which is shown in Fig. 11.1.

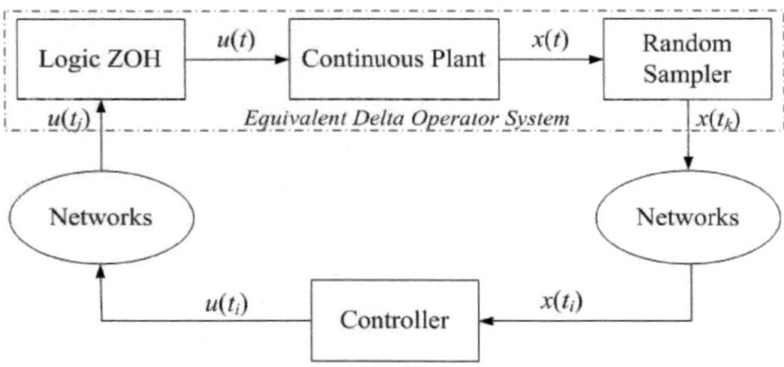

Fig. 11.1 Schematic diagram of NCS.

In this NCS framework, the control plant is a continuous-time system, which is described as follows:

$$\dot{x}(t) = Ax(t) + Bu(t) \tag{11.1}$$

where $x(t) \in R^n$ and $u(t) \in R^m$ are the state vector and control input vector in s-domain, respectively. A and B are two known constant matrices with appropriate dimensions. We assume that the original plant has been designed without considering networks. That is, if we consider a continuous-time state feedback controller, the closed-loop continuous-time system will be stable without the network connection.

The original plant can be modeled discretely with nonuniform sampling periods by a random sampler. Let $k \in Z^+$ be time step and $\{\mathtt{T}_k : \ k = 0, 1, 2, \cdots\}$ be a class of random sampling periods. The sampler samples the value of the system state $x(t)$ in the continuous-time system (11.1) at time step k with a random sampling rate $1/\mathtt{T}_k$ in sampling instant t_k. The time diagram of nonuniform sampling process is shown in Fig. 11.2.

Fig. 11.2 Time diagram of nonuniform sampling process.

According to actual situation of NCS model, the sampling period $\mathtt{T}_k$ can be random chosen in a real number interval by the random sampler. It can be obtained that at the sampling instants there exist $t_k = \sum_{i=0}^{k-1} \mathtt{T}_i$ and $t_{k+1} = t_k + \mathtt{T}_k$.

It is known that DOS has numerical advantage for fast sampling periods. Furthermore, the sampling-period $\mathtt{T}_k$ is an explicit parameter in DOS, so it is easy to observe and analyze the effect of the state feedback controller with different sampling periods. Therefore, we discretize the continuous-time system (11.1) by using delta operator approach with sampling period $\mathtt{T}_k$ at time step k, the following DOS is obtained as

$$\delta x(t_k) = A_k x(t_k) + B_k u(t_k) \tag{11.2}$$

where the delta operator $\delta x(t_k)$ is defined for $\mathtt{T}_k \neq 0$ and $k \in Z^+$ as follows

$$\delta x(t_k) = \frac{x(t_k + \mathtt{T}_k) - x(t_k)}{\mathtt{T}_k} \tag{11.3}$$

Also, A_k and B_k are two time-varying matrices which are given as below:

$$A_k = \frac{e^{A\mathtt{T}_k} - I}{\mathtt{T}_k}, \quad B_k = \frac{1}{\mathtt{T}_k} \int_0^{\mathtt{T}_k} e^{A(\mathtt{T}_k - s)} B ds.$$

The numerical advantage of the delta operator approach has been clearly shown in Chapter 1.

By considering network-induced delays and packets dropout in backward network channel, the following networked digital controller can be given as:

$$u(t_i) = K_k x(t_i), \ i \geq k \text{ and } i \in Z^+. \tag{11.4}$$

It is evident that the networked digital controller (11.4) is a conventional discrete state-feedback controller which is event-driven. When the data packets transmitted through forward network channel get to logic ZOH, $u(t_i)$ will be changed to $u(t_j)$ with $j > i$ and $j \in Z^+$. There are two sources of delays in networks: random sampler to controller $\tau_{sc}(t_k)$ and controller to logic ZOH $\tau_{cl}(t_k)$. Any controller delay can be absorbed into either $\tau_{sc}(t_k)$ or $\tau_{cl}(t_k)$, without loss of generality. For time-invariant controller, both delays can be lumped together $\tau(t_k) = \tau_{sc}(t_k) + \tau_{cl}(t_k)$. Let $0 < \tau(t_k) \leq \tau_M$, where τ_M denotes the largest step delay and M is an integer in Z^+.

We now recall the following logic ZOH established in [272] for the NCSs with packets dropout and network-induced delays.

Definition 11.1. *Logic ZOH*: Given $u(t_0)$, let $i = 0$ and $k = 0$.

1 At sampling instant t_k, ZOH changes its output to $u(t_i) = u(t_{i_k})$ for $t_k \leq t_i < t_{k+1}$. Let $i_{k+1} = i_k$.
2 During $t_k \leq t_i < t_{k+1}$, if a packet $u(t_j)$ arrives and $j > i_{k+1}$, then ZOH stores $u(t_j)$ and lets $i_{k+1} = j$.
3 Repeat Step 2 until t_k reaches the next sampling instant t_{k+1}. Let $k = k+1$ and go to Step 1.

where $i_k (k = 1, 2, 3, \cdots)$ is an integer and there exists $\{i_1, i_2, i_3, \cdots\} \subset \{1, 2, 3, \cdots\}$.

Logic ZOH is both time-driven and event-driven, and has logical capability of comparing time instants of arrived control input packets and choosing the latest one to control the process, which means that $\tau(t_{k+1}) \leq \tau(t_k) + \mathrm{T}_k$ is guaranteed at time instant t_k. If the time delay $\tau(t_k)$ is larger than τ_M, it will be thought as packets dropout. We assume that the above bounds of both packets dropout and network-induced delays are known. Let d denote the step delay of the control signal chosen by logic ZOH. It can be obtained that $\tau(t_k)$ takes values in a finite set $\{\tau_d : \ d = 1, \cdots, M\}$ and $\tau(t_k) = \sum_{l=k-d}^{k-1} \mathrm{T}_l = \tau_d$. There exists

$$t_{k-d} = t_k - \tau(t_k) = t_k - \tau_d = \sum_{l=0}^{k-1} \mathrm{T}_l - \sum_{l=k-d}^{k-1} \mathrm{T}_l = \sum_{l=0}^{k-d-1} \mathrm{T}_l$$

It is assumed that the process $\{\tau_d\}$ is independent and identically-distributed, with the probabilities given as follows:

$$\text{Prob}\ \{\tau_d = \tau_j\} = \alpha_j, \quad \text{with} \quad j \in \{1, 2, \cdots, M\}$$

where α_j is a positive scalar which satisfies $0 \leq \alpha_j \leq 1$. In another case, control inputs do not suffer delays but can be lost during transmission. Normally, feedback-controlled plants can tolerate a certain amount of packets dropout, therefore, it is valuable to determine whether the system is stable. If there is no new control input packet arrived at time step k, the controller $u(t_k) = u(t_{k-1})$ will be used in logic ZOH for the NCS (11.2). It means that if there exist packets dropout, the last time step digital controller will be used at time step k. The probability of packets dropout is given by:

$$\text{Prob}\ \{\tau_d = \tau_l\} = \beta, \quad \text{for} \quad \forall\ l > M$$

where β is a positive scalar which satisfies $0 \leq \beta \leq 1$. Hence, we have the following relation equation

$$\beta + \sum_{j=1}^{M} \alpha_j = 1, \quad \text{where} \quad j \in \{1, 2, \cdots, M\}.$$

Considering the effect of random communication delay $\tau(t_k)$, state-feedback control law $u(t_k)$ in the NCS (11.2) will be changed as:

$$\begin{aligned} u(t_k) &= u(t_{i_k}) = K_k x(t_k - \tau(t_k)) \\ x(t_k) &= \psi(t_k), \quad k \in \{-M, \cdots, 0\} \end{aligned} \tag{11.5}$$

where $\psi(t_k)$ is a vector-valued initial function. Since $t_{i_k} = t_k - (t_k - t_{i_k})$, define $\tau(t_k) = t_k - t_{i_k}$, which denotes the time-varying delay in the control signal. Obviously, there exists

$$\begin{aligned} &\tau_k \leq \tau(t_k) \leq t(i_{k+1}) - t(i_k) + \tau_{k+1}, \\ &t_k \in [t(i_k) + \tau_k,\ t(i_{k+1}) + \tau_{k+1}). \end{aligned}$$

where time-delay τ_k denotes the time from the instant $t(i_k)$ when sensor nodes sample sensor data from a plant to the instant when actuator transfers data to the plant. It can be obtained that

$$\bigcup_{k=1}^{\infty} [t(i_k) + \tau_k,\ t(i_{k+1}) + \tau_{k+1}) = [t_0,\ \infty),\ t_0 \geq 0.$$

These assumptions are reasonable and they are employed in [292].

The closed-loop NCS is changed to a delayed switched DOS which is summed as follows

$$\begin{aligned} \delta x(t_k) &= A_k x(t_k) + B_k K_k x(t_k - \tau(t_k)) \\ x(t_k) &= \psi(t_k), \quad k \in \{-M, \cdots, 0\} \end{aligned} \tag{11.6}$$

where $\tau(t_k)$ is a network-induced delays function and $\psi(t_k)$ is a vector-valued initial function.

Remark 11.2. It has been shown that delta operator $\delta x(t_k)$ in (11.3) is different from the definition in Chapter 1 for the time-varying sampling period T_k. On the other hand, the case of sampling period being zero is omitted in (11.3) for the reason of that only discrete system model is valuable for NCSs. Hence, it is an extended delta operator definition for time-varying sampling period used in this chapter.

Remark 11.3. $\{i_1,\ i_2,\ i_3,\ \cdots\}$ is a subset of $\{1,\ 2,\ 3,\ \cdots\}$, and it denotes the sequence of time points of successful data transmissions from the sensor to the actuator. Moreover, it is not required that $i_{k+1} > i_k$, $\{i_1,\ i_2,\ i_3,\ \cdots\} = \{1,\ 2,\ 3,\ \cdots\}$ means that no packet dropout occurs in the transmission. If $i_{k+1} = i_k + 1$, it implies that $h + \tau_{k+1} > \tau_k$.

Before ending this section, the following lemma will be used to develop our main results.

Lemma 11.4. *The property of delta operator: for any time-varying function* $x(t_k)$ *and* $y(t_k)$

$$\delta(x(t_k)y(t_k)) = \delta(x(t_k))y(t_k) + x(t_k)\delta(y(t_k)) + T_k\delta(x(t_k))\delta(y(t_k)),$$

where T_k is the random sampling period at time instant k.

Proof. Taking the delta operator manipulations of $x(t_k)y(t_k)$, there exists

$$\begin{aligned}\delta(x(t_k)y(t_k)) &= \frac{1}{\mathrm{T}_k}\left[x(t_k+\mathrm{T}_k)y(t_k+\mathrm{T}_k) - x(t_k)y(t_k)\right] \\ &= \frac{1}{\mathrm{T}_k}\{x(t_k+\mathrm{T}_k)y(t_k+\mathrm{T}_k) + x(t_k)y(t_k+\mathrm{T}_k) - x(t_k)y(t_k+\mathrm{T}_k) \\ &\qquad +y(t_k)x(t_k+\mathrm{T}_k) - y(t_k)x(t_k+\mathrm{T}_k) - x(t_k)y(t_k)\} \\ &= \frac{1}{\mathrm{T}_k}\{x(t_k)[y(t_k+\mathrm{T}_k) - y(t_k)] + y(t_k)[x(t_k+\mathrm{T}_k) - x(t_k)] \\ &\qquad +[x(t_k+\mathrm{T}_k) - x(t_k)][y(t_k+\mathrm{T}_k) - y(t_k)]\} \\ &= \delta(x(t_k))y(t_k) + x(t_k)\delta(y(t_k)) + \mathrm{T}_k\delta(x(t_k))\delta(y(t_k))\end{aligned}$$

This completes the proof.

11.3 Main Results

11.3.1 Stabilization of NCSs

Before proceeding further, a general stability criteria is given for delayed switched DOS in the following theorem which plays a key role in this chapter.

Theorem 11.5. *A delayed switched DOS is asymptotically stable, if there exists a Lyapunov functional in delta domain as follows:*

$$V(x(t_k)) = \xi(x(t_k)) + \zeta(x(t_k+\theta)), \quad -\tau_M \le \theta \le 0 \tag{11.7}$$

where $\xi(\cdot)$ and $\zeta(\cdot)$ are two quasi-continuous positive definite functions, such that the following conditions hold:

(1)$V(x(t_k)) \ge 0$, with equality if and only if $x(t_k) = 0$;
(2)$\delta V(x(t_k)) = \left[V(x(t_{k+1})) - V(x(t_k))\right]/T_k < 0$.

Proof. Firstly, without loss of generality refer to [86], we assume that there exists a positive constant $q > 1$ such that

$$|x(t_k+\theta)| \le q|x(t_k)|, \quad -\tau_M \le \theta \le 0$$

On the other hand, we can obtain that there exists a positive constant $p < 1$ such that

$$|x(t_k+\theta)| \ge p|x(t_k)|, \quad -\tau_M \le \theta \le 0$$

Moreover, $\xi(\cdot)$ and $\zeta(\cdot)$ are two quasi-continuous positive definite functions. Therefore, it can be obtained that $V(x(t_k))$ is a bounded functional at time step k.

Then, from condition (2) there exists

$$\delta V(x(t_k)) < 0 \Rightarrow \left[V(x(t_{k+1})) - V(x(t_k))\right] / \mathtt{T}_k < 0 \Rightarrow V(x(t_{k+1})) < V(x(t_k))$$

It can be seen that $V(x(t_k)) > 0$ is a monotone decreasing functional.

Finally, due to delta domain Lyapunov functional $V(x(t_k))$ in (11.7) is bounded and monotone decreasing, it can be obtained that there exist three nondecreasing scalar functions $\alpha(\|x(t_k)\|)$, $\beta(\|x(t_k)\|)$ and $\gamma(\|x(t_k)\|)$ such that

$$\delta V(x(t_k)) = -\gamma(\|x(t_k)\|) < 0$$
$$\beta(\|x(t_k)\|) \geq V(x(t_k)) \geq \alpha(\|x(t_k)\|) > 0$$

Hence, there exists an equation $\|x(t_k)\| = 0$, such that $\alpha(0) = 0$, $\beta(0) = 0$ and $\gamma(0) = 0$ hold, respectively. Furthermore, when $\|x(t_k)\| \to 0$, there must exist $\alpha(\|x(t_k)\|) \to 0$ such that $V(x(t_k)) \to 0$.

In the following, three steps will be used to prove this theorem clearly.

First of all, we let $\phi(t_k, x(t_k))$ be a disturbed motion which sets out from any initial state $x(t_0)$.

(i) The uniform stability of initial equilibrium state $\boldsymbol{x}(\boldsymbol{t}_0)$.

For any real number $\varepsilon > 0$, there exists another corresponding real number $\delta(\varepsilon) > 0$ such that $\beta(\delta) \leq \alpha(\varepsilon)$. For the reason that the Lyapunov functional $V(x(t_k))$ is monotone and decreasing, the inequality $V(\phi(t_k, x(t_0))) - V(x(t_0)) \leq 0$ holds for arbitrary $t_k \in [t_0, \infty)$. Therefore, for arbitrary initial time t_0 and initial state $x(t_0)$ with $\|x(t_0)\| < \delta(\varepsilon)$, we have

$$\alpha(\varepsilon) \geq \beta(\delta) \geq V(x(t_0)) \geq V(\phi(t_k, x(t_0)) \geq \alpha(\|\phi(t_k, x(t_0)\|) \quad (11.8)$$

Since $\alpha(\|x(t_k)\|)$ is a quasi-continuous and nondecreasing function with $\alpha(0) = 0$, considering inequality (11.8), we get that

$$\|\phi(t_k, x(t_0)\| \leq \varepsilon, \quad \forall\, t \geq t_0$$

It has been shown that the initial equilibrium state $x(t_k) = 0$ is uniformly stable.

(ii) The disturbed motion $\boldsymbol{\phi}(\boldsymbol{t}_k, \boldsymbol{x}(\boldsymbol{t}_0))$, which generates from a nonzero initial state $\boldsymbol{x}(\boldsymbol{t}_0)$ satisfying $\|\boldsymbol{x}(\boldsymbol{t}_0)\| \leq \boldsymbol{\delta}(\boldsymbol{\varepsilon})$, is convergent to the initial equilibrium state $\boldsymbol{x}(\boldsymbol{t}_0) = 0$.

Firstly, for any real number $\mu > 0$ and $\delta(\varepsilon) > 0$ construct another real number $\mathtt{T}(\mu, \delta) > 0$. Let t_0 be an arbitrary initial time and $0 \leq \mu \leq \|x(t_0)\|$ hold. The initial equilibrium state $x(t_0)$ satisfies $\|x(t_0)\| \leq \delta(\varepsilon)$. The delta domain Lyapunov functional $V(x(t_k))$ is bounded and $\mu > 0$, a corresponding real number $\nu(\mu) > 0$ can be chosen such that $\beta(\nu) \leq \alpha(\mu)$ holds. Let $\rho(\mu, \delta)$ be the minimal value of $\gamma(\|x(t_k)\|)$ in the interval $\nu(\mu) \le \|x(t_k)\| \le \varepsilon$. By using the quasi-continuous nondecreasing function $\gamma(\|x(t_k)\|)$, it can be given that

$$\mathtt{T}(\mu,\delta) = \beta(\delta)/\rho(\mu,\delta).$$

It has been shown clearly that the corresponding real number $\mathtt{T}(\mu,\delta)$, which is independent of initial time t_0, can be constructed for all $\mu > 0$.

Secondly, there exists a time instant t_2' satisfying $t_0 \le t_2' \le t_0 + \mathtt{T}(\mu,\delta)$ such that $\phi(t_2', x(t_0)) = \nu(\mu)$. For notational expression simplicity, we denote $t_1' = t_0 + \mathtt{T}(\mu,\delta)$. By using reduction to absurdity technique and setting $\phi(t_k, x(t_0)) > \nu(\mu)$ for all t_k satisfying $t_0 \le t_k \le t_1'$, we have the following inequality:

$$\begin{aligned} 0 < \alpha(\nu) &\le V(\phi(t_1', x(t_0))) \le V(x(t_1')) \le V(x(t_0)) - (t_1' - t_0)\rho(\mu,\delta) \\ &\le \beta(\delta) - \mathtt{T}(\mu,\delta)\rho(\mu,\delta) = \beta(\delta) - \beta(\delta) = 0 \end{aligned}$$

Obviously, the above inequality is a contradictory result. So, there exists a time instant t_2' in the interval $[t_0, t_1']$ such that $\phi(t_2', x(t_0)) = \nu(\mu)$ holds.

Thirdly, the inequality $\phi(t_k, x(t_0)) \le \mu$ will be proved for all $t_k \ge t_0 + \mathtt{T}(\mu,\delta)$. Due to $\phi(t_2', x(t_0)) = \nu(\mu)$ holds for all $t_k \ge t_2'$, the following inequality can be gotten:

$$\alpha(\|\phi(t_k, x(t_0))\|) \le V(\phi(t_k, x(t_0))) \le V(\phi(t_2', x(t_0))) \le \beta(\nu) \le \alpha(\mu)$$

From above inequality and quasi-continuous nondecreasing function $\alpha(\|x(t_k)\|)$, it is easy to see that for all $t_k \ge t_2'$ we have the following inequality

$$\|\phi(t_k, x(t_0))\| \le \mu$$

which holds for all $t_k \ge t_0 + \mathtt{T}(\mu,\delta)$ following the relation $t_0 + \mathtt{T}(\mu,\delta) \ge t_2'$. In addition, it can be obtained that $\mathtt{T}(\mu,\delta) \to \infty$ as $\mu \to \infty$.

Consequently, it has been proved that the disturbed motion $\phi(t_k, x(t_0))$ is convergent to the initial equilibrium state $x(t_k) = 0$.

***(iii)* The disturbed motion $\phi(t_k, x(t_0))$ is uniformly bounded for arbitrary initial state $x(t_0)$.**

It is given that $\alpha(\|x(t_k)\|) \to \infty$ when $\|x(t_0)\| \to \infty$. Therefore, there exists a finite real number $\varepsilon(\delta)$ such that $\beta(\delta) < \alpha(\varepsilon)$ holds for any arbitrary large finite number $\delta > 0$.

From the bounded and monotone decreasing $V(x(t_k))$ for all $t_k \in [t_0, \infty)$ and arbitrary nonzero $x(t_0)$, the following inequality can be gotten

$$\alpha(\varepsilon) > \beta(\delta) \ge V(x(t_0)) \ge V(\phi(t_k, x(t_0))) \ge \alpha\|\phi(t_k, x(t_0))\|$$

The function $\alpha(\|x(t_k)\|)$ is a nondecreasing function, hence, we have

$$\|\phi(t_k, x(t_0))\| \le \varepsilon(\delta), \quad \forall t_k \ge t_0$$

in which there is any no relation between $\varepsilon(\delta)$ and t_0.

From the above analysis, we have proved that $\phi(t_k, x(t_0))$ is uniformly bounded for arbitrary initial state $x(t_0)$.

In summary, we can get that the delayed switched DOS (11.6) with packets dropout and network-induced delays is asymptotically stable. This completes the proof.

In the following section, we will develop an LMI approach to solve the stabilization problem of NCS which has been changed into the delayed switched DOS (11.6).

Theorem 11.6. *The delayed switched DOS (11.6) is asymptotically stable for network-induced delays case, if there exist symmetric positive definite matrices $X > 0$ and $Z > 0$, real matrices Y_k, N_{1k} and N_{2k}, such that the following LMIs hold:*

$$\begin{bmatrix} \Sigma_{11} & A_k X & B_k Y_k & 0 \\ * & \Sigma_{22} & B_k Y_k - N_{1k} + N_{2k} & \tau_{dk} N_{1k} \\ * & * & -N_{2k} - N_{2k}^T & \tau_{dk} N_{2k} \\ * & * & * & -\tau_{dk} Z \end{bmatrix} < 0, \ \forall \ k \in Z^+ \tag{11.9}$$

with

$$\Sigma_{11} = (T_k - 2)X + \tau_{dk} Z, \ \Sigma_{22} = X A_k^T + A_k X + N_{1k} + N_{1k}^T$$

and $\tau_{dk} = dT_k$. In this case, the networked digital controller is given by

$$u(t_k) = Y_k X^{-1} x(t_k - \tau(t_k)). \tag{11.10}$$

Furthermore, due to the effect of logic ZOH in NCS framework, the networked digital controller for packets dropout case is obtained as $u(t_k) = u(t_{k-1})$.

Proof. Consider a delta domain Lyapunov functional which is belong to (11.7) in Theorem 11.5 as follows:

$$V(x(t_k)) = V_1(x(t_k)) + V_2(x(t_k)) \tag{11.11}$$

with

$$\begin{aligned} V_1(x(t_k)) &= x^T(t_k) P x(t_k), \\ V_2(x(t_k)) &= \textstyle\sum_{l=-d+1}^{0} \sum_{h_m=t_{k-1+l}}^{t_{k-1}} e^T(h_m) R e(h_m). \end{aligned}$$

where $e(h_m) = x(h_{m+1}) - x(h_m)$. It can be seen that $\delta x(t_k) = -e(t_k)/\mathrm{T}_k$.

By using Lemma 11.4 and taking the delta operator manipulations of $V_1(x(t_k))$ along the trajectory of the delayed switched DOS (11.6), it can be obtained that

$$\begin{aligned}\delta V_1(x(t_k)) &= \delta^T x(t_k)Px(t_k) + x^T(t_k)P\delta x(t_k) + \mathtt{T}_k\delta^T x(t_k)P\delta x(t_k)\\ &= x^T(t_k)A_k^T Px(t_k) + x^T(t_k-\tau(t_k))K_k^T B_k^T Px(t_k)\\ &\quad + x^T(t_k)PB_kK_kx(t_k-\tau(t_k)) + x^T(t_k)PA_kx(t_k)\\ &\quad + \mathtt{T}_k\delta^T x(t_k)P\delta x(t_k).\end{aligned} \tag{11.12}$$

For the positive definite real matrix P, one has that

$$\begin{aligned}0 &= -2\delta^T x(t_k)P[\delta x(t_k) - A_kx(t_k) - B_kK_kx(t_k-\tau(t_k))]\\ &= -2\delta^T x(t_k)P\delta x(t_k) + 2\delta^T x(t_k)PA_kx(t_k)\\ &\quad + 2\delta^T x(t_k)PB_kK_kx(t_k-\tau(t_k)).\end{aligned} \tag{11.13}$$

Taking the delta operator manipulations of $V_2(x(t_k))$, it can be gotten

$$\begin{aligned}\delta V_2(x(t_k)) &= \frac{1}{\mathtt{T}_k}\sum_{l=-d+1}^{0}\left[\sum_{h_m=t_{k+l}}^{t_k} e^T(h_m)Re(h_m)\right.\\ &\qquad\left. - \sum_{h_m=t_{k-1+l}}^{t_{k-1}} e^T(h_m)Re(h_m)\right]\\ &= \frac{1}{\mathtt{T}_k}\sum_{l=-d+1}^{0}\left[e^T(t_k)Re^T(t_k) - e^T(t_{k-1+l})Re^T(t_{k-1+l})\right]\\ &= d\mathtt{T}_k\delta^T x(t_k)R\delta x(t_k) - \frac{1}{\mathtt{T}_k}\sum_{l=-d+1}^{0} e^T(t_{k-1+l})Re^T(t_{k-1+l})\\ &= \tau_{dk}\delta^T x(t_k)R\delta x(t_k) - \mathtt{T}_k\sum_{l=k-d}^{k-1}\delta^T x(t_l)R\delta x(t_l)\end{aligned} \tag{11.14}$$

For the positive definite matrix R and any real matrix $M_k \in R^{2n\times n}$, there exists

$$\begin{bmatrix} M_kR^{-1}M_k^T & M_k\\ M_k^T & R\end{bmatrix} \geq 0, \quad \text{where} \quad M_k = \begin{bmatrix} M_{1k}\\ M_{2k}\end{bmatrix}.$$

Hence, letting $\chi^T(t_k) = \left[x^T(t_k)\ x^T(t_k-\tau(t_k))\right]$, we have

$$\begin{aligned}0 &\leq \mathtt{T}_k\sum_{l=k-d}^{k-1}\begin{bmatrix}\chi(t_k)\\ \delta x(t_l)\end{bmatrix}^T\begin{bmatrix} M_kR^{-1}M_k^T & M_k\\ M_k^T & R\end{bmatrix}\begin{bmatrix}\chi(t_k)\\ \delta x(t_l)\end{bmatrix}\\ &= \mathtt{T}_k\sum_{l=k-d}^{k-1}\chi^T(t_k)M_kR^{-1}M_k^T\chi(t_k) + 2\mathtt{T}_k\chi^T(t_k)M_k\sum_{l=k-d}^{k-1}\delta x(t_l)\end{aligned}$$

$$
\begin{aligned}
&+\mathrm{T}_k \sum_{l=k-d}^{k-1} \delta^T x(t_l) R \delta x(t_l) \\
&= \tau_{dk} \chi^T(t_k) M_k R^{-1} M_k^T \chi(t_k) + 2\chi^T(t_k) M_k \begin{bmatrix} I & -I \end{bmatrix} \chi(t_k) \\
&+\mathrm{T}_k \sum_{l=k-d}^{k-1} \delta^T x(t_l) R \delta x(t_l)
\end{aligned} \tag{11.15}
$$

Combining (11.12)-(11.15) results in

$$
\delta V(x(t_k)) \le \eta^T(t_k) \left(\Sigma_1 + \tau_{dk} M_k R^{-1} M_k^T \right) \eta(t_k) < 0, \tag{11.16}
$$

where

$$
\eta^T(t_k) = \left[\delta^T x(t_k) \; x^T(t_k) \; x^T(t_k - \tau(t_k)) \right],
$$

and

$$
\Sigma_1 = \begin{bmatrix} (\mathrm{T}_k - 2)P + \tau_{dk} R & PA_k & PB_k K_k \\ * & A_k^T P + PA_k + M_{1k}^T + M_{1k} & PB_k - M_{1k} + M_{2k}^T \\ * & * & -M_{2k} - M_{2k}^T \end{bmatrix}.
$$

Using Schur's complement, $\Sigma_1 + \tau_{dk} M_k R^{-1} M_k^T < 0$ can be changed to $\Sigma_2 < 0$, where

$$
\Sigma_2 = \begin{bmatrix} (\mathrm{T}_k - 2)P + \tau_{dk} R & PA_k & PB_k K_k & 0 \\ * & \Sigma_2(2,2) & PB_k - M_{1k} + M_{2k}^T & \tau_{dk} M_{1k} \\ * & * & -M_{2k} - M_{2k}^T & \tau_{dk} M_{2k} \\ * & * & * & -\tau_{dk} R \end{bmatrix},
$$

with

$$
\Sigma_2(2,2) = A_k^T P + PA_k + M_{1k}^T + M_{1k}.
$$

Pre-multiply and post-multiply $\Sigma_2 < 0$ by $\mathrm{diag}\{P^{-1}, \ P^{-1}, \ P^{-1}, \ P^{-1}\}$, respectively. Then, letting $P^{-1} = X$, $K_k P^{-1} = Y_k$, $P^{-1} M_{1k} P^{-1} = N_{1k}$, $P^{-1} M_{2k} P^{-1} = N_{2k}$ and $P^{-1} R P^{-1} = Z$, inequality $\Sigma_2 < 0$ is equivalent to LMIs (11.9). Hence, we have obtained the two inequalities: $V(x(t_k)) \ge 0$ and $\delta V(x(t_k)) < 0$. According to Theorem 11.5, the closed-loop switched DOS (11.6) is asymptotically stable. This completes the proof.

Remark 11.7. We choose an ordinary method as the one in [272], which is not the main contribution in this chapter. It is well known that there are so many methods to deal with time delays both in continuous and discrete systems but not in switched DOSs. It is noted that P and R are two time invariant positive definite matrices. If let P and R be time variant it will get better results. However, this restriction is necessary for the proof, for the reason of that it is hard to make sure that $\delta V(x(t_k)) < 0$ holds for time-varying random sampling periods by using delta operator approach.

Remark 11.8. Following the LMI method for DOS with time-varying delays [196], the following Lyapunov functional in delta domain can be also taken as:

$$V_3(x(t_k)) = \mathtt{T}_k \sum_{l=t_{k-1-d}}^{t_{k-1}} x^T(l)Qx(l) \tag{11.17}$$

where Q is a time invariant positive definite matrix. Taking the delta operator manipulations of $V_3(x(t_k))$ along the trajectory of system (11.6) at time instant t_k, there exists

$$\delta V_3(x(t_k)) = \sum_{l=t_{k-d}}^{t_k} x^T(l)Qx(l) - \sum_{l=t_{k-1-d}}^{t_{k-1}} x^T(l)Qx(l)$$

It can be obtained that $t_{k-1-d} = t_{k-1} - \tau(t_{k-1}) = t_k - \mathtt{T}_{k-1} - \tau(t_{k-1})$ and $\mathtt{T}_{k-1} + \tau(t_{k-1}) \geq \tau(t_k)$ by the effect of logic ZOH in NCS framework. There exists $t_{k-1-d} \leq t_{k-d}$. It can be gotten that $\delta V_3(x(t_k)) \leq x^T(t_k)Qx(t_k)$. There is no need to introduce such additional term of $V_3(x(t_k))$, since $\|Q\|$ can be set sufficiently small in a positive definite sense in the result, which is the same as the one in [272].

For Theorem 11.6, LMIs (11.9) consists of a class of infinite LMIs for the reason of that $\{\mathtt{T}_k : k = 0, 1, \cdots, \infty\}$ is an infinite set. Therefore, we should give a set of finite LMIs to verify LMIs (11.9).

Let Y_k, N_{1k} and N_{2k} be replaced by Y, N_1 and N_2 in LMIs (11.9), which is changed to

$$\begin{bmatrix} \Gamma_{11} & A_kX & B_kY & 0 \\ * & \Gamma_{22} & B_kY - N_1 + N_2 & \tau_M N_1 \\ * & * & -N_2 - N_2^T & \tau_M N_2 \\ * & * & * & -\tau_M Z \end{bmatrix} < 0, \quad \forall\, k \in Z^+ \tag{11.18}$$

with

$$\Gamma_{11} = (\mathtt{T}_M - 2)X + \tau_M Z, \ \Gamma_{22} = XA_k^T + A_kX + N_1 + N_1^T.$$

It is easy to see that if LMIs (11.18) hold then LMIs (11.9) hold for the reasons of that there are more unknown matrix variables. In the following, we will verify the feasibility of infinite LMIs (11.18) by verifying a class of finite LMIs. Let $A_k = \Phi(\mathtt{T}_k)$, $B_k = \Psi(\mathtt{T}_k)$ and $K_k = K$, the closed-loop delayed switched DOS (11.6) can be rewritten as

$$\delta x(t_k) = \Phi(\mathtt{T}_k)x(t_k) + \Psi(\mathtt{T}_k)Kx(t_k - \tau(t_k)) \tag{11.19}$$

Letting $i, j, h = 1, 2. \cdots, n$ and $l = 1, 2. \cdots, m$, define two matrix sets as follows

$$\bar{\mathcal{A}} = \{\bar{\Phi} \in R^{n \times n} : \ \bar{a}_{ij} = q_{ij} \text{ or } \bar{a}_{ij} = r_{ij}\}$$

$$\bar{\mathcal{B}} = \left\{\bar{\Psi} \in R^{n\times m} : \ \bar{b}_{hl} = p_{hl} \text{ or } \bar{b}_{hl} = s_{hl}\right\}$$

where $\bar{a}_{ij}$ is the element of $\bar{\Phi}$ in the ith row and the jth column, $\bar{b}_{hl}$ is the element of $\bar{\Psi}$ in the hth row and the lth column; $q_{ij} = \min_{\mathrm{T}_k\in[\mathrm{T}_m,\mathrm{T}_M]} a_{ij}(\mathrm{T}_k)$ and $r_{ij} = \max_{\mathrm{T}_k\in[\mathrm{T}_m,\mathrm{T}_M]} a_{ij}(\mathrm{T}_k)$ are the minimum and the maximum elements of $\bar{\Phi}$ in the ith row and the jth column, $p_{hl} = \min_{\mathrm{T}_k\in[\mathrm{T}_m,\mathrm{T}_M]} b_{hl}(\mathrm{T}_k)$ and $s_{hl} = \max_{\mathrm{T}_k\in[\mathrm{T}_m,\mathrm{T}_M]} b_{hl}(\mathrm{T}_k)$ are the minimum and the maximum elements of $\bar{\Psi}$ in the ith row and the jth column. Let $\tau_M = \max\{d\mathrm{T}_k\}$, $K_k = K$, $N_{1k} = N_1$ and $N_{2k} = N_2$ in Theorem 11.6. We have the following verification theorem.

Theorem 11.9. *The closed-loop delayed switched DOS (11.19) is asymptotically stable, if there exist symmetric positive definite matrices $X > 0$ and $Z > 0$, real matrices Y, N_1 and N_2, such that the following LMIs hold:*

$$\Sigma(\bar{\Phi}, \bar{\Psi}) = \begin{bmatrix} \Xi_{11} & \bar{\Phi}X & \bar{\Psi}Y & 0 \\ * & \Xi_{22} & \bar{\Psi}Y - N_1 + N_2 & \tau_M N_1 \\ * & * & -N_2 - N_2^T & \tau_M N_2 \\ * & * & * & -\tau_M Z \end{bmatrix} < 0, \tag{11.20}$$

with

$$\Xi_{11} = (T_M - 2)X + \tau_M Z, \ \Xi_{22} = X\bar{\Phi}^T + \bar{\Phi}X + N_1 + N_1^T.$$

Proof. Define two interval matrix sets as follows:

$$\mathcal{A} = Co\bar{\mathcal{A}} = \left\{\hat{\Phi} = (\hat{a}_{ij}) : \ q_{ij} \le \hat{a}_{ij} \le r_{ij}\right\}$$

$$\mathcal{B} = Co\bar{\mathcal{B}} = \left\{\hat{\Psi} = (\hat{b}_{hl}) : \ p_{hl} \le \hat{b}_{hl} \le s_{hl}\right\}$$

Obviously, we have $\{\Phi(\mathrm{T}_k)|\mathrm{T}_k \in [\mathrm{T}_m, \mathrm{T}_M]\} \subset \mathcal{A}$ and $\{\Psi(\mathrm{T}_k)|\mathrm{T}_k \in [\mathrm{T}_m, \mathrm{T}_M]\} \subset \mathcal{B}$, respectively. Due to $\mathcal{A} = Co\bar{\mathcal{A}}$ and $\mathcal{B} = Co\bar{\mathcal{B}}$, for $\forall\ \hat{\Phi} \in \mathcal{A}$ and $\hat{\Psi} \in \mathcal{B}$, there exist

$$\begin{aligned} \theta_\imath \ge 0 \ \text{ and } \ \bar{\Phi} \in \mathcal{A}, \quad \imath = 1, 2, \cdots, L; \\ \vartheta_\jmath \ge 0 \ \text{ and } \ \bar{\Psi} \in \mathcal{B}, \quad \jmath = 1, 2, \cdots, H. \end{aligned}$$

such that

$$\hat{\Phi} = \sum_{\imath=\imath}^{L} \theta_\imath \bar{\Phi}, \quad \imath = 1, 2, \cdots, L \ \text{ and } \ \hat{\Psi} = \sum_{\jmath=1}^{L} \vartheta_\jmath \bar{\Psi}, \quad \jmath = 1, 2, \cdots, H.$$

Consider the following DOS

$$\delta x(t_k) = \hat{\Phi}x(t_k) + \hat{\Psi}Kx(t_k - \tau(t_k)) \tag{11.21}$$

Constructing Lyapunov functional as the same as the one in Theorem 11.6 and following its proof, we can get

$$\begin{bmatrix} (\mathtt{T}_M-2)X+\tau_M Z & \hat{\Phi}X & \hat{\Psi}Y & 0 \\ * & \Omega_{22} & \hat{\Psi}Y-N_1+N_2 & \tau_M N_1 \\ * & * & -N_2-N_2^T & \tau_M N_2 \\ * & * & * & -\tau_M Z \end{bmatrix} < 0 \tag{11.22}$$

with

$$\Omega_{22}=X\hat{\Phi}^T+\hat{\Phi}X+N_1+N_1^T$$

which can be rewritten as

$$\sum_{i=1}^{L}\theta_i\sum_{j=1}^{H}\vartheta_j\Sigma(\bar{\Phi},\bar{\Psi})<0 \quad \text{with} \quad \theta_i\geq 0,\ \vartheta_j\geq 0,$$

Hence, LMI $\Sigma(\bar{\Phi},\bar{\Psi})<0$ is hold. It has been gotten that the DOS (11.21) is asymptotically stable. From the following relation

$$\{\Phi(\mathtt{T}_k)|\mathtt{T}_k\in[\mathtt{T}_m,\mathtt{T}_M]\}\subset\mathcal{A} \quad \text{and} \quad \{\Psi(\mathtt{T}_k)|\mathtt{T}_k\in[\mathtt{T}_m,\mathtt{T}_M]\}\subset\mathcal{B} \tag{11.23}$$

Lastly, it has been obtained that the closed-loop delayed switched DOS (11.6) is asymptotically stable.

Remark 11.10. In order to verify that the closed-loop delayed switched DOS (11.6) is asymptotically stable, the number of LMIs have to be verified is $LH=2^{n^2}\cdot 2^{nm}=2^{n(n+m)}$, where n and m is the dimension of $\Phi(\mathtt{T}_k)$ and $\Psi(\mathtt{T}_k)$, respectively.

11.3.2 NCS with Packets Dropout

In the present section, we investigate a special NCS framework, in which some data packets can arrive to logic ZOH straightway for example in some local area networks. Since networks transfer distance is short enough and networks load is not heavy, that is most of data packets transmitted without any time delays. On the other hand, the other data packets which do not get instantly will be taken for loss, which is formulated in a probability context as

$$\text{Prob}\ \{Loss\}=\gamma, \quad \text{for} \quad 0\leq\gamma\leq 1.$$

The networked digital controller is obtained as follows:

$$u(t_k)=K_k x(t_k), \quad \text{with } k\in Z^+. \tag{11.24}$$

If there exist packets dropout in network channel, for the effect of logic ZOH, networked digital controller is changed as $u(t_k) = u(t_{k-1})$. Hence, the closed-loop NCS is expressed as

$$\delta x(t_k) = (A_k + B_k K_k)x(t_k) \tag{11.25}$$

The following theorem provides us the stabilization condition of the NCS (11.25) for the networked digital controller (11.24) design.

Theorem 11.11. *The closed-loop NCS (11.25) is asymptotically stable, if there exist symmetric positive definite matrix $X > 0$, a real matrix Y_k, such that the following LMIs hold:*

$$\begin{bmatrix} (T_k - 2)X & A_k X + B_k Y_k \\ * & A_k X + B_k Y_k + X A_k^T + Y_k^T B_k^T \end{bmatrix} < 0, \quad \forall\, k \in Z^+ \tag{11.26}$$

In the case of no data packets dropout, networked digital controller is given by $K_k = Y_k X^{-1}$. Furthermore, for packets dropout case the networked digital controller is given as $u(t_k) = u(t_{k-1})$ due to the effect of logic ZOH in NCS framework.

Proof. Construct a simple Lyapunov functional in delta domain as follows:

$$V_0(x(t_k)) = x^T(t_k)Px(t_k).$$

Using Lemma 11.4 and taking delta operator manipulations of $V_0(x(t_k))$ along the trajectory of the closed-loop NCS (11.25), we can obtain:

$$\begin{aligned} \delta V_0(x(t_k)) &= \delta^T x(t_k)Px(t_k) + x^T(t_k)P\delta x(t_k) + \mathtt{T}_k \delta^T x(t_k)P\delta x(t_k) \\ &= x^T(t_k)(A_k + B_k K_k)^T Px(t_k) + x^T(t_k)P(A_k + B_k K_k)x(t_k) \\ &\quad + \mathtt{T}_k \delta^T x(t_k)P\delta x(t_k). \end{aligned} \tag{11.27}$$

For the positive definite real matrix P, one has that

$$\begin{aligned} 0 &= -2\delta^T x(t_k)P[\delta x(t_k) - (A_k + B_k K_k)x(t_k)] \\ &= -2\delta^T x(t_k)P\delta x(t_k) + 2\delta^T x(t_k)P(A_k + B_k K_k)x(t_k). \end{aligned} \tag{11.28}$$

Combining (11.27) and (11.28) results in

$$\delta V_0(x(t_k)) \leq \xi^T(t_k)\Xi_1\xi(t_k) < 0, \tag{11.29}$$

where

$$\xi^T(t_k) = \begin{bmatrix} \delta^T x(t_k) \; x^T(t_k) \end{bmatrix},$$

and

$$\Xi_1 = \begin{bmatrix} (\mathtt{T}_k - 2)P & PA_k + PB_k K_k \\ * & PA_k + PB_k + A_k^T P + B_k^T P \end{bmatrix}.$$

Pre-multiplying and post-multiplying $\Xi_1 < 0$ by matrix diag$\{P^{-1},\ P^{-1}\}$, respectively, letting $P^{-1} = X$, $K_k P^{-1} = Y_k$, the inequality $\Xi_1 < 0$ is equivalent to LMIs (11.26).

We will prove that $x(t_k) \to 0$ as $k \to \infty$. Due to LMIs (11.26) hold, there exists

$$\delta V_0(x(t_k)) = \frac{1}{\mathtt{T}_k}\left[x^T(t_k+\mathtt{T}_k)Px(t_k+\mathtt{T}_k) - x^T(t_k)Px(t_k)\right] < 0$$

It can be seen that for $x(t_k) \neq 0$ the following inequality holds

$$x^T(t_k+\mathtt{T}_k)Px(t_k+\mathtt{T}_k) - x^T(t_k)Px(t_k) < 0,$$

which implies that the Lyapunov functional $x^T(t_k)Px(t_k)$ in delta domain is a strictly decreasing function. Thus, we obtain

$$x^T(t_k)Px(t_k) \to 0, \quad \text{when } k \to \infty$$

Since P is a positive definite constant matrix, we have

$$x(t_k) \to 0, \quad \text{when } k \to \infty$$

Therefore, the feasible state feedback control law obtained in Theorem 11.11 asymptotically stabilizes the closed-loop NCS (11.25). This completes the proof.

The verification theorem is given as follows.

Theorem 11.12. *The closed-loop delayed switched DOS (11.25) is asymptotically stable, if there exist a symmetric positive definite matrix $X > 0$, a real matrix Y, such that the following LMIs hold:*

$$\begin{bmatrix} (T_M - 2)X & \bar{\Phi}X + \bar{\Psi}Y \\ * & X\bar{\Phi}^T + Y^T\bar{\Psi}^T + \bar{\Phi}X + \bar{\Psi}Y \end{bmatrix} < 0 \tag{11.30}$$

The proof of the verification theorem can be obtained by following the proof process of Theorem 11.9.

11.3.3 Invariant Sampling Periods Case

In this section, we will give some corollaries on DOSs with invariant sampling periods for comparing with the results in [272] and Chapter 4.

In the following, we firstly consider the constant sampling case for the continuous system (11.1) by using delta operator approach. That is for $\forall\ k \in 0, 1, 2, \cdots$, there exist $t_k = k\mathtt{T}_s$ and $\mathtt{T}_k = \mathtt{T}_s$, where $\mathtt{T}_s$ is a constant sampling period such that the following DOS is obtained:

$$\delta x(t_k) = A_s x(t_k) + B_s u(t_k) \tag{11.31}$$

where

$$A_s = \frac{e^{A\mathtt{T}_s} - I}{\mathtt{T}_s}, \quad B_s = \frac{1}{\mathtt{T}_s}\int_0^{\mathtt{T}_s} e^{A(\mathtt{T}_s - \sigma)} B d\sigma.$$

It is easy to see that A_s and B_s are two known constant matrices.

Following the process of data packets transmitted in network channel, for considering network-induced delays case the control input $u(t_k)$ in system (11.31) with time-varying delay $\tau(t_k)$ is shown as:

$$u(t_k) = Kx(t_k - \tau(t_k)). \tag{11.32}$$

where K is a constant state feedback control gain matrix. The linear switched DOS (11.6) is simplified into a general DOS as follows:

$$\begin{aligned} \delta x(t_k) &= A_s x(t_k) + B_s K x(t_k - \tau(t_k)) \\ x(t_k) &= \psi(t_k), \quad k \in \{-M, \cdots, 0\} \end{aligned} \tag{11.33}$$

where $\psi(t_k)$ is a vector-valued initial function.

The following corollary presents the corresponding simplified results.

Corollary 11.13. *The closed-loop NCS (11.33) is asymptotically stable, if there exist symmetric positive definite matrices $X > 0$ and $Z > 0$, some real matrices Y, N_1 and N_2, such that the following LMI holds:*

$$\begin{bmatrix} (T_s - 2)X + \tau_M Z & A_s X & B_s Y & 0 \\ * & \Theta_{22} & B_s Y - N_1 + N_2 & \tau_M N_1 \\ * & * & -N_2 - N_2^T & \tau_M N_2 \\ * & * & * & -\tau_M Z \end{bmatrix} < 0. \tag{11.34}$$

with

$$\Theta_{22} = X A_s^T + A_s X + N_1 + N_1^T$$

where τ_M is the largest delay. In network-induced delays case, the networked digital controller is given by

$$u(t_k) = Y X^{-1} x(t_k - \tau(t_k)).$$

In addition, the networked digital controller for packets dropout case is obtained as $u(t_k) = u(t_{k-1})$.

The proof of above corollary is easily obtained by following the same process as the one of Theorem 11.6.

It is easy to see that when $\mathtt{T}_s = 1$, i.e. $t_k = k$, the DOS (11.33) can be rewritten as

$$x(k+1) = (A_s + I)x(k) + B_s K x(k - \tau(k)) \tag{11.35}$$

$$x(k) = \psi(k), \quad k \in \{-M, \cdots, 0\}$$

where $\psi(k)$ is a vector-valued initial function.

It can be seen that the discrete-time system (11.35) is the same as the one investigated in [272], which only considered an especial case of (11.33). It is worthy saying that the obtained results in this chapter are provided with more extensive than the ones in [272] which only traditional shift operator and uniform sampling periods are used.

In the following, in order to show that the LMI method used in Theorem 11.6 is less conservative, the same method will be used to a class of uncertain DOS with time-varying time delays which has been investigated in Chapter 4. We consider the following uncertain DOS with time delays

$$\delta x(t_k) = (A + \Delta A(t))x(t_k) + (A_d + \Delta A_d(t_k))x(t_k - d(t_k)) + Bu(t_k) \quad (11.36)$$

where $x(t_k) \in R^n$ is the state variable; $u(t_k) \in R^m$ is control input; A, A_d and B are three known constant matrices of appropriate dimensions; the time delay function $d(t_k)$ is a time-varying function that satisfies $0 \leq d(t_k) \leq d_M$, with $d_M = n_M \mathrm{T}_s$, n_M is a known positive and finite integer. Linear fractional parametric uncertainties $\Delta A(t_k)$ and $\Delta A_d(t_k)$ are time-varying matrices, which are defined as follows:

$$\left[\Delta A(t_k)\ \Delta A_d(t_k)\right] = D\hat{F}(t_k)\left[E_1\ E_2\right], \quad (11.37)$$

$$\hat{F}(t_k) = F(t_k)\left[I - GF(t_k)\right]^{-1} \quad (11.38)$$

where D, G, E_1 and E_2 are known constant real matrices with appropriate dimensions. $F(t_k)$ is an unknown time-varying matrix satisfying $F^T(t_k)F(t_k) \leq I$. It is assumed that the matrix $[I - GF(t_k)]^{-1}$ is invertible for any $F(t_k)$ and $I - G^TG > 0$.

The following control law is employed to deal with the problem of stabilization via state feedback:

$$u(t_k) = Kx(t_k) \quad (11.39)$$

where K is the state feedback controller gain to be determined such that the closed-loop system is asymptotically stable for any $0 \leq d(t_k) \leq d_M$. Consider the delayed DOS (11.36) associated with control law (11.39), then the resulted closed-loop system can be expressed as follows:

$$\delta x(t_k) = (A + BK + \Delta A(t_k))x(t_k) + (A_d + \Delta A_d(t_k))x(t_k - d(t_k)). \quad (11.40)$$

The following corollary is given by using the Lyapunov functional in Theorem 11.6.

Corollary 11.14. *Considering the closed-loop uncertain DOS (11.40), if there exist symmetric positive definite matrices $X > 0$, $W > 0$ and $Z > 0$,* *real matrices Y, N_1 and N_2, as well as a positive scalar $\alpha > 0$ such that the following LMI holds*

$$\begin{bmatrix} \Psi_{11} & AX+BY & A_dX & 0 & 0 & \alpha D \\ * & \Psi_{22} & A_dX - N_1 + N_2^T & d_M N_1 & XE_1^T & \alpha D \\ * & * & -W - N_2^T - N_2 & d_M N_2 & XE_2^T & 0 \\ * & * & * & -d_M Z & 0 & 0 \\ * & * & * & * & -\alpha I & \alpha G \\ * & * & * & * & * & -\alpha I \end{bmatrix} < 0, \quad (11.41)$$

with

$$\Psi_{11} = (T_s - 2)X + d_M Z,$$
$$\Psi_{22} = (AX + BY)^T + AX + BY + N_1^T + N_1 + W.$$

Then, the closed-loop uncertain DOS (11.40) is robustly asymptotically stable. Moreover, a suitable stabilizing state feedback controller (11.39) can be chosen by $u(t_k) = YX^{-1}x(t_k)$.

Proof. Combine the Lyapunov functionals in (11.11) and (11.17), it can be obtained that

$$V(x(t_k)) = x^T(t_k)Px(t_k) + \mathtt{T}_s \sum_{l=t_{k-1-d}}^{t_{k-1}} x^T(l)Qx(l)$$
$$+ \sum_{l=-d+1}^{0} \sum_{h_m=t_{k-1+l}}^{t_{k-1}} e^T(h_m)Re(h_m) \quad (11.42)$$

where $e(h_m) = x(h_m + \mathtt{T}_s) - x(h_m)$ which is the same as the one in the proof of Theorem 11.6. It is easy to see that $\delta x(t_k) = -e(t_k)/\mathtt{T}_s$.

Following the proof of Theorem 11.6 to deal with Lyapunov functional (11.42) and the LMI method in [196] to deal with the uncertainty parameter term in (11.37), the following LMI is obtained

$$\begin{bmatrix} \Phi_{11} & P(A+BK) & PA_d & 0 & 0 & \varepsilon^{-1}PD \\ * & \Phi_{22} & PA_d - M_1 + M_2^T & d_M M_1 & \varepsilon E_1^T & \varepsilon^{-1}PD \\ * & * & -Q - M_2^T - M_2 & d_M M_2 & \varepsilon E_2^T & 0 \\ * & * & * & -d_M R & 0 & 0 \\ * & * & * & * & -I & G \\ * & * & * & * & * & -I \end{bmatrix} < 0. (11.43)$$

with

$$\Phi_{11} = (\mathtt{T}_s - 2)P + d_M R,$$
$$\Phi_{22} = (A + BK)^T P + P(A + BK) + M_1^T + M_1 + Q.$$

Pre-multiplying and post-multiplying (11.43) by the diagonal matrix

$$\mathrm{diag}\{P^{-1},\ P^{-1},\ P^{-1},\ P^{-1},\ \varepsilon I,\ \varepsilon I,\}$$

and letting $\varepsilon^2 = \alpha$, $X = P^{-1}$, $Y = KX = KP^{-1}$, $W = P^{-1}QP^{-1}$, $Z = P^{-1}RP^{-1}$, $N_1 = P^{-1}M_1P^{-1}$ and $N_2 = P^{-1}M_2P^{-1}$, inequality (11.43) is equivalent to LMI (11.41). This completes the proof.

11.4 Numerical Examples

In this section, we consider some examples for numerical implementation and compare the results with the existing methods.

Example 11.15. In the first example, we give a rotating base pendulum system model to illustrate the main results obtained in Theorem 11.5 and Theorem 11.6. Consider the linearized state-space model of motion about the upward unstable equilibrium position of a rotating base pendulum, which has been described in [190]. The structure of the Furuta Pendulum is shown in Fig. 11.3.

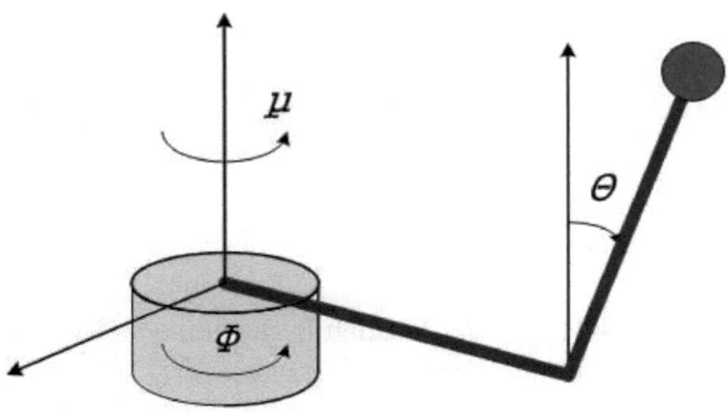

Fig. 11.3 Furuta pendulum Model.

Pendulum angle $\boldsymbol{\theta}$ is defined to be zero in the upright position with clockwise rotation considered positive. The location of the zero position for the arm angle $\boldsymbol{\Phi}$ can be chosen arbitrarily, and the counterclockwise direction of rotation $\boldsymbol{\mu}$ is termed positive. The mass of pendulum rod $m = 0.025$ Kg, the mass of pendulum ball $M = 0.004$ Kg, the length of pendulum rod $l = 0.241$ m, the length of horizontal arm $r = 0.152$ m, the moment of inertia of central shaft $J = 1.21 \times 10^{-3}$ Kg·m^2, the gravitational constant $g = 9.81$ m/s^2. Let $x^T(t) = \begin{bmatrix} \boldsymbol{\theta}\ \dot{\boldsymbol{\theta}}\ \boldsymbol{\Phi}\ \dot{\boldsymbol{\Phi}} \end{bmatrix}$ and

$$a = \left(M + \frac{m}{3}\right) l^2,\ b = J + (M + m) r^2,$$
$$c = \left(M + \frac{m}{2}\right) rl,\ d = \left(M + \frac{m}{2}\right) gl.$$

Therefore, there exist

$$\frac{bd}{ab-c^2}=63.25,\ \frac{-c}{ab-c^2}=-33.31,$$
$$\frac{-c}{ab-c^2}=-520.72,\ \frac{a}{ab-c^2}=804.13.$$

The continuous linear rotating base pendulum system model, is given by:

$$\dot{x}(t)=\begin{bmatrix}0 & 1\,0\,0\\ 63.25 & 0\,0\,0\\ 0 & 0\,0\,1\\ -33.31 & 0\,0\,0\end{bmatrix}x(t)+\begin{bmatrix}0\\ -520.72\\ 0\\ 804.13\end{bmatrix}u(t),$$

which can be changed to a linear switched DOS with packets dropout and network-induced delays by networks. The sampling period is a random constant $\mathtt{T}_k$ for $k \in Z^+$. Letting $\mathtt{T}_k = (0.003 + \varrho_k)$ s with $0 \le \varrho_k \le 0.002$, the constant $\mathtt{T}_k$ is given at each time step k in Fig. 11.4.

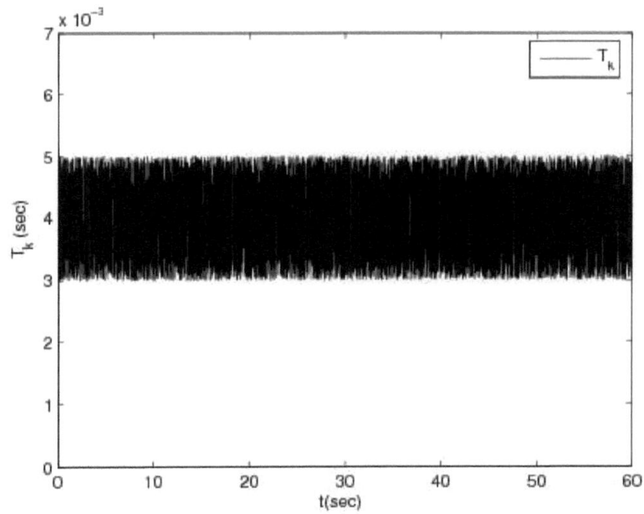

Fig. 11.4 Sampling periods $\mathtt{T}_k$.

Let the time delay steps exist in $d \in \{1, 2, 3, 4, 5\}$. The distribution of time delays and packets dropout is given in Fig. 11.5. as follows:

By using Matlab LMI Control Toolbox, solve LMIs (11.20) in Theorem 11.9. We have that LMIs (11.9) in Theorem 11.6 is solvable. Solve LMIs (11.9), we obtain the control in the switched model $u(t_k) = K_k x(t_k)$, which is computed at every time step k. Letting $x_1(0) = x_2(0) = x_3(0) = x_4(0) = 0$,

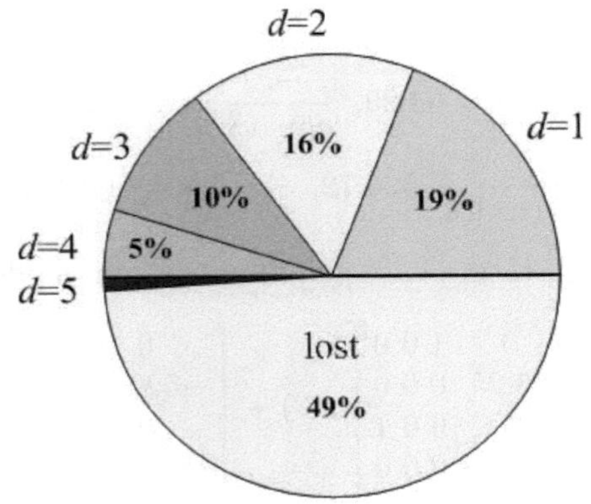

Fig. 11.5 Distribution of time delays and packets dropout.

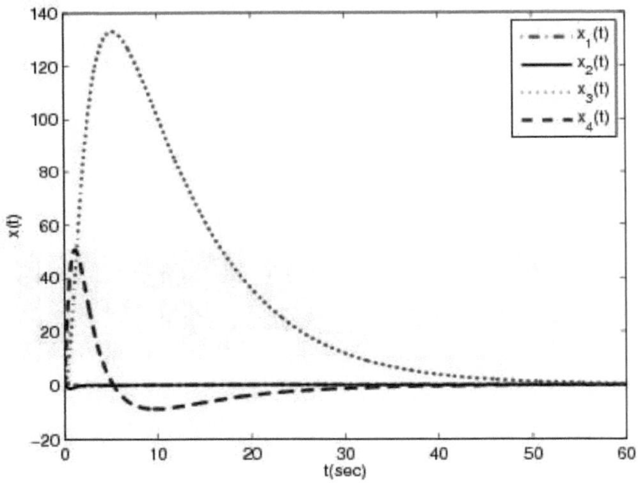

Fig. 11.6 The state response of $x(t)$.

the state trajectories of the closed-loop switched DOS caused by the obtained networked digital controller are shown in Fig. 11.6.

The control input of the closed-loop switched DOS with initial condition $u(0) = 0$ is given in Fig. 11.7.

It is clear from above two figures that the time-varying sample DOS is asymptotically stable. Hereby, the method of Theorem 11.5 and Theorem 11.6 in this chapter is effective in NCSs.

In the following example, we will provide an inverted pendulum example to demonstrate the effectiveness of the proposed method in Theorem 11.11.

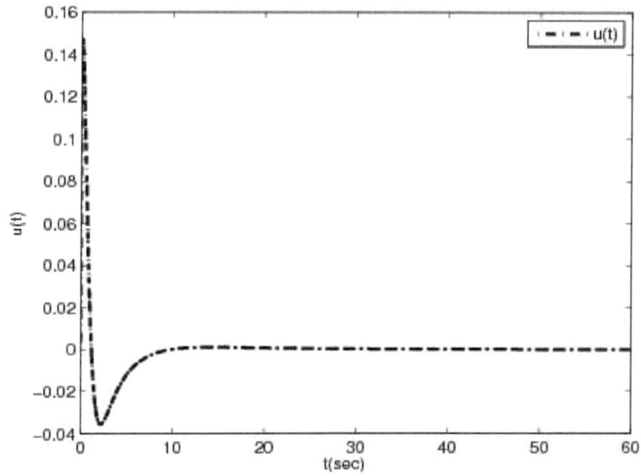

Fig. 11.7 The control input $u(t)$.

Example 11.16. The proposed method will be applied to design a controller for an inverted pendulum system to illustrate the results developed. Consider the inverted pendulum on a cart such as in [22]. The physical structure is shown in Fig. 11.8.

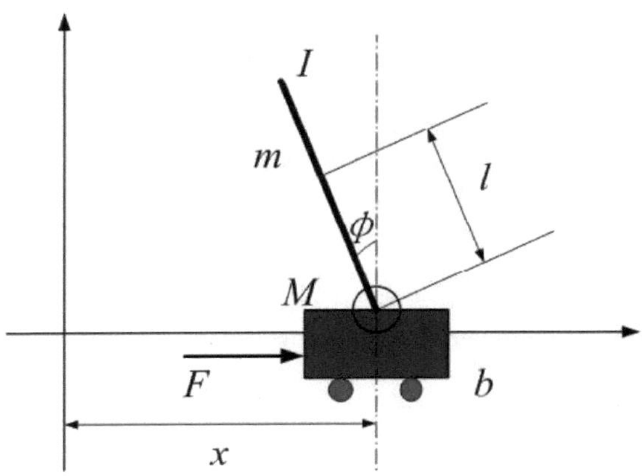

Fig. 11.8 Inverted pendulum Model.

where M is the mass of the cart, m is the mass of the pendulum rod, b is the friction coefficient of the cart, l is the length of the pendulum rod, F is the force acting on the cart, $\boldsymbol{x}$ is the horizontal displacement of the cart,

ϕ is the angle between the pendulum rod and the vertical. The surface is friction free. Let state variables be $\boldsymbol{x}$, $\dot{\boldsymbol{x}}$, ϕ and $\dot{\phi}$ which corresponds to the horizontal position, horizontal velocity of the cart, angle and angle velocity of the pole respectively. We are interested only in the linear system about the equilibrium point at the origin. The equations of motion derived by using Newton's Second Law can be obtained as follows:

$$\begin{aligned}
\dot{\boldsymbol{x}} &= \dot{\boldsymbol{x}} \\
\ddot{\boldsymbol{x}} &= \frac{-(I+ml^2)b}{I(M+m)+Mml^2}\dot{\boldsymbol{x}} + \frac{m^2gl^2}{I(M+m)+Mml^2}\phi + \frac{(I+ml^2)b}{I(M+m)+Mml^2}u \\
\dot{\phi} &= \dot{\phi} \\
\ddot{\phi} &= \frac{-mlb}{I(M+m)+Mml^2}\dot{\boldsymbol{x}} + \frac{mgl(M+m)}{I(M+m)+Mml^2}\phi + \frac{ml}{I(M+m)+Mml^2}u
\end{aligned}$$

where $M = 1.096$ Kg, $m = 0.109$ Kg, $b = 0.1$ N/m/sec, $l = 0.25$ m and $I = 0.034$ kg·m^2.

In the following, let state variables be $\boldsymbol{x} = x_1(t)$, $\dot{\boldsymbol{x}} = x_2(t)$, $\phi = x_3(t)$ and $\dot{\phi} = x_4(t)$, we can obtained as

$$\dot{x}(t) = \begin{bmatrix} 0 & 1 & 0 & 0 \\ 0 & -0.0883167 & 0.629317 & 0 \\ 0 & 0 & 0 & 1 \\ 0 & -0.235655 & 27.8285 & 0 \end{bmatrix} x(t) + \begin{bmatrix} 0 \\ 0.883167 \\ 0 \\ 2.35655 \end{bmatrix} u(t) \qquad (11.44)$$

which can be changed to a linear switched DOS only with packets dropout by investigating the case of control pass networks. The probability of packets dropout is 50%. It is shown in Fig. 11.9 that the sampling period $\mathtt{T}_k$ is a random constant at each time step k which is given in the interval [0.01s, 0.09s].

By using Matlab LMI Control Toolbox, solve LMIs (11.26) in Theorem 11.12. We get LMIs (11.26) in Theorem 11.11 is solvable. Solve LMIs (11.26) at every time step k in Theorem 11.12, we obtain the control in the switched model $u(t_k) = K_k x(t_k)$.

For initial condition: $x_1(0) = x_2(0) = x_3(0) = x_4(0) = 1$, the simulation results of the state response of closed-loop switched DOS in this example is given in Fig. 11.10.

Furthermore, Fig. 11.11 shows the simulation results of the control input for initial condition $u(0) = 0$.

From these simulation results, it is clear that the designed networked state feedback digital controller ensures the asymptotic stability of the closed-loop NCS with packet dropout.

In the next example, we will demonstrate the results obtained in Corollary 11.14 is less conservative than the ones in Chapter 4.

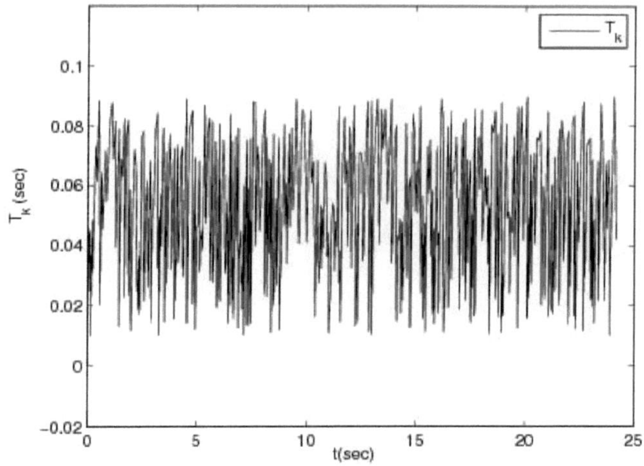

Fig. 11.9 The sampling periods $\mathtt{T}_k$.

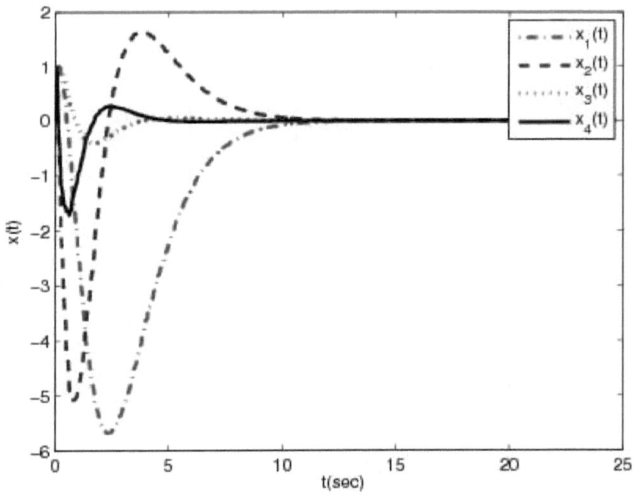

Fig. 11.10 The state response of $x(t)$.

Example 11.17. The LMI results in Corollary 11.14 will be applied to design a robust state feedback controller for truck trailer delay system. Letting sampling period $\mathtt{T}_s = 0.01$, the continuous system is sampled to the following robust DOS which is the same as the one in Chapter 4.

$$\begin{aligned}\delta x(t_k) = {} & (A + \Delta A(t_k))\, x(t_k) + (A_d + \Delta A_d(t_k))\, x(t_k - d(t_k)) \\ & + Bu(t_k)\end{aligned} \tag{11.45}$$

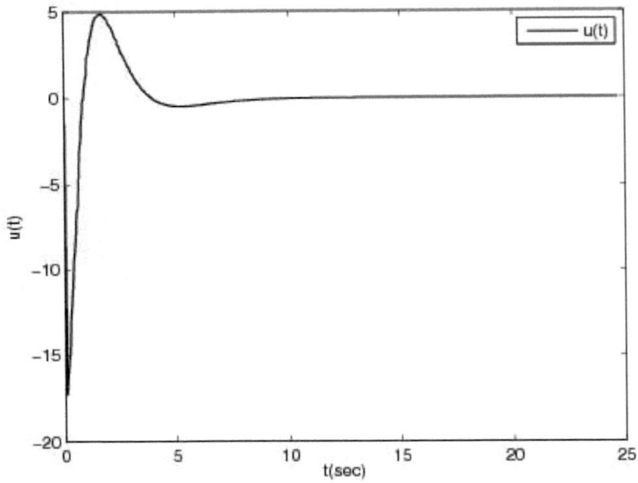

Fig. 11.11 The control input $u(t)$.

with

$$A=\begin{bmatrix}0.5109 & 0 & 0\\ -0.5109 & 0 & 0\\ 0.5212 & -4 & 0\end{bmatrix},\ B=\begin{bmatrix}-1.4338\\ 0.0052\\ -0.0053\end{bmatrix},\ a_d=\begin{bmatrix}0.219\\ -0.219\\ 0.2234\end{bmatrix},\ h=\begin{bmatrix}1\\ -1\\ -1\end{bmatrix},$$

$$A_d=\begin{bmatrix}a_d & 0_{3\times 2}\end{bmatrix},\ H=\begin{bmatrix}0.5091h & 0_{3\times 2}\end{bmatrix},\ H_d=\begin{bmatrix}0.2182h & 0_{3\times 2}\end{bmatrix},$$
$$F(t_k)=\mathrm{diag}\left\{\sin(t_k)\ \cos(t_k)\ \sin(t_k)\right\},\ G=G_d=0.05I_{3\times 3},\ W=0.01I_{3\times 3}.$$

In order to compare the designed state feedback controller for the robust DOS (11.45), we choose the same large time delay $d_M = 2$ as the one in Chapter 4. Then, using Matlab LMI Control Toolbox to solve LMI (11.41) in Corollary 11.14, the solution is obtained as:

$$X=\begin{bmatrix}9.7014 & 2.7267 & -1.9335\\ 2.7267 & 1.9096 & 5.0637\\ -1.9335 & 5.0637 & 81.4255\end{bmatrix},\ Z=\begin{bmatrix}0.4266 & 0.1469 & -0.2368\\ 0.1469 & 0.1364 & 0.4713\\ -0.2368 & 0.4713 & 11.0952\end{bmatrix},$$

$$N_1=\begin{bmatrix}-0.0778 & -0.0152 & 0.2740\\ -0.0324 & -0.0157 & 0.1320\\ 0.0114 & -0.0122 & -0.3611\end{bmatrix},\ N_2=\begin{bmatrix}0.0742 & 0.0098 & -0.3507\\ 0.0304 & 0.0122 & -0.1805\\ -0.0134 & 0.0015 & 0.1688\end{bmatrix},$$

$$W=\begin{bmatrix}11.9315 & 3.3372 & 1.6423\\ 3.3372 & 1.2917 & 2.4390\\ 1.6423 & 2.4390 & 19.1767\end{bmatrix},\ Y=\begin{bmatrix}15.1925 & 1.4206 & -5.5903\end{bmatrix}.$$

and $\alpha = 28.0266$. Therefore, the robust state feedback controller is given as:

$$u(t_k) = YX^{-1}x(t_k) = \begin{bmatrix} 2.6028 & -3.5379 & 0.2132 \end{bmatrix} x(t_k).$$

In order to contrast the control effect, we first give the states response and control input of the same closed-loop DOS in Chapter 4 as follows

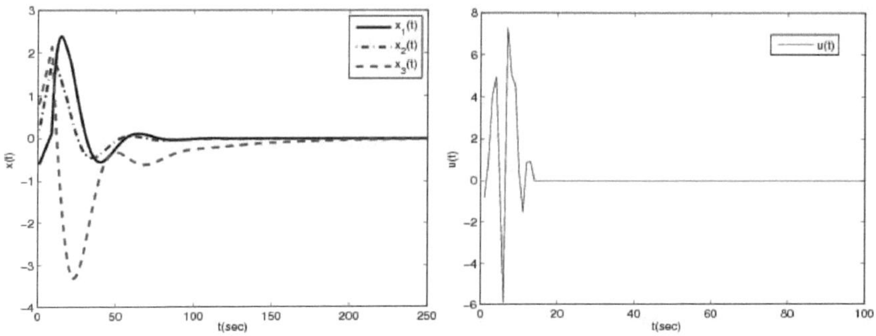

Fig. 11.12 The states and input in Chapter 4.

Then, the states response and control input of the closed-loop DOS (11.45) are given as in 11.13.

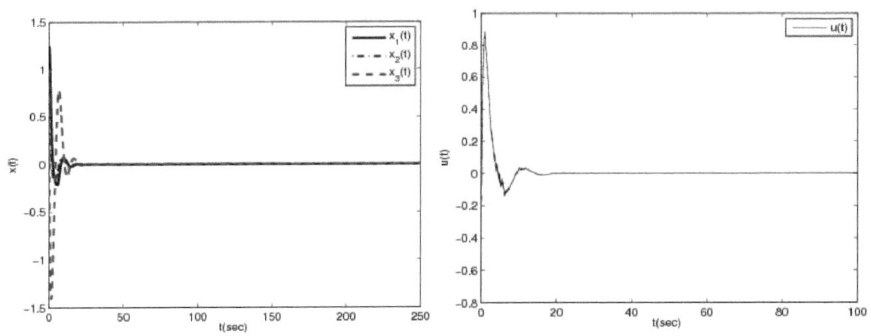

Fig. 11.13 The states and input of (11.45).

It is clear from the above contrastive simulation results that the convergence speed of state response in Fig. 11.13 of this example is much faster than the one in Fig. 11.12 which comes from Chapter 4. From these contrastive simulation results, it has been proved that the LMI method used in Theorem 11.6 of this chapter is less conservative than that of Chapter 4.

11.5 Summary

In this chapter, a new linear delayed switched DOS model has been proposed to describe the NCS with both packets dropout and network-induced delays. A general delta domain Lyapunov stability criteria has been given for switched DOSs with time delays. We have presented a new method for the stabilization of NCSs with nonuniform sampling periods based on delta domain Lyapunov functional method and LMI technique. A verification theorem has been given to show the solvability of the stabilization conditions by solving a class of finite LMIs. Furthermore, a NCS with packets dropout for local area networks case has also been investigated. A new sufficient condition for the solvability of a robust stabilization problem was also presented for comparing the results in Chapter 4 by using delta operator approach. Three numerical examples were given to illustrate the effectiveness and potential for the developed techniques.

Chapter 12
Stabilization of NCSs by Using DOKF

12.1 Introduction

Control systems in which information from the sensors and controllers is sent over an electronic communication network are called NCSs. The general theory for NCSs have attracted much research interest in recent years, such as [73, 108, 167] for the network-induced delay issue, [97, 244, 304] for packets dropout in digital communication networks, [153, 200, 205] for the H_∞ control performance of NCSs, [157, 166] for Markovian characterization of NCSs, [155, 156, 306] for predictive NCSs, [94, 187] for sampling strategy of NCSs. For limited bandwidth networks, a nonuniform sampling strategy may be achieved better performance than a uniform sampling strategy [220]. The nonuniform sampled-data NCSs have been extensively studied such as in [201, 241, 287]. A better idea is to shorten network-induced delay and reduce packet dropout by taking into account the system QoS criterion, i.e., the network throughput. As a method of remote control, a QoS-based remote control scheme for a NCS has been presented in [39, 138]. Fortunately, Kalman filter firstly presented by Kalman in [132] is an efficient recursive filter that estimates the state of a linear dynamic system from a series of noisy measurements. In [281], a continuous Kalman filter connected with sampler has been used in NCSs. Delta operator method is a reasonable approach than traditional shift operator in sampling continuous-time systems for short sampling period [53, 82, 170, 137]. A delta operator Riccati difference equation was gained for discrete-time Kalman filter in [202], in which the detailed DOKF algorithm had not been given.

In this chapter, we let transmission time is a switch multiple of sampling-period in overall dynamical system. We let the model transition be governed by a Markov chain as [41]. A finite memory buffer is used to implement the architecture of an estimator such that the delivered packets have a finite maximum delay as [204]. The logic ZOH is an efficient method to compare the time steps of the arrived data packets and choose the latest one to control

H. Yang et al.: Analysis and Synthesis of Delta Operator Systems, LNCIS 430, pp. 203–230.
springerlink.com © Springer-Verlag Berlin Heidelberg 2012

the process [272]. Furthermore, guaranteed cost control is a useful method in the systems which is not only stable but also guarantees an adequate level of performance [28, 55, 267, 294]. However, to the best of our knowledge, there have been few papers on guaranteed cost control for a Markov jump DOS which is used to describe NCS by using DOKF in Chapter 7 to estimate state variable approach, which motivates us to make an effort in this chapter.

The main contributions of this chapter are summarized as below:

i A new type of Lyapunov functional is given for Markov jump DOSs.
ii A class of DOKF arithmetic with small sampling periods is used for NCSs.
iii Both a smart sensor and a logic actuator are designed to solve the problems in NCSs.

12.2 Problem Statement

In this chapter, we consider the NCS model with packet dropout and network-induced delay both in the forward and the backward channels shown in Fig. 12.1. In the NCS framework, the continuous-time system can be modeled discretely with nonuniform sampling periods by a random sampler.

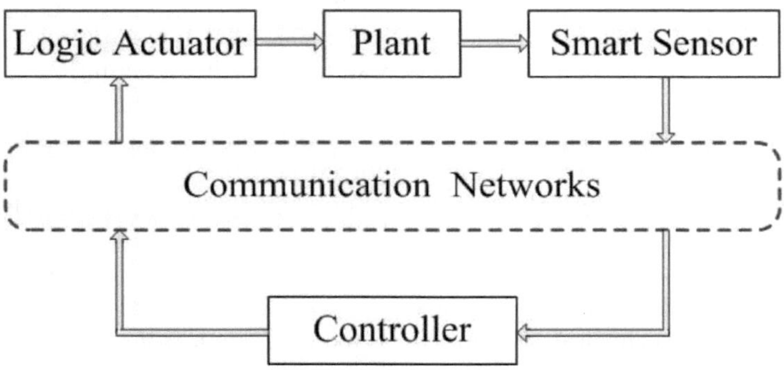

Fig. 12.1 Schematic diagram of NCS.

We address an estimation problem for linear plants in NCSs. The continuous-time plant under consideration is

$$\dot{x}(t) = A_s x(t) + B_s u(t) + \omega(t) \tag{12.1}$$
$$y(t) = C_s x(t) + v(t) \tag{12.2}$$

where $x(t) \in R^n$ is the state variable, $u(t) \in R^p$ is the control variable, $y(t) \in R^m$ is the measurement output, $A_s \in r^{n\times n}$ is system matrix, $B_s \in r^{n\times p}$ is the input matrix, $C_s \in r^{m\times n}$ is the measurement matrix, $\omega(t) \in R^n$ is the input noise, $\upsilon(t) \in R^m$ is the measurement noise. The following assumptions are standard.

Assumption 12.1. *System (A, B) is controllable*

Assumption 12.2. *For all constants $k > 0$ and $\tau > 0$, $\omega(t) \in R^n$ and $\upsilon(t) \in R^m$ are the uncorrelated white noises with zero mean and two ordinary covariance matrices Q_s and R_s*

$$\begin{aligned} \boldsymbol{E}[\omega(t)\omega^T(\tau)] &= Q_s\delta(t-\tau) \\ \boldsymbol{E}[\upsilon(t)\upsilon^T(\tau)] &= R_s\delta(t-\tau) \\ \boldsymbol{E}[\omega(t)\upsilon^T(\tau)] &= 0 \end{aligned}$$

where $\boldsymbol{E}(\cdot)$ denotes the expectation and $\delta(t)$ is the Kronecker Delta.

Assumption 12.3. *The initial state $x(0)$ is independent of $\omega(t) \in R^n$ and $\upsilon(t) \in R^m$*

$$\begin{aligned} \hat{x}(0) &= \boldsymbol{E}[x(0)] \\ P(0) &= \boldsymbol{E}[(\hat{x}(0) - x(0))(\hat{x}(0) - x(0))^T] \end{aligned}$$

The estimation scheme is motivated by the burgeon of smart sensors, which incorporate both computation and communication units. A smart sensor is co-located with the plant, Fig. 12.2.

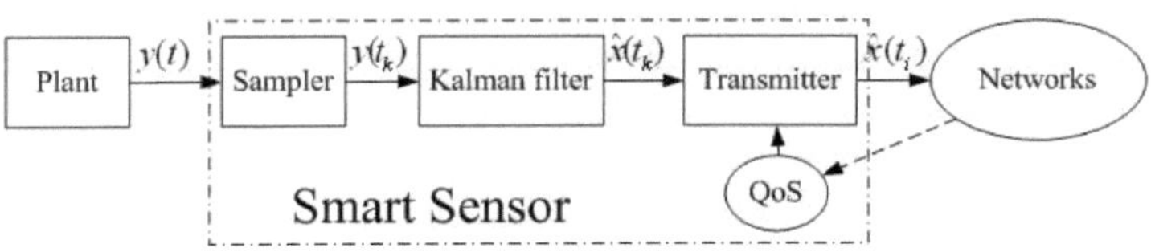

Fig. 12.2 Schematic diagram of smart sensor.

In the smart sensor, there exist a sampler to discretize continuous-time system model, a Kalman filter to estimate the plant state and a transmitter to sent data packets to a remote controller with considering the QoS of networks.

When output signal of system (12.1)-(12.2) pass through the sampler, the sampler samples the value of the system output $y(t)$ with a invariable sampling rate $1/\mathtt{T}_s$ where $\mathtt{T}_s$ is a sampling period. Letting $k \in Z^+$ be time step and the time sequence $t_k = k\mathtt{T}_s$ as shown in Fig 12.3, discretize continuous-time system (12.1)-(12.2) by using delta operator approach, the following DOS is obtained as

$$\delta x(t_k) = Ax(t_k) + Bu(t_k) + D\omega(t_k) \tag{12.3}$$

$$y(t_k) = Cx(t_k) + \upsilon(t_k) \tag{12.4}$$

in which the delta operator was defined by:

$$\delta x(t_k) = \begin{cases} dx(t)/dt, & \mathrm{T}_s = 0 \\ \frac{x(t_k+\mathrm{T}_s)-x(t_k)}{\mathrm{T}_s}, & \mathrm{T}_s \neq 0 \end{cases}$$

And A, B, C and D are constant matrices which are defined as below:

$$A = \frac{e^{A_s\mathrm{T}_s} - I}{\mathrm{T}_s}, \ B = \frac{1}{\mathrm{T}_s}\int_0^{\mathrm{T}_s} e^{A_s(\mathrm{T}_s-\tau)}B_s d\tau,$$
$$C = C_s, \ D = \frac{1}{\mathrm{T}_s}\int_0^{\mathrm{T}_s} e^{A_s(\mathrm{T}_s-\tau)}d\tau.$$

There exist $\lim_{\mathrm{T}_s\to 0} A = A_s$, $\lim_{\mathrm{T}_s\to 0} B = B_s$ and $\lim_{\mathrm{T}_s\to 0} D = I$, respectively.

The uncorrelated white noises $\omega(t_k) \in R^n$ and $\upsilon(t_k) \in R^m$ are given as follows

$$\omega(t_k) \sim (0, Q_s), \ Q = Q_s$$
$$\upsilon(t_k) \sim (0, R_s), \ R = R_s$$

where Q and R are two ordinary covariance matrices.

The state feedback control method is an efficient control arithmetic, however, it is hard to be used straightway because of that we can not obtain system states in a practical NCS. Thanks to Kalman filter which has been proved an efficient technique to estimate system states. In most of NCS such in [281], continuous-time Kalman filter instead of discrete-time case has been used to accurately estimate the plant state. However, it can not be applied in realistic NCSs for the application of digital computers. Consequently, we have to apply discrete-time Kalman filter to estimate system states in NCSs.

By utilizing the designed Kalman filter for DOSs (12.3)-(12.4) in Chapter 7, the estimate state $\hat{x}(t_k)$ in sampling instant t_k can be gotten and satisfied for delta operator state estimate system:

$$\delta\hat{x}(t_k) = A\hat{x}(t_k) + Bu(t_k) \tag{12.5}$$
$$y(t_k) = C\hat{x}(t_k) \tag{12.6}$$

Therefore, we can use estimate state feedback control method in the following process.

In NCS framework, data is transmitted in atomic units called packets. The communication scheduler on the transmitter of the smart sensor are modeled by considering a QoS measure which can decide on the intermittent time, i.e., the time interval between one data sending and the next.

Let $i \in Z^+$ be another time step and $\{\Gamma_i : i = 0, 1, 2, \cdots\}$ be a class of random intermittent time. It can be obtained that the sampling instants

$t_k = t_i$. That is to say there exists $t_{i+1} = t_i + \Gamma_i$. According to actual situation of NCS model, the the intermittent time Γ_i can be random chosen in a real number interval by the transmitter. The time diagram of sampling process is shown in Fig. 12.3.

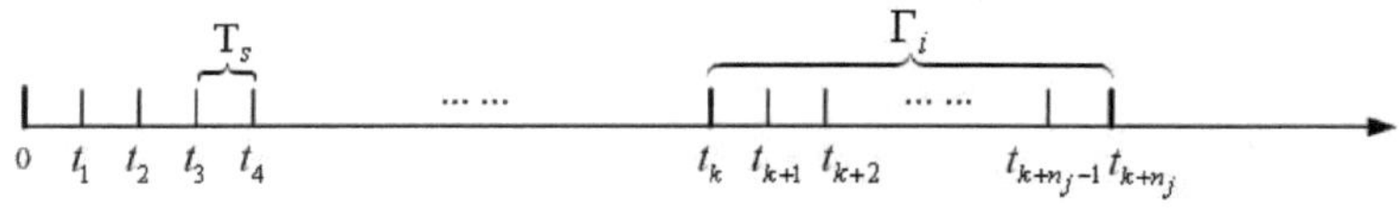

Fig. 12.3 Time diagram of sampling and transmit process.

As is shown in Fig. 12.3, there exists

$$t_i = t_k = k\mathrm{T}_s = \Gamma_0 + \Gamma_1 + \Gamma_2 + \cdots + \Gamma_{i-1} = \textstyle\sum_{l=0}^{i-1} \Gamma_l,$$

the intermittent time is also denoted as $\Gamma_i = n_j \mathrm{T}_s$ with $n_m \leq n_j \leq n_m$, where n_m and n_M are two known positive and finite integers, n_j is a time-varying finite integer. The overall dynamical system can be described in terms of switching between different models from a finite set of DOSs (12.5)-(12.6). In [41], the modal transition is governed by a Markov chain, the control system is then modeled as an Markov jump linear system with each state representing a different operational mode. That is, integer n_j at each intermittent time Γ_i satisfies the corresponding Markov jump process: the system mode $\gamma_i = l$ is a time homogeneous Markov chain taking values in a finite set $S = \{1,\ 2,\ \cdots, s\}$ with stationary transition probabilities from mode h at time i to mode l at time $i+1$

$$Pr(\gamma_{i+1} = l|\ \gamma_i = h) = \pi_{hl} \tag{12.7}$$

The Markov process transition probability matrix Π_z is defined by

$$\Pi_z = \begin{bmatrix} \pi_{11} & \pi_{12} & \cdots & \pi_{1s} \\ \pi_{21} & \pi_{22} & \cdots & \pi_{2s} \\ \vdots & \vdots & \ddots & \vdots \\ \pi_{s1} & \pi_{s2} & \cdots & \pi_{ss} \end{bmatrix}, \quad \text{with} \quad \pi_{ij} \geq 0, \quad \sum_{j=1}^{s} \pi_{ij} = 1. \tag{12.8}$$

In a corresponding continuous-time Markovian jump system, let (η_t) be a discrete-state Markov process with values in a finite set $S = \{1,\ 2,\ \cdots, s\}$. The transition probability from mode h at time t to the mode l at time $t + \Delta t,\ i,\ j \in S$, is

$$Pr(\eta_{t+\Delta t} = l|\ \eta_t = h) = \begin{cases} \vartheta_{hl}\Delta t + o(\Delta t), & h \neq l, \\ 1 + \vartheta_{hh}\Delta t + o(\Delta t), & h = l, \end{cases} \tag{12.9}$$

where $\Delta t > 0$, $\lim_{\Delta t \to 0} o(\Delta t)/\Delta t = 0$. The continuous Markov process transition rate matrix Λ_s is defined by

$$\Lambda_s = \begin{bmatrix} \vartheta_{11} & \vartheta_{12} & \cdots & \vartheta_{1s} \\ \vartheta_{21} & \vartheta_{22} & \cdots & \vartheta_{2s} \\ \vdots & \vdots & \ddots & \vdots \\ \vartheta_{s1} & \vartheta_{s2} & \cdots & \vartheta_{ss} \end{bmatrix}, \text{ with } \vartheta_{ij} \geq 0, \ \vartheta_{ii} = -\sum_{j=1, j\neq i}^{s} \vartheta_{ij}. \quad (12.10)$$

Letting $\Pi_s(t)$ be the transition probability matrix of continuous Markov chain, there exists

$$\Pi_z = \Pi_s(1) = e^{\Lambda_s}, \ \Lambda_s = \lim_{\Delta t \to 0} \frac{\Pi_s(\Delta t) - \Pi_s(0)}{\Delta t} = \Pi_s'(0), \ \Pi_s(\Delta t) = e^{\Lambda_s \Delta t}.$$

Furthermore, we give the following delta operator defined at time step t_i is given as

$$\delta \hat{x}(t_i) = \frac{\hat{x}(t_i + \Gamma_i) - \hat{x}(t_i)}{\Gamma_i}, \quad \forall \ \Gamma_i \neq 0 \quad (12.11)$$

The dynamics of the NCS (12.5)-(12.6) can be approximated by the following Markov jump linear DOS:

$$\delta \hat{x}(t_i) = A(\eta_i)\hat{x}(t_i) + B(\eta_i)u(t_i), \quad (12.12)$$

where $\hat{x}(t_i) \in R^n$ and $u(t_i) \in R^p$ are the state vector and control input vector. where $A(\eta_i)$ and $B(\eta_i)$ are two time-varying matrices which are defined as below:

$$A(\eta_i) = \frac{e^{A_s \Gamma_i} - I}{\Gamma_i}, \quad B(\eta_i) = \frac{1}{\Gamma_i} \int_0^{\Gamma_i} e^{A_s(\Gamma_i - \theta)} B_s d\theta.$$

where (η_i) is a discrete-state Markov process with values in a finite set $S = \{1, \ 2, \ \cdots, s\}$. The Markov process transition rate matrix Λ_δ is defined by

$$\Lambda_\delta = \frac{\Pi_s(\mathtt{T}_s) - \Pi_s(0)}{\mathtt{T}_s - 0} = \frac{e^{\Lambda_s \mathtt{T}_s} - I}{\mathtt{T}_s} = \begin{bmatrix} \lambda_{11} & \lambda_{12} & \cdots & \lambda_{1s} \\ \lambda_{21} & \lambda_{22} & \cdots & \lambda_{2s} \\ \vdots & \vdots & \ddots & \vdots \\ \lambda_{s1} & \lambda_{s2} & \cdots & \lambda_{ss} \end{bmatrix}. \quad (12.13)$$

Because of the use of QoS information, an appropriate intermittent time Γ_i within the interval $[t_i, t_{i+1})$ can be chosen such that for each input delay τ_i is not larger than intermittent time Γ_i, i.e., $\tau_i \in \{0, \mathtt{T}_s, 2\mathtt{T}_s, \cdots, n_j\mathtt{T}_s\}$. The effect of the intermittent time policy and QoS information is incorporated into NCSs by assigning the appropriate intermittent time to its corresponding maximum allowable equivalent delay bound $\eta(QoS)$ for each state. A logic actuator is co-located with the plant, Fig. 12.4.

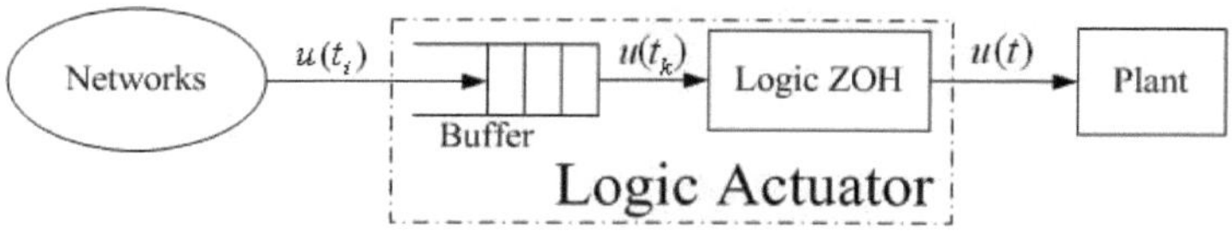

Fig. 12.4 Schematic diagram of Logic Actuator.

The main feature of the actuator is a buffer located at logic actuator which delays the application of a control signal until the maximum allowable equivalent delay bound is reached. The the maximum allowable equivalent delay bound is determined at the controller and is paired with the appropriate control signal in a single packet. Control signals which arrive after the maximum allowable equivalent delay time bound are taken for lost. For packets dropout case, the last control signal $u(t_{i-1})$ at time step t_{i-1} will be used at time step t_i due to the effect of logic ZOH in the logic actuator.

The goal of this chapter is to develop a procedure to design a memoryless estimate state-feedback guaranteed cost controller of the form

$$u^*(t_i) = K_h\hat{x}(t_i), \quad \text{when } \eta_i = h \tag{12.14}$$

Let maximum allowable equivalent delay bound $\eta(QoS) = \bar{\tau}$. Considering the maximum allowable equivalent delay bound $\bar{\tau}$ in Markov jump DOS (12.12), the control input $u(t_i)$ is shown as:

$$u(t_i) = \begin{cases} K_h\hat{x}(t_i), & t_i \in [k\mathrm{T}_s + \bar{\tau}, k\mathrm{T}_s + \Gamma_i) \\ K_{0h}\hat{x}(t_{i-1}), & t_i \in [k\mathrm{T}_s, k\mathrm{T}_s + \bar{\tau}) \end{cases} \tag{12.15}$$

For notational simplicity, in the sequel, for $\eta_i = h \in S$, the closed-loop estimate state Markov jump DOS (12.12) is given as follows:

$$\begin{aligned} \delta\hat{x}(t_i) &= A_h\hat{x}(t_i) + \Phi_h K_h\hat{x}(t_i) + \Psi_h K_{0h}\hat{x}(t_{i-1}) \\ \hat{x}(t_{-1}) &= \varphi(t_0), \ (\eta_{-1} = \eta_-) \quad \text{for } i = 0 \end{aligned} \tag{12.16}$$

with

$$A_h = \frac{e^{A_s\Gamma_i} - I}{\Gamma_i}, \quad \Phi_h = \frac{1}{\Gamma_i}\int_0^{\Gamma_i - \bar{\tau}} e^{A_s\theta}B_s d\theta, \quad \Psi_h = \frac{1}{\Gamma_i}\int_{\Gamma_i - \bar{\tau}}^{\Gamma_i} e^{A_s\theta}B_s d\theta$$

and $\varphi(t_0)$ is a vector-valued initial function, the integer variable $n_h \in \{n_m, n_{m+1}, \cdots, n_M\}$ is the current state of the MJLS. Each state in the Markov chain represents the system operating under specific values of sampling interval $\Gamma_i = n_h\mathrm{T}_s$. The duration in a state depends on the sampling interval $\Gamma_i \in \{n_m\mathrm{T}_s, n_{m+1}\mathrm{T}_s, \cdots, n_M\mathrm{T}_s\}$.

Associated with Markovian jump DOS (12.16) is the following quadratic cost function

$$J = \mathbf{E}\left\{\sum_{i=0}^{\infty} \Gamma_i \left[\hat{x}^T(t_i)Q\hat{x}(t_i) + u^T(t_i)Ru(t_i)\middle|\, \hat{x}(t_0), \eta_0\right]\right\} \tag{12.17}$$

where $Q \in R^{n\times n}$ and $R \in R^{p\times p}$ are two given positive-definite matrices.

Here, the objective of this chapter is to develop a procedure to design a state feedback controller $u(t_i)$ for the system (12.16) and cost function (12.17) such that the resulting closed-loop system is asymptotically stable and the closed-loop value of the cost function (12.17) satisfies $J \leq J^*$, where J^* is some specified constant.

Definition 12.1. The delta operator of the stochastic process $\{x(t_i),\ \eta_i,\ t_i \geq 0\}$, acting on $V(x(t_i), \eta_i)$ at the point $\{x(t_i),\ \eta_i = h\}$, can be expressed as follows:

$$\sigma(V(x(t_i), \eta_i)) = \frac{\mathbf{E}\left[V(x(t_{i+1}), \eta_{i+1})\middle|\, x(t_0), \eta_0\right] - V(x(t_i), \eta_i)}{\Gamma_i}. \tag{12.18}$$

In the following, we let the mode at time t_i be h, that is $\eta_i = h$. Recall that at time t_{i+1}, the system may jump to any mode $\eta_{i+1} = l$.

Definition 12.2. The Markovian jump DOS (12.40) is stochastic robustly asymptotically stable, if the following conditions for a delta domain Lyapunov function $V(x(t_i))$:

i $V(x(t_i), h) \geq 0$, with equality if and only if $x(t_i) = 0$;

ii $\sigma V(x(t_i), h) = \{\mathbf{E}[V(x(t_{i+1}), l)|\, x(t_0), \eta_0] - V(x(t_i), h)\}/\Gamma_i < 0$,

hold such that for every initial state $(x(t_0), \eta_0)$, the following condition:

$$\mathbf{E}\left\{\sum_{i=0}^{\infty} \Gamma_i \left\|\hat{x}(t_i)\right\|^2 \middle|\, \hat{x}(t_0), \eta_0\right\} < \infty \tag{12.19}$$

is satisfied.

Definition 12.3. For Markovian jump DOS (12.16) and cost function (12.17), if there exist a control law $u^*(t)$ and a positive J^* such that Markovian jump DOS (12.16) is asymptotically stable and the closed-loop value of the cost function (12.17) satisfies $J \leq J^*$, then J^* is said to be a guaranteed cost and $u^*(t)$ is said to be a guaranteed cost control law of Markovian jump DOS (12.16) and cost function (12.17).

Remark 12.4. It is well known that the smaller the sampling period the better the control effect in a discrete system by using delta operator approach. The numerical advantage of the delta operator approach has been clearly shown in [196]. Hence, in order to suit for small sampling period $\mathtt{T}_s$ we use DOS to replace traditional discrete system and it is necessary to study Kalman filter arithmetic for DOSs, which is a key problem in this chapter.

Remark 12.5. It has been shown that the operator $\delta x(t_i)$ in (12.5) is different from the above delta operator definition $\sigma(\cdot)$ for the time-varying sampling period Γ_i in (12.18). On the other hand, the case of sampling period being zero is omitted in (12.5) for the reasons that only discrete system model is valuable for NCSs.

Remark 12.6. The Markov process transition rate matrix Λ_s is a special case of Λ_δ, since there exists

$$\lim_{\mathtt{T}_s\to 0}\Lambda_\delta=\lim_{\mathtt{T}_s\to 0}\frac{e^{\Lambda_s\mathtt{T}_s}-I}{\mathtt{T}_s}=\lim_{\mathtt{T}_s\to 0}\frac{d(e^{\Lambda_s\mathtt{T}_s}-I)/d\mathtt{T}_s}{d(\mathtt{T}_s)/d\mathtt{T}_s}=\lim_{\mathtt{T}_s\to 0}\Lambda_s e^{\Lambda_s\mathtt{T}_s}=\Lambda_s.$$

Furthermore, the transition rate matrix Λ_δ also covers the transition probability matrix Π_z which is relative to Π_s for $\Pi_z=\Pi_s(1)$. Therefore, it can be said that continuous-time Markov jump process and discrete-time Markov jump process are the special case of Markov jump process in delta operator domain when sampling period $\mathtt{T}_s\to 0$ and $\mathtt{T}_s=1$, respectively.

Remark 12.7. With Assumptions 12.1, 12.2 and 12.3 hold, the continuous-time Kalman filter equations for continuous system (12.1)-(12.2) are given as

$$\begin{aligned}K_s&=P(t)C_s^TR_s^{-1}\\ \dot{\hat{x}}(t)&=A_s\hat{x}(t)+B_su(t)+K_s(y(t)-C_s\hat{x}(t))\\ \dot{P}(t)&=-P(t)C_s^TR_sC_sP(t)+A_sP(t)+P(t)A_s^T+Q_s\end{aligned}$$

The method of deriving the continuous-time Kalman filter is shown in [212].

Before ending this section, the following lemmas will be used to prove our main results.

Lemma 12.8. *The property of delta operator: for any time-varying function $x(t_k)$ and $y(t_k)$*

$$\delta(x(t_k)y(t_k))=\delta(x(t_k))y(t_k)+x(t_k)\delta(y(t_k))+T_k\delta(x(t_k))\delta(y(t_k)),$$

where T_k is the random sampling period at time instant k.

Proof. Taking the delta operator manipulations of $x(t_k)y(t_k)$, there exists

$$\begin{aligned}\delta(x(t_k)y(t_k))&=\frac{1}{\mathtt{T}_k}\left[x(t_k+\mathtt{T}_k)y(t_k+\mathtt{T}_k)-x(t_k)y(t_k)\right]\\ &=\frac{1}{\mathtt{T}_k}\{x(t_k+\mathtt{T}_k)y(t_k+\mathtt{T}_k)+x(t_k)y(t_k+\mathtt{T}_k)-x(t_k)y(t_k+\mathtt{T}_k)\\ &\qquad+y(t_k)x(t_k+\mathtt{T}_k)-y(t_k)x(t_k+\mathtt{T}_k)-x(t_k)y(t_k)\}\end{aligned}$$

$$
\begin{aligned}
&= \frac{1}{\mathtt{T}_k}\{x(t_k)[y(t_k+\mathtt{T}_k)-y(t_k)]+y(t_k)[x(t_k+\mathtt{T}_k)-x(t_k)] \\
&\qquad +[x(t_k+\mathtt{T}_k)-x(t_k)][y(t_k+\mathtt{T}_k)-y(t_k)]\} \\
&= \delta(x(t_k))y(t_k)+x(t_k)\delta(y(t_k))+\mathtt{T}_k\delta(x(t_k))\delta(y(t_k))
\end{aligned}
$$

This completes the proof.

Lemma 12.9. *[212] The discrete-time Kalman filter is summarized as follows.*

1 For all integers k, $\tau > 0$ and the ordinary covariance matrices $\{Q_k\}$ and $\{R_k\}$, the dynamic discrete-time system is given by the following equations:

$$x_k = F_{k-1}x_{k-1} + G_{k-1}u_{k-1} + \omega_{k-1} \tag{12.20}$$

$$y_k = H_k x_k + \upsilon_k \tag{12.21}$$

$$
\begin{aligned}
E[\omega_k\omega_j^T] &= Q_k\delta_{k-j} \\
E[\upsilon_k\upsilon_j^T] &= R_k\delta_{k-j} \\
E[\omega_k\upsilon_j^T] &= 0
\end{aligned}
$$

2 The Kalman filter is initialized as follows

$$
\begin{aligned}
\hat{x}_0^+ &= E[x_0] \\
P_0^+ &= E[(x_0-\hat{x}_0^+)(x_0-\hat{x}_0^+)^T]
\end{aligned}
$$

3 The discrete-time Kalman filter is given by the following equations, which are computed for each time step $k = 1, 2, \cdots$:

$$
\begin{aligned}
P_k^- &= F_{k-1}P_{k-1}^+F_{k-1}^T + Q_{k-1} \\
K_k &= P_k^-H_k^T(H_kP_k^-H_k^T+R_k)^{-1} \\
&= P^+H_k^TR_k^{-1} \\
\hat{x}_k^- &= F_{k-1}\hat{x}_{k-1}^+ + G_{k-1}u_{k-1} = \textit{a priori state estimate} \\
\hat{x}_k^+ &= \hat{x}_k^- + K_k(y_k - H_k\hat{x}_k^-) = \textit{a posteriori state estimate} \\
P_k^+ &= (I-K_kH_k)P_k^-(I-K_kH_k)^T + K_kR_kK_k^T \\
&= [(P_k^-)^{-1} + H_k^TR_k^{-1}H_k]^{-1} \\
&= (I-K_kH_k)P_k^-
\end{aligned}
$$

12.3 Main Results

12.3.1 DOKF for NCSs

DOKF is a valuable and significant part of this chapter. In order to show the effectiveness of DOKF used in the estimate state DOS (12.5), we firstly consider the following DOKF lemma in Chapter 7 in this subsection.

Lemma 12.10. *For DOS (12.3)-(12.4). A single algorithm of DOKF is given as follows:*

$$K = P(t_k)C^T(CP(t_k)C^T T_s + R)^{-1} \tag{12.22}$$

$$\begin{aligned}\delta\hat{x}(t_k) &= A\hat{x}(t_k) + Bu(t_k) + K[y(t_k) - C\hat{x}(t_k)] \\ &\quad - T_s KC[A\hat{x}(t_k) + Bu(t_k)]\end{aligned} \tag{12.23}$$

$$\begin{aligned}\delta P(t_k) &= -KCP(t_k) + AP(t_k) + P(t_k)A^T + T_s A(I - KCT_s)P(t_k)A^T \\ &\quad + Q - T_s[AKCP(t_k) + KCP(t_k)A^T]\end{aligned} \tag{12.24}$$

which is initialized with

$$\begin{aligned}\hat{x}(0) &= E[x(0)] \\ P(0) &= E[(\hat{x}(0) - x(0))(\hat{x}(0) - x(0))^T]\end{aligned}$$

and then it operates recursively performing a single cycle each time a new set of measurements becomes available. Each iteration propagates the estimate from the time the last measurement was obtained to the current time.

The estimate state feedback controller is given as

$$u(t_k) = K\hat{x}(t_k) \tag{12.25}$$

can be gotten. The quadratic cost function (12.17) can be rewritten as follows

$$J = T_s \sum_{k=0}^{\infty} [\hat{x}^T(t_k)Q\hat{x}(t_k) + u^T(t_k)Ru(t_k)] \tag{12.26}$$

where $Q \in R^{n\times n}$ and $R \in R^{p\times p}$ are two given positive-definite matrices.

Theorem 12.11. *The closed-loop NCS (12.5) is asymptotically stable, if there exist symmetric positive definite matrix $X > 0$, a real matrix Y, such that the following LMI holds:*

$$\Xi = \begin{bmatrix} \Xi(1,1) & XA^T + Y^T B^T & T_s XQ & T_s Y^T R \\ * & (T_s - 2)X & 0 & 0 \\ * & * & -T_s Q & 0 \\ * & * & * & -T_s R \end{bmatrix} < 0 \tag{12.27}$$

with

$$\Xi(1,1) = AX + BY + XA^T + YB_k^T.$$

The digital controller is given by

$$u(t_k) = K\hat{x}(t_k) = YX^{-1}\hat{x}(t_k) \tag{12.28}$$

and the corresponding upper bound of the closed-loop quadratic cost function J is

$$J^* = \hat{x}^T(t_0)X^{-1}\hat{x}(t_0). \tag{12.29}$$

Proof. By using Schur's complement, $\Xi < 0$ can be rewritten as $\Xi_0 < 0$ with

$$\Xi_0 = \begin{bmatrix} \Xi_0(1,1) & XA^T + Y^T B^T \\ * & (\mathtt{T}_s - 2)X \end{bmatrix} < 0 \tag{12.30}$$

with

$$\Xi_0(1,1) = AX + BY + XA^T + YB_k^T + \mathtt{T}_s XQX + \mathtt{T}_s Y^T RY.$$

Construct a simple Lyapunov functional in delta domain as follows:

$$V(\hat{x}(t_k)) = \hat{x}^T(t_k)P\hat{x}(t_k).$$

Using Lemma 12.8 and taking the delta operator manipulations of $V(\hat{x}(t_k))$ along the trajectory of closed-loop system (12.5), we can obtain:

$$\begin{aligned}\delta V(\hat{x}(t_k)) &= \delta^T\hat{x}(t_k)P\hat{x}(t_k) + \hat{x}^T(t_k)P\delta\hat{x}(t_k) + \mathtt{T}_s\delta^T\hat{x}(t_k)P\delta\hat{x}(t_k) \\ &= \hat{x}^T(t_k)(A+BK)^T P\hat{x}(t) + \hat{x}^T(t_k)P(A+BK)\hat{x}(t_k) \\ &\quad + \mathtt{T}_s\delta^T\hat{x}(t)P\delta\hat{x}(t_k). \end{aligned} \tag{12.31}$$

For the positive definite real matrix P, one has that

$$\begin{aligned} 0 &= -2\delta^T\hat{x}(t_k)P[\delta\hat{x}(t_k) - (A+BK)\hat{x}(t_k)] \\ &= -2\delta^T\hat{x}(t_k)P\delta\hat{x}(t_k) + 2\delta^T\hat{x}(t_k)P(A+BK)\hat{x}(t_k). \end{aligned} \tag{12.32}$$

Combining (12.31) and (12.32), letting

$$\delta V(x(t_k)) < -\mathtt{T}_s\hat{x}^T(t_k)(Q + K^T RK)\hat{x}(t_k) \le 0 \tag{12.33}$$

results in

$$\delta V(x(t_k)) + \mathtt{T}_s\hat{x}^T(t_k)(Q + K^T RK)\hat{x}(t_k) \le \xi^T(t_k)\Xi_1\xi(t_k) < 0, \tag{12.34}$$

where

$$\xi^T(t_k) = \begin{bmatrix} x^T(t_k) & \delta^T x(t_k) \end{bmatrix},$$

$$\Xi_1 = \begin{bmatrix} PA + PB + A^T P + B^T P + \mathtt{T}_s Q + \mathtt{T}_s K^T RK & A^T P + K^T B^T P \\ * & (\mathtt{T}_s - 2)P \end{bmatrix}.$$

Pre-multiplying and post-multiplying $\Xi_1 < 0$ by $diag\{P^{-1},\ P^{-1}\}$, respectively, letting $P^{-1} = X$, $KP^{-1} = Y$, the inequality $\Xi_1 < 0$ is equivalent to LMI (12.27). Due to LMI (12.27) holds, there exists

$$\delta V(\hat{x}(t_k)) = \frac{1}{\mathtt{T}_s}\left[\hat{x}^T(t_k+\mathtt{T}_s)P\hat{x}(t_k+\mathtt{T}_s) - \hat{x}^T(t_k)P\hat{x}(t_k)\right] < 0$$

It can be seen that for $\hat{x}(t_k) \neq 0$ the following inequality holds

$$\hat{x}^T(t_k+\mathtt{T}_s)P\hat{x}(t_k+\mathtt{T}_s) - \hat{x}^T(t_k)P\hat{x}(t_k) < 0,$$

which implies that the Lyapunov functional $\hat{x}^T(t_k)P\hat{x}(t_k)$ in delta domain is a strictly decreasing function. Thus, we can obtain

$$\hat{x}^T(t_k)P\hat{x}(t_k) \to 0, \quad \text{when } k \to \infty$$

Since $P > 0$ is a constant matrix, we have

$$\hat{x}(t_k) \to 0, \quad \text{when } k \to \infty$$

Therefore, the feasible state feedback control law obtained in Theorem 12.11 asymptotically stabilizes the closed-loop NCS (12.5).

On the other hand, summing both sides of the inequality (12.33) on k from 0 to ∞, we have

$$\sum_{i=0}^{\infty} \Gamma_i \left[\hat{x}^T(t_i)Q\hat{x}(t_i) + u^T(t_i)Ru(t_i)\right] \leq V(x(t_0))$$

That can be rewritten as

$$J \leq \hat{x}^T(t_0)X^{-1}\hat{x}(t_0)$$

This completes the proof.

Theorem 12.12. *Consider system (12.16) with cost function (12.17). If the following optimization problem*

$$\min_{X,Y,\gamma} \quad \gamma$$
$$s.t. \begin{cases} (i) \quad LMI \quad (12.27) \\ (ii) \begin{bmatrix} -\gamma & \hat{x}^T(t_0) \\ \hat{x}(t_0) & -X \end{bmatrix} < 0 \end{cases} \tag{12.35}$$

has a positive solution set (X, Y, γ), then the control law (12.28) is an optimal robust guaranteed cost control law which ensures the minimization of the guaranteed cost (12.29) for DOS (12.5).

Proof. By Theorem 12.11, (i) in (12.35) is clear. Also, we can choose a positive constant γ such that

$$\gamma > \hat{x}^T(t_0)X^{-1}\hat{x}(t_0), \tag{12.36}$$

It follows from Schur's complement that (ii) in (12.35) are equivalent to inequation (12.36). Hence, it follows from (12.29) that $J^* \leq \gamma$. Thus, the minimization of γ implies the minimization of the guaranteed cost for DOS (12.5). This completes the proof.

12.3.2 Stabilization of NCSs

The following theorem provides us the stabilization condition of NCS (12.16) and it plays a key role in the networked digital controller design.

Theorem 12.13. *The closed-loop NCS (12.16) is asymptotically stable, if there exist symmetric positive definite matrices* $X_h > 0$, $Z_h > 0$ *and* $\bar{X}_l = \sum_{l=1}^{s} \lambda_{hl} X_l > 0$, $\forall\ n_h \in Z^+$ *a real matrix* Y_h, Y_{0h}, N_{1h} *and* N_{2h}, *such that the following LMI holds:*

$$\Sigma_h = \begin{bmatrix} \Sigma_h(1,1) & \Sigma_h(1,2) & \Sigma_h(1,3) & \Gamma_i N_{1h} & \Gamma_i X_h Q & \Gamma_i Y_h^T R \\ * & \Sigma_h(2,2) & \Psi_h Y_{0h} & 0 & 0 & 0 \\ * & * & \Sigma_h(3,3) & \Gamma_i N_{2h} & 0 & 0 \\ * & * & * & -\Gamma_i Z_h & 0 & 0 \\ * & * & * & * & -\Gamma_i Q & 0 \\ * & * & * & * & * & -\Gamma_i R \end{bmatrix} < 0,$$
$$\forall\ h, l \in S \quad (12.37)$$

with

$$\Sigma_h(1,1) = \tau_s \bar{X}_l / \Gamma_i + A_h X_h + \Phi_h Y_h + X_h A_h^T + Y_h^T \Phi_h^T + N_{1h}^T + N_{1h},$$
$$\Sigma_h(1,2) = \tau_s \bar{X}_l + X_h \mathcal{A}_h^T + Y_h^T \Phi_h^T,\ \ \Sigma_h(1,3) = \Psi_h Y_{0h} - N_{1h}^T + N_{2h},$$
$$\Sigma_h(2,2) = \tau_s \Gamma_i \bar{X}_l + (\Gamma_i - 2) X_h + \Gamma_i Z_h,\ \ \Sigma_h(3,3) = -N_{2h}^T - N_{2h}.$$

A suitable stabilizing guaranteed cost estimate state feedback networked digital controller can be chosen by

$$u^*(t_i) = K_h \hat{x}(t_i) = Y_h X_h^{-1} \hat{x}(t_i) \quad (12.38)$$

and the corresponding upper bound of the closed-loop quadratic cost function J *is*

$$J^* = \hat{x}^T(t_0) X_\kappa^{-1} \hat{x}(t_0) + \varphi^T(t_0) X_\kappa^{-1} \varphi(t_0) \quad (12.39)$$

with $X_\kappa = \min\{X_h, h \in S\}$. *Furthermore, for packets dropout case the networked digital controller is obtained as* $u(t_i) = u(t_{i-1})$ *due to the effect of the Logic ZOH in NCS framework.*

Proof. By using Schur's complement, $\Sigma_h < 0$ can be rewritten as $\Sigma_{0h} < 0$ with

$$\Sigma_{0h} = \begin{bmatrix} \Sigma_{0h}(1,1) & \mathtt{T}_s \bar{X}_l + X_h \mathcal{A}_h^T + Y_h^T \Phi_h^T & \Psi_h Y_{0h} - N_{1h}^T + N_{2h} & \Gamma_i N_{1h} \\ * & \mathtt{T}_s \Gamma_i \bar{X}_l + (\Gamma_i - 2) X_h + \Gamma_i Z_h & \Psi_h Y_{0h} & 0 \\ * & * & -N_{2h}^T - N_{2h} & \Gamma_i N_{2h} \\ * & * & * & -\Gamma_i Z_h \end{bmatrix}$$

with

$$\begin{aligned}\Sigma_{0h}(1,1) = {} & \mathtt{T}_s \bar{X}_l / \Gamma_i + A_h X_h + \Phi_h Y_h + X_h A_h^T + Y_h^T \Phi_h^T + N_{1h}^T + N_{1h} \\ & + \Gamma_i X_h Q X_h + \Gamma_i Y_h^T R Y_h.\end{aligned}$$

Letting $\mathcal{A}_h = A_h + \Phi_h K_h$ and $\mathcal{B}_h = \Psi_h K_{0h}$, the closed-loop NCS (12.16) can be gotten

$$\delta \hat{x}(t_i) = \mathcal{A}_h \hat{x}(t_i) + \mathcal{B}_h \hat{x}(t_{i-1}) \tag{12.40}$$

Setting $\hat{e}(t_i) = \hat{x}(t_i) - \hat{x}(t_{i-1})$, construct stochastic Lyapunov-Krasovskii functional in delta domain as follows:

$$V(\hat{x}(t_i), h) = \hat{x}^T(t_i) P_h \hat{x}(t_i) + \hat{e}^T(t_i) \mathcal{R} \hat{e}(t_i) \tag{12.41}$$

with

$$V_1(\hat{x}(t_i), h) = \hat{x}^T(t_i) P_h \hat{x}(t_i), \quad V_2(\hat{x}(t_i), h) = \hat{e}^T(t_i) \mathcal{R} \hat{e}(t_i).$$

By letting $\mathbf{E}(P_l) = \mathtt{T}_s \sum_{l=1}^{s} \lambda_{hl} P_l + P_h$ and taking stochastic delta operator manipulations of $V(\hat{x}(t_i), h)$ along the trajectory of system (12.40), we can obtain:

$$\begin{aligned}\sigma V_1(\hat{x}(t_i), h) = {} & \frac{1}{\Gamma_i} \left\{ \mathbf{E} \left[\hat{x}^T(t_{i+1}) P_l \hat{x}(t_{i+1}) \middle| \hat{x}(t_0), \eta_0 \right] - \hat{x}^T(t_i) P_h \hat{x}(t_i) \right\} \\ = {} & \frac{\mathtt{T}_s}{\Gamma_i} \sum_{l=1}^{s} \lambda_{hl} \hat{x}^T(t_{i+1}) P_l \hat{x}(t_{i+1}) + \frac{1}{\Gamma_i} \left[\hat{x}^T(t_{i+1}) P_h \hat{x}(t_{i+1}) - \hat{x}^T(t_i) P_h \hat{x}(t_i) \right] \\ = {} & \mathtt{T}_s \Gamma_i \sum_{l=1}^{s} \lambda_{hl} \delta^T(\hat{x}(t_i)) P_l \delta(\hat{x}(t_i)) + \mathtt{T}_s \sum_{l=1}^{s} \lambda_{hl} \delta^T(\hat{x}(t_i)) P_l \hat{x}(t_i) \\ & + \hat{x}^T(t_i) \mathcal{A}_h^T P_h \hat{x}(t_i) + \mathtt{T}_s \sum_{l=1}^{s} \lambda_{hl} \hat{x}^T(t_i) P_l \delta(\hat{x}(t_i)) \\ & + \frac{\mathtt{T}_s}{\Gamma_i} \sum_{l=1}^{s} \lambda_{hl} \hat{x}^T(t_i) P_l \hat{x}(t_i) + \Gamma_i \delta^T(\hat{x}(t_i)) P_h \delta(\hat{x}(t_i)) + \hat{x}^T(t_i) P_h \mathcal{A}_h \hat{x}(t_i) \\ & + \hat{x}^T(t_{i-1}) \mathcal{B}_h^T P_h \hat{x}(t_i) + \hat{x}^T(t_i) P_h \mathcal{B}_h \hat{x}(t_{i-1})\end{aligned} \tag{12.42}$$

For the positive definite real matrix P_h, one has that

$$\begin{aligned} 0 &= -2\delta^T\hat{x}(t_i)P_h\left[\delta\hat{x}(t_i) - \mathcal{A}_h\hat{x}(t_i) - \mathcal{B}_h\hat{x}(t_{i-1})\right] \\ &= -2\delta^T\hat{x}(t_i)P_h\delta\hat{x}(t_i) + 2\delta^T\hat{x}(t_i)P_h\mathcal{A}_h\hat{x}(t_i) \\ &\quad +2\delta^T\hat{x}(t_i)P_h\mathcal{B}_h\hat{x}(t_{i-1}) \end{aligned} \tag{12.43}$$

Taking the stochastic delta operator manipulations of $V_2(\hat{x}(t_i), h)$, there exists

$$\begin{aligned} \sigma V_2(\hat{x}(t_i), h) &= \frac{1}{\Gamma_i}\left[\hat{e}^T(t_{i+1})\mathcal{R}\hat{e}(t_{i+1}) - \hat{e}^T(t_i)\mathcal{R}\hat{e}(t_i)\right] \\ &= \Gamma_i\delta^T\hat{x}(t_i)\mathcal{R}\delta\hat{x}(t_i) - \hat{e}^T(t_i)\mathcal{R}\hat{e}(t_i)/\Gamma_i \end{aligned} \tag{12.44}$$

For the positive definite matrix $\mathcal{R}$ and any real matrix $M_h \in R^{2n\times n}$, we have

$$\begin{bmatrix} M_h\mathcal{R}^{-1}M_h^T & M_h \\ M_h^T & \mathcal{R} \end{bmatrix} \geq 0, \quad \text{where} \quad M_h = \begin{bmatrix} M_{1h} \\ M_{2h} \end{bmatrix}.$$

Hence, letting $\eta^T(t_i) = \left[x^T(t_i)\ x^T(t_{i-1})\right]$ and $\psi(t_i) = \hat{e}(t_i)/\Gamma_i$, one has that

$$\begin{aligned} 0 &\leq \Gamma_i\begin{bmatrix} \eta(t_i) \\ \psi(t_i) \end{bmatrix}^T \begin{bmatrix} M_h\mathcal{R}^{-1}M_h^T & M_h \\ M_h^T & \mathcal{R} \end{bmatrix} \begin{bmatrix} \eta(t_i) \\ \psi(t_i) \end{bmatrix} \\ &= \Gamma_i\eta^T(t_k)M_h\mathcal{R}^{-1}M_h^T\eta(t_k) + 2\eta^T(t_k)M_h\left[I \ \ -I\right]\eta(t_k) \\ &\quad +\frac{1}{\Gamma_i}\hat{e}^T(t_i)\mathcal{R}\hat{e}(t_i) \end{aligned} \tag{12.45}$$

From Definition 12.2, there exists $\sigma V(\hat{x}(t_i), h) < 0$. Combining (12.42)-(12.45) results in

$$\sigma V(\hat{x}(t_i), h) \leq \xi^T(t_i)\left(\Sigma_{1h} + \Gamma_i M_h\mathcal{R}^{-1}M_h^T\right)\xi(t_i) < 0, \tag{12.46}$$

where

$$\xi^T(t_i) = \left[\hat{x}^T(t_i)\ \delta^T\hat{x}(t_i)\ \hat{x}^T(t_{i-1})\right],$$

$$\Sigma_{1h} = \begin{bmatrix} \Sigma_{1h}(1,1) & \mathtt{T}_s\sum_{l=1}^{s}\lambda_{hl}P_l + \mathcal{A}_h^TP_h & P_h\mathcal{B}_h - M_{1h}^T + M_{2h} \\ * & \Sigma_{1h}(2,2) & P_h\mathcal{B}_h \\ * & * & -M_{2h}^T - M_{2h} \end{bmatrix}$$

with

$$\Sigma_{1h}(1,1) = \frac{\mathtt{T}_s}{\Gamma_i}\sum_{l=1}^{s}\lambda_{hl}P_l + P_h\mathcal{A}_h + \mathcal{A}_h^TP_h + M_{1h}^T + M_{1h},$$

$$\Sigma_{1h}(2,2) = \mathtt{T}_s\Gamma_i\sum_{l=1}^{s}\lambda_{hl}P_l + (\Gamma_i - 2)P_h + \Gamma_i\mathcal{R}.$$

By using Schur's complement and letting $\sum_{l=1}^{s} \lambda_{hl} P_l = \bar{P}_l$, the matrix inequation $\Sigma_{1h} + \Gamma_i M_h \mathcal{R}^{-1} M_h^T < 0$ can be changed to $\Sigma_{2h} < 0$, where

$$\Sigma_{2h} = \begin{bmatrix} \Sigma_{2h}(1,1) & \mathtt{T}_s \bar{P}_l + \mathcal{A}_h^T P_h & P_h \mathcal{B}_h - M_{1h}^T + M_{2h} & \Gamma_i M_{1h} \\ * & \mathtt{T}_s \Gamma_i \bar{P}_l + (\Gamma_i - 2) P_h + \Gamma_i \mathcal{R} & P_h \mathcal{B}_h & 0 \\ * & * & -M_{2h}^T - M_{2h} & \Gamma_i M_{2h} \\ * & * & * & -\Gamma_i \mathcal{R} \end{bmatrix}$$

with

$$\Sigma_{2h}(1,1) = \mathtt{T}_s \bar{P}_l / \Gamma_i + P_h \mathcal{A}_h + \mathcal{A}_h^T P_h + M_{1h}^T + M_{1h}.$$

Pre-multiplying and post-multiplying $\Sigma_{2h} < 0$ by the diagonal matrix $diag\{P_h^{-1},\ P_h^{-1},\ P_h^{-1},\ P_h^{-1}\}$, respectively, letting $X_h = P_h^{-1}$, $Y_h = K_h P_h^{-1}$, $Y_{0h} = K_{0h} P_h^{-1}$, $\bar{X}_l = P_h^{-1} \bar{P}_l P_h^{-1} = \sum_{l=1}^{s} \lambda_{hl} X_l$, $P_h^{-1} M_{1h} P_h^{-1} = N_{1k}$, $P_h^{-1} M_{2h} P_h^{-1} = N_{2h}$ and $P_h^{-1} R P_h^{-1} = Z_h$, there exists $\Sigma_{3h} < 0$ which is equal to $\Sigma_{2h} < 0$ as follows

$$\Sigma_{3h} = \begin{bmatrix} \Sigma_{3h}(1,1) & \mathtt{T}_s \bar{X}_l + X_h \mathcal{A}_h^T + Y_h^T \Phi_h^T & \Psi_h Y_{0h} - N_{1h}^T + N_{2h} & \Gamma_i N_{1h} \\ * & \mathtt{T}_s \Gamma_i \bar{X}_l + (\Gamma_i - 2) X_h + \Gamma_i Z_h & \Psi_h Y_{0h} & 0 \\ * & * & -N_{2h}^T - N_{2h} & \Gamma_i N_{2h} \\ * & * & * & -\Gamma_i Z_h \end{bmatrix}$$

with

$$\Sigma_3(1,1) = \mathtt{T}_s \bar{X}_l / \Gamma_i + A_h X_h + \Phi_h Y_h + X_h A_h^T + Y_h^T \Phi_h^T + N_{1h}^T + N_{1h}.$$

Comparing with Σ_{3h} and Σ_{0h}, it also can be obtained that

$$\Sigma_{3h} < \text{diag}\{-\Gamma_i X_h Q X_h - \Gamma_i Y_h^T R Y_h,\ 0,\ 0,\ 0\} \leq 0$$

Therefore, from $\Sigma_{3h} < 0$ it can be obtained that $\sigma V(\hat{x}(t_i), h) < 0$. Let $\lambda_0 = \min\{\lambda_{\min}(-\Sigma_h), h \in S\}$, then $\lambda_0 > 0$ due to $\Sigma_{3h} < 0$. Then, from (12.46) we obtain that for any $i > 0$, there exists

$$\sigma V(\hat{x}(t_i), h) \leq -\lambda_0 \psi^T(t_i) \psi(t_i) < -\lambda_0 x^T(t_0) x(t_0) < 0$$

and

$$\mathbf{E}\left[V(\hat{x}(t_{i+1}), l) | \, \hat{x}(t_0), \eta_0\right] \leq V(\hat{x}(t_i), h) - \lambda_0 \Gamma_i x^T(t_0) x(t_0) \quad (12.47)$$

Setting $i = 0$ and $i = 1$ in (12.47) yields

$$\mathbf{E}\left[V(\hat{x}(t_1), \eta_1) | \, \hat{x}(t_0), \eta_0\right] \leq V(\hat{x}(t_0), \eta_0) - \lambda_0 \Gamma_0 x^T(t_0) x(t_0) \quad (12.48)$$

and

$$\mathbf{E}\left[V(\hat{x}(t_2), \eta_2) | \, \hat{x}(t_0), \eta_0\right] \leq V(\hat{x}(t_1), \eta_1) - \lambda_0 \Gamma_1 x^T(t_0) x(t_0) \quad (12.49)$$

Taking expectation $\mathbf{E}[\cdot|\,\hat{x}(t_0),\eta_0]$ on both sides of (12.49), with the aid of (12.48), leads to

$$\begin{aligned}\mathbf{E}\left[V(\hat{x}(t_2),\eta_2)|\,\hat{x}(t_0),\eta_0\right] &\leq V(\hat{x}(t_0),\eta_0)\\ &\quad-\lambda_0\sum_{j=0}^{1}\mathbf{E}\left\{\Gamma_j x^T(t_j)x(t_j)|\,\hat{x}(t_0),\eta_0\right\}\end{aligned}\tag{12.50}$$

Then, one can continue the iterative procedure (12.50) about i from 0 to ∞, we have

$$\begin{aligned}&V(\hat{x}(t_0),\eta_0)-\lambda_0\sum_{j=0}^{\infty}\mathbf{E}\left\{\Gamma_j x^T(t_j)x(t_j)|\,\hat{x}(t_0),\eta_0\right\}\\ &\geq\lim_{K\to 0}\mathbf{E}\left[V(\hat{x}(t_K),\eta_K)|\,\hat{x}(t_0),\eta_0\right]=0\end{aligned}$$

From above inequation, we know that

$$\sum_{j=0}^{\infty}\mathbf{E}\left\{\Gamma_j x^T(t_j)x(t_j)|\,\hat{x}(t_0),\eta_0\right\}\geq\frac{1}{\lambda_0}V(\hat{x}(t_0),\eta_0)<\infty$$

From Definition 12.1, Markovian jump DOS (12.16) is stochastic robustly asymptotically stable.

On the other hand, pre-multiplying and post-multiplying $\Sigma_0<0$ by the diagonal matrix

$$diag\left\{X_h^{-1},\ X_h^{-1},\ X_h^{-1},\ X_h^{-1}\right\}$$

we can give that

$$\Sigma_2<diag\{-\Gamma_i Q-\Gamma_i K_h^T R K_h,\ 0,\ 0,\ 0\}\leq 0$$

Hence, it can be obtained

$$\sigma V(\hat{x}(t_i),h)=\xi^T(t_i)\Sigma_{2h}\xi(t_i)<-\hat{x}^T(t_i)\Gamma_i(Q+K_h^T R K_h)\hat{x}(t_i)<0\tag{12.51}$$

Summing both sides of the inequality (12.51) about i from 0 to ∞, we have

$$\mathbf{E}\left\{\sum_{i=0}^{\infty}\Gamma_i\left[\hat{x}^T(t_i)Q\hat{x}(t_i)+u^T(t_i)Ru(t_i)|\,\hat{x}(t_0),\eta_0\right]\right\}\leq V(\hat{x}(t_0),\eta_0)$$

That is

$$J\leq\hat{x}^T(t_0)P_{\eta_0}\hat{x}(t_0)+\varphi^T(t_0)P_{\eta_-}\varphi(t_0)$$

Setting $P_\kappa=\max\{P_h,h\in S\}$ and letting $X_\kappa=P_\kappa^{-1}$, we get

$$J\leq\hat{x}^T(t_0)X_\kappa^{-1}\hat{x}(t_0)+\varphi^T(t_0)X_\kappa^{-1}\varphi(t_0)$$

It means that $u(t_i) = K_h\hat{x}(t_i)$ is one of guaranteed cost control laws of (12.16). It follows from Definition 12.2, and the theorem is proofed.

Remark 12.14. If we let intermittent time $\Gamma_i \to 0$, the result of stochastic delta operator for $V(x(t_i), h)$ will be equal to the ones of weak infinitesimal operator which was employed in [211]. The weak infinitesimal operator $\Im_t^x[\cdot]$ of the stochastic process $\{x(t),\ \eta_t,\ t \geq 0\}$, acting on $V(x,\ t,\ \eta_t)$ at the point $\{x,\ t,\ \eta_t = i\}$, is given by

$$\Im_t^x[V] = \frac{\partial V}{\partial t} + \dot{x}^T(t)\frac{\partial V}{\partial x}\bigg|_{\eta_t=i} + \sum_{j=1}^{s} \lambda_{ij} V(t,\ x,\ i,\ j). \qquad (12.52)$$

Therefore, it can be said that weak infinitesimal operator is an especial case of the stochastic delta operator defined in (12.18). We have gotten a less conservative result by using delta operator approach.

It is easy to see that Theorem 12.11 presents a method of designing a estimate state feedback guaranteed cost controller. The following theorem presents a method of selecting a controller minimizing the upper bound of the guaranteed cost function (12.17).

Theorem 12.15. *Consider system (12.16) with cost function (12.17). If the following optimization problem*

$$\min_{X_h, X_l, Y_h, \alpha, \beta} \quad \{\alpha + \beta\}$$

$$s.t. \begin{cases} (i) \quad LMI\ (12.37), \quad \forall\ h, l \in S \\ (ii) \begin{bmatrix} -\alpha & x^T(t_0) \\ x(t_0) & -X_h \end{bmatrix} < 0, \quad \forall\ h \in S \\ (iii) \begin{bmatrix} -\beta & \varphi^T(t_0) \\ \varphi(t_0) & -X_h \end{bmatrix} < 0, \quad \forall\ h \in S \end{cases} \qquad (12.53)$$

has a positive solution set $(X_h, X_l, Y_h, \alpha, \beta)$, then the control law (12.38) is an optimal robust guaranteed cost control law which ensures the minimization of the guaranteed cost (12.17) for the Markov jump DOS (12.16).

Proof. By Theorem 12.11, (i) in (12.53) is clear. Also, we can choose two positive constants α and β such that

$$\alpha > \hat{x}^T(t_0)X_\kappa^{-1}\hat{x}(t_0), \quad \text{for } \kappa \in S \qquad (12.54)$$

$$\beta > \varphi^T(t_0)X_\kappa^{-1}\varphi(t_0), \quad \text{for } \kappa \in S \qquad (12.55)$$

Due to $X_\kappa = \min\{X_h, h \in S\}$, i.e., $X_\kappa^{-1} = \max\{X_h^{-1}, h \in S\}$, inequations (12.54) and (12.55) are equivalent to

$$\alpha > \hat{x}^T(t_0)X_h^{-1}\hat{x}(t_0), \quad \forall\ h \in S \qquad (12.56)$$

$$\beta > \varphi^T(t_0)X_h^{-1}\varphi(t_0), \quad \forall\ h \in S \qquad (12.57)$$

It follows from Schur's complement that (ii) and (iii) in (12.53) are equivalent to inequations (12.56) and (12.57). Hence, it follows from (12.39) that

$$J^* \leq \alpha + \beta$$

Thus, the minimization of $\alpha + \beta$ implies the minimization of the guaranteed cost for Markov jump DOS (12.16). This completes the proof.

12.4 Numerical Example

In the following example, we will provide an inverted pendulum example to demonstrate the effectiveness of the proposed methods in this chapter.

Example 12.16. The proposed method will be applied to design a controller for an inverted pendulum system to illustrate the results developed. Consider the inverted pendulum on a cart. The physical structure is shown in Fig. 12.5, where M is the mass of the cart, m is the mass of the pendulum

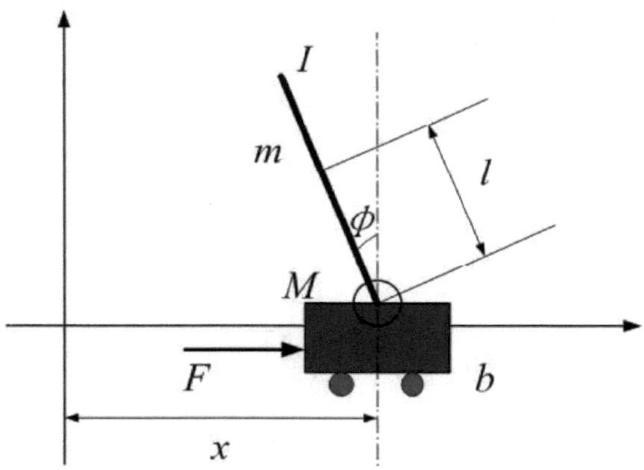

Fig. 12.5 Inverted pendulum Model.

rod, b is the friction coefficient of the cart. l is the length of the pendulum rod; F is the force acting on the cart, $\boldsymbol{x}$ is the horizontal displacement of the cart, ϕ is the angle between the pendulum rod and the vertical. The surface is friction free. The inverted pendulum is driven by a electric motor, hence, only acceleration input can be used in this pendulum system model. Let g be the gravitational constant. Let state variables be $\boldsymbol{x}$, $\dot{\boldsymbol{x}}$, ϕ and $\dot{\phi}$ which corresponds to the horizontal position, horizontal velocity of the cart, angle and angle velocity of the pole, respectively. The equation of motion can be obtained as follows:

$$\left(I + ml^2\right)\ddot{\phi} - mgl\phi = ml\ddot{\boldsymbol{x}}$$

For the swing link with uniform distribution quality, we have the following equation of inertia as

$$I = \frac{1}{3}ml^2$$

Based on above two equations, result in

$$\left(\frac{1}{3}ml^2 + ml^2\right)\ddot{\phi} - mgl\phi = ml\ddot{\boldsymbol{x}}$$

which is changed to

$$\ddot{\phi} = \frac{3g}{4l}\phi + \frac{3}{4l}\ddot{\boldsymbol{x}}$$

where $g = 9.81$ m/s^2 and $l = 0.25$ m. Let state variables be $\boldsymbol{x} = x_1(t)$, $\dot{\boldsymbol{x}} = x_2(t)$, $\phi = x_3(t)$, $\dot{\phi} = x_4(t)$ and $u(t) = \ddot{\boldsymbol{x}}$, we can obtained as

$$\dot{x}(t) = A_s x(t) + B_s u(t) + \omega(t) \qquad (12.58)$$
$$y(t) = C_s x(t) + \upsilon(t) \qquad (12.59)$$

with

$$A_s = \begin{bmatrix} 0 & 1 & 0 & 0 \\ 0 & 0 & 0 & 0 \\ 0 & 0 & 0 & 1 \\ 0 & 0 & 29.4 & 0 \end{bmatrix}, \quad B_s = \begin{bmatrix} 0 \\ -1 \\ 0 \\ 3 \end{bmatrix}, \quad C_s = \begin{bmatrix} 1 & 0 & 0 & 0 \\ 0 & 0 & 1 & 0 \end{bmatrix}.$$

the simulation parameters of guaranteed cost function are given as:

$$Q = \begin{bmatrix} 1000 & 0 & 0 & 0 \\ 0 & 10 & 0 & 0 \\ 0 & 0 & 200 & 0 \\ 0 & 0 & 0 & 1 \end{bmatrix}, \quad R = \begin{bmatrix} 1 \end{bmatrix}.$$

Discretize continuous system (12.58)-(12.59) by using delta operator approach with sampling period $\mathtt{T}_s = 0.01$s and utilizing the designed Kalman filter in Theorem 12.10, we have the following delta operator state estimate system:

$$\delta\hat{x}(t_k) = A\hat{x}(t_k) + Bu(t_k) \qquad (12.60)$$
$$y(t_k) = C\hat{x}(t_k) \qquad (12.61)$$

with

$$A = \begin{bmatrix} 0 & 1 & 0 & 0 \\ 0 & 0 & 0 & 0 \\ 0 & 0 & 0.14700 & 1.0005 \\ 0 & 0 & 29.4144 & 0.1470 \end{bmatrix}, \quad B = \begin{bmatrix} -0.005 \\ -1.000 \\ 0.0150 \\ 3.0015 \end{bmatrix}, \quad C = \begin{bmatrix} 1 & 0 & 0 & 0 \\ 0 & 0 & 1 & 0 \end{bmatrix}.$$

By using Matlab command mincx of LMI-toolbox to solve convex optimization problem (12.27) in Theorem 12.15, we obtain the solutions as follows:

$$\gamma = 30.6516,$$
$$X = \begin{bmatrix} 5.6469 & -10.0379 & 1.7850 & -2.4817 \\ -10.0379 & 36.3224 & -11.0969 & 23.2302 \\ 1.7850 & -11.0969 & 7.2714 & -24.1940 \\ -2.4817 & 23.2302 & -24.1940 & 93.2001 \end{bmatrix},$$
$$Y = \begin{bmatrix} -16.6166 & 94.1784 & -45.3352 & 124.8564 \end{bmatrix}.$$

the control input is given as

$$u(t_k) = YX^{-1}\hat{x}(t_k) = \begin{bmatrix} -0.5102 & -1.0496 & -17.8227 & -3.0389 \end{bmatrix} \hat{x}(t_k) \quad (12.62)$$

Furthermore, Fig. 12.6 shows the simulation results of the control input for initial condition $u(0) = 0$.

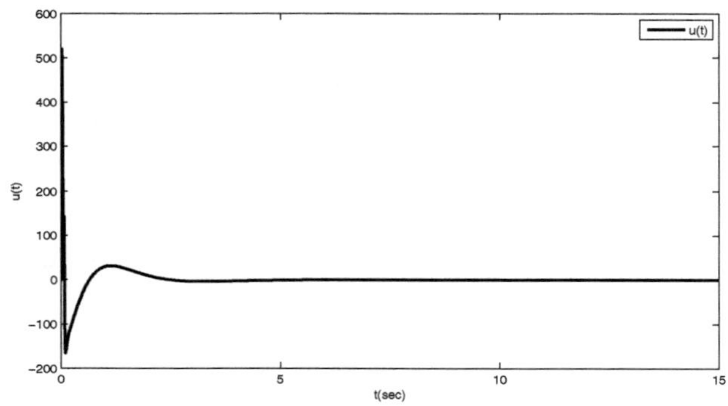

Fig. 12.6 The control input $u(t)$.

For initial condition: $x_1(0) = x_2(0) = x_3(0) = x_4(0) = 0$, the simulation results of the state response of closed-loop DOS (12.58)-(12.59) is given in Fig. 12.7, where $x(t)$ is denoted by $x_1(t)$, $x_2(t)$, $x_3(t)$ and $x_4(t)$.

The simulation results of the estimate state response of closed-loop DOS (12.58)-(12.59) is given for initial condition $\hat{x}_1(0) = \hat{x}_2(0) = \hat{x}_3(0) = \hat{x}_4(0) = 0$ in Fig. 12.8, where $\hat{x}(t)$ are denoted by $xe_1(t)$, $xe_2(t)$, $xe_3(t)$ and $xe_4(t)$.

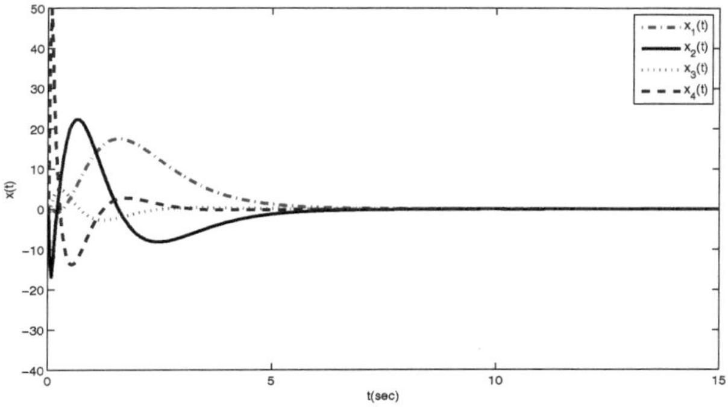

Fig. 12.7 The state response of $x(t)$.

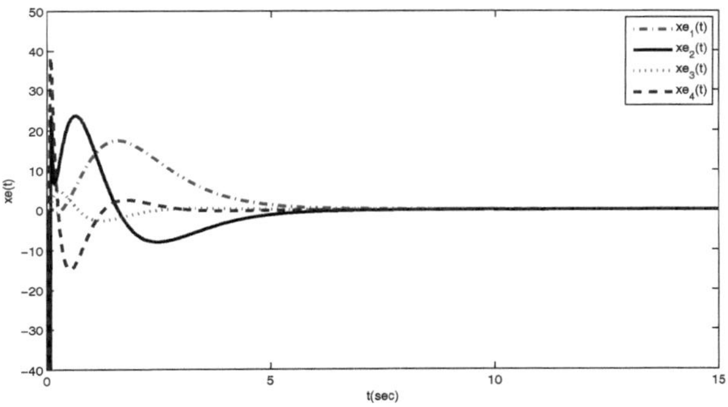

Fig. 12.8 The state response of $\hat{x}(t)$.

It can be gotten that the errors of the state variables for horizontal position, angle of the pole in Fig. 12.9, where $x_1(t)-\hat{x}_1(t)$ and $x_3(t)-\hat{x}_3(t)$ are denoted by $e_1(t)$ and $e_3(t)$.

the errors of the state variables for horizontal velocity of the cart, angle velocity of the pole in Fig. 12.10, where $x_2(t)-\hat{x}_2(t)$ and $x_4(t)-\hat{x}_4(t)$ are denoted by $e_2(t)$ and $e_4(t)$.

It is clear from the above contrastive simulation results that the estimate state vector is similar to the real state vector in this inverted pendulum model. Hence, a real state vector in DOS can be replaced by an estimate state vector using the DOKF obtained in Theorem 12.10.

In the following, we will consider the effectiveness of the methods in Theorem 12.13 and Theorem 12.15 for the case of control signals passing through communication networks.

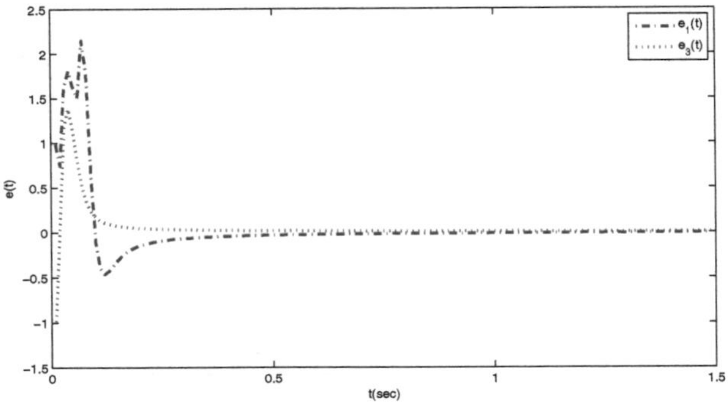

Fig. 12.9 The state response errors of $x_1(t) - \hat{x}_1(t)$ and $x_3(t) - \hat{x}_3(t)$.

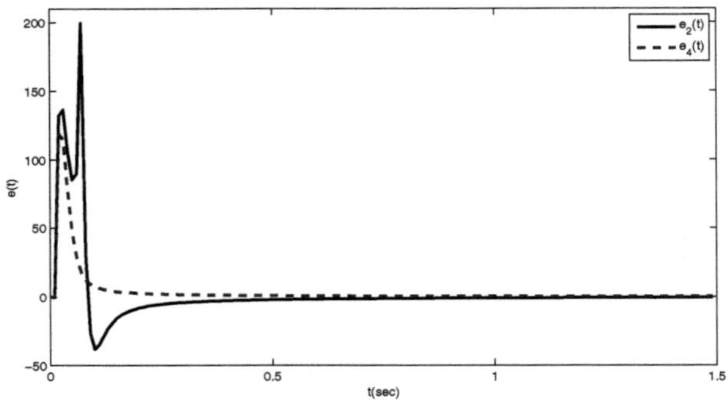

Fig. 12.10 The state response errors of $x_2(t) - \hat{x}_2(t)$ and $x_4(t) - \hat{x}_4(t)$.

The nonuniform transmission periods Γ_i is a Markovian jump process which is given as $\Gamma_i \in \{0.5\text{s},\ 0.6\text{s},\ 0.7\text{s}\}$. The Markovian jump process transition probability matrix Π_z is defined by

$$\Pi_z = \begin{bmatrix} 0.5 \; 0.3 \; 0.2 \\ 0.3 \; 0.4 \; 0.3 \\ 0.2 \; 0.3 \; 0.5 \end{bmatrix}.$$

One of the possible realizations of Markovian jumping mode is given in Fig. 12.11.

Passing through communication networks, the closed-loop estimate state Markov jump DOS is given as follows:

$$\delta\hat{x}(t_i) = A_1\hat{x}(t_i) + \Phi_1 K_1\hat{x}(t_i) + \Psi_1 K_{01}\hat{x}(t_{i-1}) \tag{12.63}$$

$$\delta\hat{x}(t_i) = A_2\hat{x}(t_i) + \Phi_2 K_2\hat{x}(t_i) + \Psi_2 K_{02}\hat{x}(t_{i-1}) \tag{12.64}$$

$$\delta\hat{x}(t_i) = A_3\hat{x}(t_i) + \Phi_3 K_3\hat{x}(t_i) + \Psi_3 K_{03}\hat{x}(t_{i-1}) \tag{12.65}$$

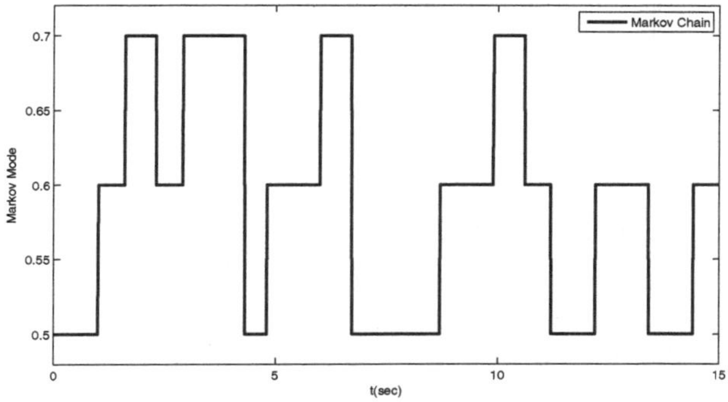

Fig. 12.11 Random Markovian jump mode.

with

$$A_1 = \begin{bmatrix} 0 & 1 & 0 & 0 \\ 0 & 0 & 0 & 0 \\ 0 & 0 & 13.1121 & 2.7626 \\ 0 & 0 & 81.2197 & 13.1121 \end{bmatrix}, \quad \Phi_1 = \begin{bmatrix} -0.1936 \\ -0.8800 \\ 0.9142 \\ 5.9618 \end{bmatrix}, \quad \Psi_1 = \begin{bmatrix} -0.0564 \\ -0.1200 \\ 0.4238 \\ 2.3259 \end{bmatrix},$$

$$A_2 = \begin{bmatrix} 0 & 1 & 0 & 0 \\ 0 & 0 & 0 & 0 \\ 0 & 0 & 19.9287 & 3.9709 \\ 0 & 0 & 116.7444 & 19.9287 \end{bmatrix}, \quad \Phi_2 = \begin{bmatrix} -0.2430 \\ -0.9000 \\ 1.4237 \\ 8.5926 \end{bmatrix}, \quad \Psi_2 = \begin{bmatrix} -0.0570 \\ -0.1000 \\ 0.6098 \\ 3.3201 \end{bmatrix},$$

$$A_3 = \begin{bmatrix} 0 & 1 & 0 & 0 \\ 0 & 0 & 0 & 0 \\ 0 & 0 & 30.3743 & 5.8594 \\ 0 & 0 & 172.2667 & 3.03743 \end{bmatrix}, \quad \Phi_3 = \begin{bmatrix} -0.2926 \\ -0.9143 \\ 2.1993 \\ 12.6907 \end{bmatrix}, \quad \Psi_3 = \begin{bmatrix} -0.0574 \\ -0.0857 \\ 0.9001 \\ 4.8876 \end{bmatrix}.$$

The transition rate matrix Λ_s for the Markovian jump DOS (12.63)-(12.65) is given by

$$\Lambda_\delta = (\Pi_z - I)/\mathtt{T}_s = \begin{bmatrix} -50 & 30 & 20 \\ 30 & -60 & 30 \\ 20 & 30 & -50 \end{bmatrix}.$$

By using Matlab command mincx of LMI-toolbox to solve convex optimization problem (12.27) in Theorem 12.15, we obtain the solutions as follows:

$$\alpha = 44.5277, \quad \beta = 0.0218,$$

$$X_1 = \begin{bmatrix} 0.0962 & -0.1031 & 0.0090 & -0.0089 \\ -0.1031 & 0.3606 & -0.0799 & 0.2104 \\ 0.0090 & -0.0799 & 0.1499 & -0.7372 \\ -0.0089 & 0.2104 & -0.7372 & 3.7624 \end{bmatrix},$$

$$X_2 = \begin{bmatrix} 0.0643 & -0.0724 & -0.0021 & 0.0328 \\ -0.0724 & 0.3421 & -0.0148 & -0.1749 \\ -0.0021 & -0.0148 & 0.0895 & -0.4652 \\ 0.0328 & -0.1749 & -0.4652 & 2.6517 \end{bmatrix},$$

$$X_3 = \begin{bmatrix} 0.0462 & -0.0462 & 0.0009 & -0.0002 \\ -0.0462 & 0.2741 & -0.0229 & -0.0929 \\ 0.0009 & -0.0229 & 0.1002 & -0.5186 \\ -0.0002 & -0.0929 & -0.5186 & 2.9020 \end{bmatrix},$$

$$Z = \begin{bmatrix} 0.0005 & -0.0001 & 0.0002 & -0.0012 \\ -0.0001 & 0.0010 & -0.0005 & 0.0021 \\ 0.0002 & -0.0005 & 0.0006 & -0.0029 \\ -0.0012 & 0.0021 & -0.0029 & 0.0142 \end{bmatrix},$$

$$N_{11} = \begin{bmatrix} 0.0010 & -0.0002 & 0.0005 & -0.0023 \\ -0.0002 & 0.0020 & -0.0009 & 0.0042 \\ 0.0005 & -0.0009 & 0.0012 & -0.0058 \\ -0.0023 & 0.0042 & -0.0058 & 0.0284 \end{bmatrix},$$

$$N_{21} = \begin{bmatrix} -0.0010 & 0.0002 & -0.0005 & 0.0023 \\ 0.0002 & -0.0020 & 0.0009 & -0.0042 \\ -0.0005 & 0.0009 & -0.0012 & 0.0058 \\ 0.0023 & -0.0042 & 0.0058 & -0.0284 \end{bmatrix},$$

$$N_{12} = \begin{bmatrix} 0.0009 & -0.0005 & 0.0005 & -0.0024 \\ -0.0005 & 0.0023 & -0.0012 & 0.0053 \\ 0.0005 & -0.0012 & 0.0014 & -0.0069 \\ -0.0024 & 0.0053 & -0.0069 & 0.0335 \end{bmatrix},$$

$$N_{22} = \begin{bmatrix} -0.0007 & 0.0004 & -0.0004 & 0.0020 \\ 0.0004 & -0.0018 & 0.0010 & -0.0043 \\ -0.0004 & 0.0010 & -0.0012 & 0.0056 \\ 0.0020 & -0.0043 & 0.0056 & -0.0272 \end{bmatrix},$$

$$N_{13} = \begin{bmatrix} 0.0007 & -0.0002 & 0.0004 & -0.0017 \\ -0.0002 & 0.0014 & -0.0007 & 0.0030 \\ 0.0004 & -0.0007 & 0.0009 & -0.0042 \\ -0.0017 & 0.0030 & -0.0042 & 0.0203 \end{bmatrix},$$

$$N_{23} = \begin{bmatrix} -0.0007 & 0.0002 & -0.0004 & 0.0017 \\ 0.0002 & -0.0014 & 0.0007 & -0.0030 \\ -0.0004 & 0.0007 & -0.0009 & 0.0042 \\ 0.0017 & -0.0030 & 0.0042 & -0.0203 \end{bmatrix},$$

$$Y_1 = \begin{bmatrix} -0.1049 & 0.6011 & -0.2080 & 0.6523 \end{bmatrix},$$

$$Y_{01} = 10^{-3} \cdot \begin{bmatrix} -0.4379 & -0.5993 & -0.0501 & 0.0654 \end{bmatrix},$$

$$Y_2 = \begin{bmatrix} -0.0546 & 0.6736 & -0.0540 & -0.3408 \end{bmatrix},$$

$$Y_{02} = 10^{-3} \cdot \begin{bmatrix} -0.5448 & -0.3822 & -0.1466 & 0.6748 \end{bmatrix},$$

$$Y_3 = \begin{bmatrix} -0.0110 & 0.5643 & -0.0652 & -0.2346 \end{bmatrix},$$

$$Y_{03} = 10^{-3} \cdot \begin{bmatrix} -0.2939 & -0.1784 & -0.0875 & 0.4065 \end{bmatrix}.$$

the control input is given as

$$\begin{aligned}
K_1 &= \begin{bmatrix} -0.0120 & -0.0778 & -15.2771 & -2.8156 \end{bmatrix}, \\
K_{01} &= \begin{bmatrix} -0.0193 & -0.0250 & -0.1589 & -0.0298 \end{bmatrix}, \\
K_2 &= \begin{bmatrix} -0.0451 & -0.1073 & -15.0361 & -2.7729 \end{bmatrix}, \\
K_{02} &= \begin{bmatrix} -0.0334 & -0.0313 & -0.1683 & -0.0309 \end{bmatrix}, \\
K_3 &= \begin{bmatrix} -0.0507 & -0.0920 & -14.6543 & -2.7027 \end{bmatrix}, \\
K_{03} &= \begin{bmatrix} -0.0234 & -0.0189 & -0.0982 & -0.0180 \end{bmatrix}.
\end{aligned}$$

For initial condition: $x_1(0) = x_2(0) = x_3(0) = x_4(0) = 0$, the simulation results of the state response of closed-loop DOS in this example is given in Fig. 12.12. Furthermore, Fig. 12.13 shows the simulation results of the control input for initial condition $u(0) = 0$.

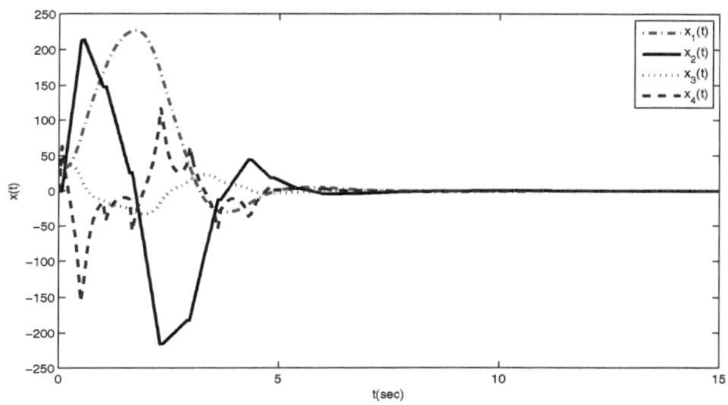

Fig. 12.12 The state response of $x(t)$.

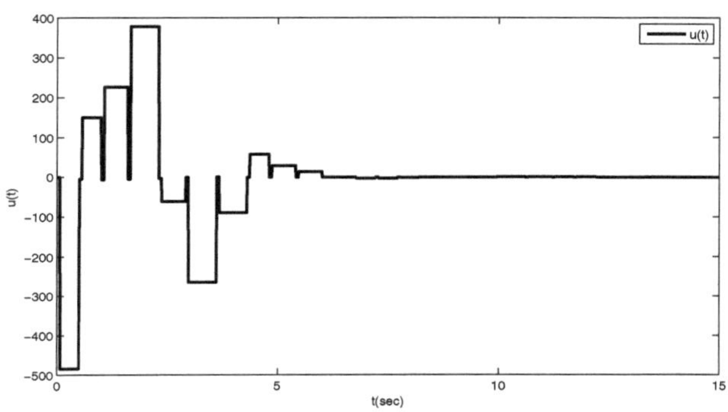

Fig. 12.13 The control input $u(t)$.

From these simulation results, it is clear that the designed networked state feedback digital controller ensures the asymptotic stability of the closed-loop NCS with packet dropout.

12.5 Summary

In this chapter, a new approach has been proposed to study the modeling and control problems for a NCS with both network-induced delay and packet-dropout. The plant in the NCS framework is a continuous-time system which is sampled with uniform sampling periods and transmitted with nonuniform transmission periods by considering QoS of communication networks. A discrete Kalman filter arithmetic for DOSs has been introduced to estimate system state variable. A kind of Markov jump linear DOS has been proposed to describe the NCS. Through constructing a new type of stochastic Lyapunov functional in delta domain, some sufficient conditions for the existence of a guaranteed cost controller, which depend on each transmission time of the modes, have been derived in terms of the familiar format of LMIs. A numerical example has been given to illustrate the effectiveness and potential for the developed techniques.

Chapter 13
Stability Analysis of High Frequency NCSs

13.1 Introduction

For the reason of that high frequency sampling will increase the possibility of network congestion, high frequency constricts of NCSs is more affected by time delays and packets dropout than low frequency constricts. Therefore, designing a high frequency system to describe NCSs and satisfying stability performance by using delta operator approach is a key task in this topic. To deal with finite frequency requirements, both weighting functions and frequency gridding has been utilized in [150] and [286], respectively. KYP lemma establishes equivalence between such frequency domain inequality for a transfer function and a LMI for its state space realization [198]. It originates from Popov's criterion [191]. The Popov's frequency condition is equivalent to existence of a certain simple Lyapunov function form [133, 282]. Based on the famous Shannon sampling theory, sampling frequency is much larger than system frequency. Goodwin introduced delta operator method to deal with high frequency sampling in [82]. A number of results on delta operator method have been reported in literatures, such as [215, 285, 284]. The generalized KYP lemma in Chapter 9 has given a direct treatment of the finite frequency performance, thus it is can avoid computational burden and guarantee gain property performances simultaneously as in [115, 118, 119].

Network-induced delays and packets dropout are two main problems in NCSs [68, 69]. The general theory for the stability of NCSs has attracted much research interest in recent years, for example [48, 73, 78, 97]. It is not reasonable to combine the delay from sensor to controller and the delay from controller to actuator together in NCSs [136]. In this chapter, a networked predictive control scheme based on Kalman filter is employed to compensate for network-induced delays and packets dropout in feedback channel. This kind of networked predictive control schemes has been investigated in [155, 156, 261]. Moreover, we develop the sampled-data control theory to deal with the network in forward channel. Motivated by the widespread use of

H. Yang et al.: Analysis and Synthesis of Delta Operator Systems, LNCIS 430, pp. 231–253.
springerlink.com © Springer-Verlag Berlin Heidelberg 2012

NCSs, the sampled-data control theory has been well-developed in the last two decades, see [34, 77, 287, 220]. It is possible to use the tools of linear sampled-data systems for analysis and design linear NCSs [61].

The main contributions of this chapter are summarized as below:

i The high frequency NCS is described as a DOS with another nonuniform sampling pattern and a high frequency constraint.
ii By developing the high frequency generalized KYP lemma, improved stability conditions in presented in terms of LMIs.
iii a networked predictive scheme is used in feedback channel and the sampled-data control theory is used in forward channel.

13.2 Problem Statement

The high frequency NCS framework diagram considered in this chapter is shown in Fig. 13.1.

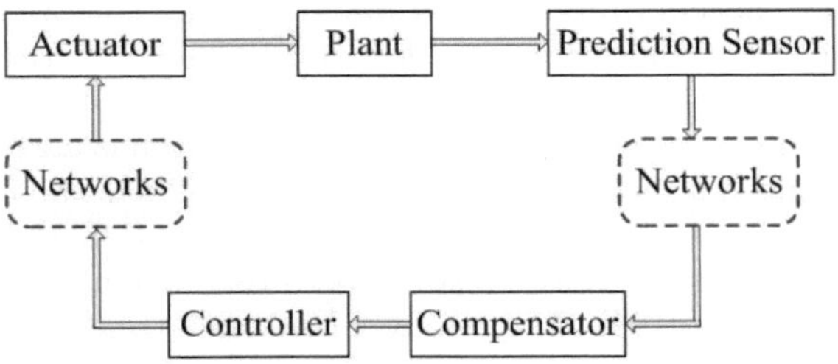

Fig. 13.1 Schematic diagram of NCS framework.

Throughout the chapter, the following assumptions are made for the high frequency NCS framework:

Assumption 13.1. *The plant with a high frequency constraint is completely controllable and completely observable. Both the prediction sensor and the controller are time driven, the actuator is event driven. The networks in feedback channel is different from the ones in forward channel.*

Assumption 13.2. *A predictive and compensative packets method will be used to deal with the packets dropout and delays in feedback channel. The upper bound of the packets dropout and delay is not greater than M multiples of the sampling period, M is a positive integer.*

Assumption 13.3. *In forward channel, packets dropout and delays are assumed as a time-varying delays. The upper bound of the delays is not greater than N multiples of the sampling period, N is a positive integer. These assumption is reasonable and also employed in [292].*

The state-space model of the plant is described as a continuous system with a high frequency constraint $|\omega| \geq \varpi$, where ω is the frequency of the continuous model and ϖ is a real number. The continuous system is given as follows:

$$\dot{x}(t) = A_s x(t) + B_s u(t) + w(t) \tag{13.1}$$
$$y(t) = C_s x(t) + v(t) \tag{13.2}$$

with $x(t) \in R^n$ is the plant state, $u(t) \in R^p$ is the control input, $y(t) \in R^m$ is the measurement output, $A_s \in R^{n\times n}$ is system matrix, $B_s \in R^{n\times p}$ is the input matrix, $C_s \in R^{m\times n}$ is the measurement matrix, $w(t) \in R^n$ is the input noise, $v(t) \in R^m$ is the measurement noise. For all constants $t > 0$ and $\ell > 0$, the uncorrelated white noises $w(t) \in R^n$ and $v(t) \in R^m$ with zero mean and two ordinary covariance matrices Q_s and R_s satisfy

$$\mathbf{E}[w(t)w^T(\ell)] = Q_s\delta(t-\ell)$$
$$\mathbf{E}[v(t)v^T(\ell)] = R_s\delta(t-\ell)$$
$$\mathbf{E}[w(t)v^T(\ell)] = 0$$

where $\delta(t)$ is the Kronecker delta function; The initial state $x(0)$ is independent on $\omega(t) \in R^n$ and $v(t) \in R^m$

$$\hat{x}(0) = \mathbf{E}[x(0)]$$
$$P(0) = \mathbf{E}[(\hat{x}(0) - x(0))(\hat{x}(0) - x(0))^T].$$

The stability region of system (13.1)-(13.2) with the high frequency constraint $|\omega| \geq \varpi$ is shadowed in Fig. 13.2.

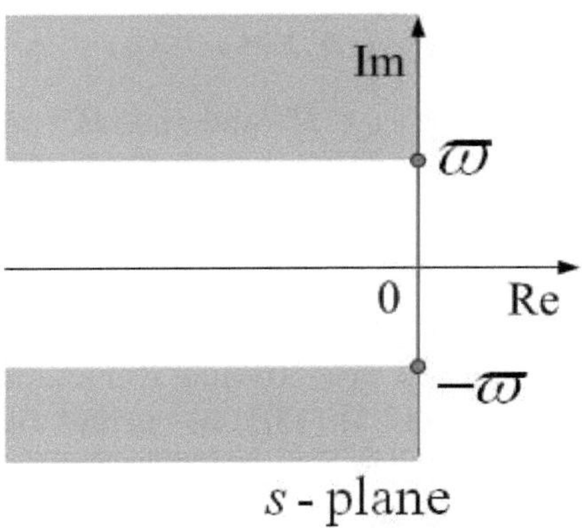

Fig. 13.2 Stability regions for high frequency constraint in s-plane.

The prediction sensor co-located with the plant in Fig. 13.1 is shown in Fig. 13.3. It is seen from Fig. 13.3 that the prediction sensor consists of three

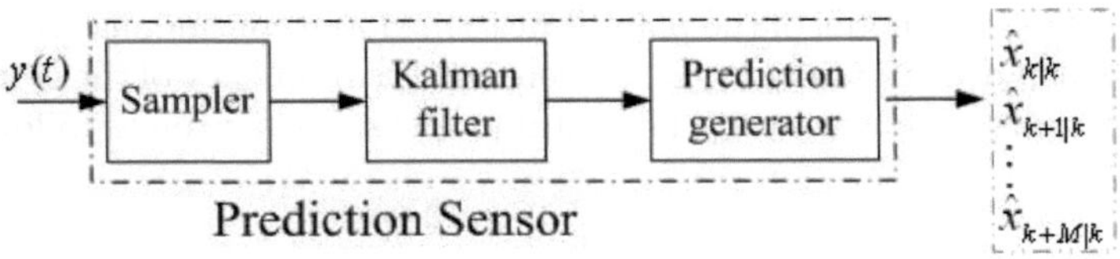

Fig. 13.3 Schematic diagram of Prediction Sensor in Fig. 13.1.

parts: a sampler to sample the output signal of plant, a Kalman filter to estimate the plant state and a prediction generator to predict a set of data and transmit it at the same time.

The NCS under consideration is restricted in a high frequency constraint. Based on the famous Shannon sampling theory, we have to choose a small enough sampling period $\mathtt{T}_s$ to fit for the high frequency constraint. That is, the sampler samples the value of the system output $y(t)$ with a fast sampling rate $1/\mathtt{T}_s$ when output signal passes through the sampler in the prediction sensor. By sampling system (13.1)-(13.2), the following discrete system is obtained as

$$x_k = A_z x_{k-1} + B_z u_{k-1} + \Lambda w_{k-1} \tag{13.3}$$
$$y_k = C_z x(t_k) + v(t_k) \tag{13.4}$$

where $x_k \in R^n$ is the plant state, $u_k \in R^p$ is the control input, $y_k \in R^m$ is the measurement output. The matrices in system (13.3)-(13.4) are computed as follows:

$$A_z = \exp(A_s \mathtt{T}_s),\ B_z = \exp(A_s \mathtt{T}_s - I) A_s^{-1} B_s,$$
$$\Lambda = \exp(A_s \mathtt{T}_s - I) A_s^{-1},\ C_z = C_s.$$

The uncorrelated white noises $w_k \in R^n$ and $v_k \in R^m$ are given as follows:

$$w_k \sim (0, Q_z),\ Q_z = \mathtt{T}_s Q_s$$
$$v_k \sim (0, R_z),\ R_z = R_s/\mathtt{T}_s$$

where Q_z and R_z are two ordinary covariance matrices.

The following lemma shows a discrete-time Kalman filter which is used to estimate the state of system (13.3)-(13.4) at current time and predict a set of estimated state for further time.

Lemma 13.1. *[212] The discrete-time Kalman filter is summarized as follows.*

i For the ordinary covariance matrices Q_z and R_z, the dynamic discrete-time system is given by the following equations:

$$x_k = F x_{k-1} + G u_{k-1} + w_{k-1}$$

$$
\begin{aligned}
y_k &= Hx_k + v_k \\
\boldsymbol{E}[w_k w_j^T] &= Q_z \delta_{k-j} \\
\boldsymbol{E}[v_k v_j^T] &= R_z \delta_{k-j} \\
\boldsymbol{E}[w_k v_j^T] &= 0
\end{aligned}
$$

ii The Kalman filter is initialized as follows

$$
\begin{aligned}
\hat{x}_0^+ &= \boldsymbol{E}[x_0] \\
P_0^+ &= \boldsymbol{E}[(x_0 - \hat{x}_0^+)(x_0 - \hat{x}_0^+)^T]
\end{aligned}
$$

iii The discrete-time Kalman filter is given by the following equations, which are computed for each time step $k = 1, 2, \cdots$:
1) measurement update

$$
\begin{aligned}
K_k &= P_k^- H^T (HP_k^- H^T + R_z)^{-1} = P_k^+ H^T R_z^{-1} \\
\hat{x}_k^+ &= \hat{x}_k^- + K_k (y_k - H\hat{x}_k^-) = \textit{a posteriori state estimate} \\
P_k^+ &= (I - K_k H) P_k^- (I - K_k H)^T + K_k R_z K_k^T \\
&= [(P_k^-)^{-1} + H^T R_z^{-1} H]^{-1} \\
&= (I - K_k H) P_k^-
\end{aligned}
$$

2) time update

$$
\begin{aligned}
P_k^- &= F P_{k-1}^+ F^T + Q_z \\
\hat{x}_k^- &= F \hat{x}_{k-1}^+ + G u_{k-1} = \textit{a priori state estimate}
\end{aligned}
$$

By using the discrete-time Kalman filter in Lemma 13.1 for system (13.3)-(13.4), we have that

$$
\begin{aligned}
K_k &= P_{k|k-1} C_z^T (C_z P_{k|k-1} C_z^T + R_z)^{-1} \\
P_{k|k-1} &= A_z P_{k-1|k-1} A_z^T + Q_z \\
P_{k|k} &= (I - K_k C_z) P_{k|k-1} \\
\hat{x}_{k|k-1} &= A_z \hat{x}_{k-1|k-1} + B_z u_{k-1} \\
\hat{x}_{k|k} &= \hat{x}_{k|k-1} + K_k (y_k - C_z \hat{x}_{k|k-1})
\end{aligned}
$$

The network delay at time instant $k\mathrm{T}_s$ in feedback channel is denoted as $d_k = m_k \mathrm{T}_s$, where m_k is a random integer. We assume that the upper bound of the delay is $d_{\max} = M\mathrm{T}_s$. That is, there exists

$$
0 \le d_k = m_k \mathrm{T}_s \le M\mathrm{T}_s = d_{\max}
$$

If the network delay d_k is larger than $d_{\max}$, we suppose that the data packet is lost at time instant $k\mathrm{T}_s$. That is, M steps is enough as the most prediction steps in the prediction generator. The following set of data is predicted at time instant t_{k-m_k} by the prediction generator:

$$
\begin{aligned}
\hat{x}_{k-m_k|k-m_k} &= \hat{x}_{k-d|k-m_k-1} + K_k(y_{k-m_k} - C_z\hat{x}_{k-m_k|k-m_k-1}) \\
\hat{x}_{k-m_k+1|k-m_k} &= A_z\hat{x}_{k-m_k|k-m_k} + B_z u_{k-m_k} \\
&\vdots \\
\hat{x}_{k|k-m_k} &= A_z\hat{x}_{k-1|k-M} + B_z u_{k-1} \\
&\vdots \\
\hat{x}_{k-m_k+M|k-m_k} &= A_z\hat{x}_{k-m_k+M-1|k-m_k} + B_z u_{k-m_k+M-1}
\end{aligned}
$$

which results in

$$
\begin{aligned}
\hat{x}_{k|k-m_k} &= A_z^{m_k-1}(A_z - K_{k-m_k}C_z)\hat{x}_{k-m_k|k-m_k} \\
&\quad + \sum_{i=1}^{m_k} A^{m_k-i}B_z u_{k-m_k+i-1} + A_z^{m_k-1}K_{k-m_k}y_{k-m_k}
\end{aligned}
$$

On the controller side, the following prediction state set

$$
\left\{\, \hat{x}_{k|k}\ \hat{x}_{k|k-1}\ \cdots\ \hat{x}_{k|k-m_k+1}\ \hat{x}_{k|k-m_k}\ \cdots\ \hat{x}_{k|k-M+1}\ \hat{x}_{k|k-M} \,\right\} \tag{13.5}
$$

or one of it's subsets must be obtained in the delay compensator. Only the latest prediction sequence is kept. The network delay compensator chooses the prediction value from the latest prediction sequence. From the above, it is shown that in the case of no network delay in feedback channel, the input of the controller is the output of the sensor. In the case of a delay $d_k = m_k\mathtt{T}_s$, the input of the controller is the m_kth-step ahead control prediction received in current sampling period. If the data packet is lost within current sampling period, the control input should be the i_kth-step ahead control prediction for the current time which is received in the sampling period. Assume that a prediction state $\hat{x}_{k|k-i_k}$ in (13.5) is available to be chosen as the latest prediction state of x_k at time instant $k\mathtt{T}_s$, the output estimate state of the network delay compensator is

$$
\hat{x}(t_k) = \hat{x}_{k|k-i_k} \tag{13.6}
$$

where $\hat{x}(t_k)$ denotes the estimate state at time $t_k = k\mathtt{T}_s$. This kind of networked predictive control schemes has been presented in [156] and [155]. In this chapter, we consider the predictive control scheme with a high frequency constraint for the continuous system (13.1)-(13.2). It has been shown that the sampling period $\mathtt{T}_s$ is must be very small for the reason of the high frequency constraint based on the Shannon sampling theory. With the small sampling period $\mathtt{T}_s$, the delta operator approach is better than the traditional shift operator approach for sampling continuous systems [82]. By the fore-mentioned analysis with the estimate state $\hat{x}(t_k)$ in (13.6), the delta operator of $\hat{x}(t_k)$ in this chapter is defined as

$$\delta\hat{x}(t_k) = \frac{\hat{x}(t_k + \mathrm{T}_s) - \hat{x}(t_k)}{\mathrm{T}_s}$$

System (13.1)-(13.2) with the high frequency constraint $|\omega| \geq \varpi$ overpass feedback channel is changed to the following DOS:

$$\delta\hat{x}(t_k) = A\hat{x}(t_k) + Bu(t_k) \tag{13.7}$$

where

$$A = \frac{e^{A_s \mathrm{T}_s} - I}{\mathrm{T}_s}, \ B = \frac{1}{\mathrm{T}_s}\int_0^{\mathrm{T}_s} e^{A_s(\mathrm{T}_s - \tau)}B_s d\tau.$$

The high frequency constraint $\vartheta \leq |\theta| \leq \pi$ for $\vartheta \in [0, \pi]$, where $\vartheta + 2n\pi = \varpi \mathrm{T}_s$ and n is a appropriate integer. That is, the high frequency constraint $|\omega| \geq \varpi$ in s-plane is changed to $\vartheta \leq |\theta| \leq \pi$ in δ-plane. The stability region of the DOS (13.7) with the high frequency $\vartheta \leq |\theta| \leq \pi$ in δ-plane is shadowed in Fig. 13.4.

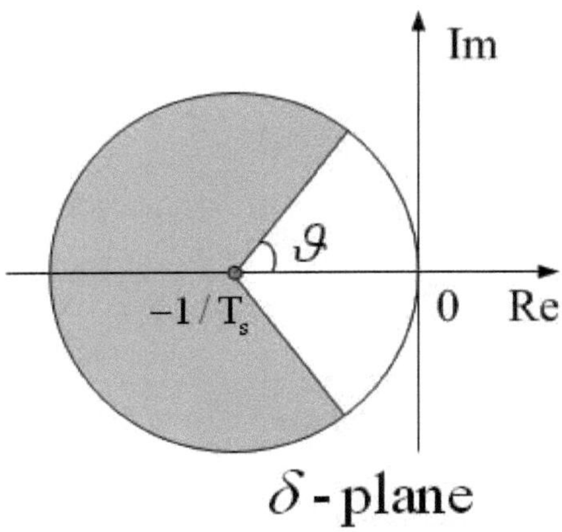

Fig. 13.4 Stability regions for high frequency constraint in δ-plane.

By the results of [292], the packets dropout and time delays in feedback channel can be assumed as nominal delays denoted as $h_k = n_k \mathrm{T}_s$, where n_k is a random integer. Moreover, we assume that the upper bound of the nominal delays is not greater than $h_{\max} = N\mathrm{T}_s$. That is, there exists

$$0 \leq h_k = n_k \mathrm{T}_s \leq N\mathrm{T}_s = h_{\max}$$

Considering the nominal delays $h_k = n_k \mathtt{T}_s$ in feedback channel as input time-varying delays, system (13.7) is rewritten as the following DOS with input delays

$$\delta\hat{x}(t_k) = A\hat{x}(t_k) + Bu(t_k - h_k) \tag{13.8}$$

where the networked control signal $u(t_k)$ is piecewise constants with the following form

$$u(t_k - h_k) = F\hat{x}(t_k - h_k), \ \forall\ t_k \in [t_k - h_k, t_k) \tag{13.9}$$

for the appropriately dimensioned matrix gain F. System (13.8) and controller (13.40) can be written as the following DOS with time-varying delays

$$\delta\hat{x}(t_k) = A\hat{x}(t_k) + BF\hat{x}(t_k - h_k) \tag{13.10}$$

Rearrange system (13.10) as follows

$$\begin{aligned}\delta\hat{x}(t_k) &= A\hat{x}(t_k) + BF\hat{x}(t_k) - BF\hat{x}(t_k) + BF\hat{x}(t_k - h_k)\\ &= (A + BF)\hat{x}(t_k) - BF \cdot \mathtt{T}_s \sum_{i=1}^{n_k} \delta\hat{x}(t_k - i\mathtt{T}_s)\end{aligned}$$

which can be rewritten as

$$\delta\hat{x}(t_k) = A_F\hat{x}(t_k) + Bw(t_k) \tag{13.11}$$

$$z(t_k) = C\hat{x}(t_k) + Dw(t_k) \tag{13.12}$$

where $A_F = A + BF$, $C = FA$, $D = FB$, $z(t_k) = F\delta\hat{x}(t_k)$ and

$$w(t_k) = -\mathtt{T}_s F \sum_{i=1}^{n_k} \delta\hat{x}(t_k - i\mathtt{T}_s) = -\mathtt{T}_s F \sum_{i=k-n_k}^{k-1} \delta\hat{x}(t_i)$$

Define a distributed-delay operator $\Delta_F : z \mapsto w$ by

$$w = \Delta_F z \Leftrightarrow w(t_k) = -\mathtt{T}_s \sum_{i=k-n_k}^{k-1} z(t_i) \tag{13.13}$$

This makes it easy to design the NCS (13.11)-(13.12) for the robustness of time delays in forward channel. The essential idea of this approach is to transform the delay system to be analyzed in such a way that it becomes a feedback interconnection of a linear time invariant plant and the distributed-delay operator Δ_F. Then an upper of the induced-gain operator Δ_F is calculated, which allows us to apply small gain theorem for the stability analysis.

13.3 Main Results

The NCS (13.11)-(13.12) can be changed to a linear time invariant system with an uncertain time-varying time delay in feedback loop. The closed-loop NCS is descried by Fig. 13.5 where $\delta = (e^{j\theta} - 1)/\mathtt{T}_s$ satisfying $\vartheta \leq |\theta| \leq \pi$. It

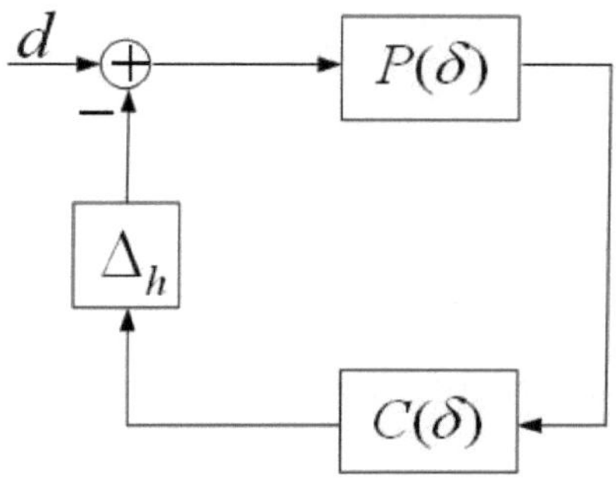

Fig. 13.5 System with a time-delay operator in feedback loop.

is shown in Fig. 13.5 that Δ_h represents a time-delay operator which perturb the network control system, $P(\delta)$ represents the plant and $C(\delta)$ represents the controller. A simple criteria of stability for the NCS descried by Fig. 13.5 is given in the following theorem.

Theorem 13.2. *The closed-loop NCS descried by Fig. 13.5 is stable for any time-varying delays defined by*

$$\Delta_h z = z(t_k - h_k),\ 0 \leq h_k \leq h_{\max}$$

where $z(t_k)$ is given in (13.12) and (13.13), if there exists

$$\left|\frac{P(\delta)C(\delta)}{1+P(\delta)C(\delta)}\right| < \frac{1}{h_{\max}}\left|T_s + \frac{1}{\delta}\right|$$

where $P(\delta)$ and $C(\delta)$ with $\delta = (e^{j\theta} - 1)/T_s$ in the frequency constraint $\vartheta \leq |\theta| \leq \pi$.

Proof. The time-delay operator Δ_h can be transformed to the distributed-delay operator Δ_F by a direct feed through an error path and a delay error, which means that $\Delta_F = (\Delta_h - 1)\cdot(\mathtt{T}_s\delta + 1)/\delta$. The closed-loop NCS descried by Fig. 13.5 is equivalent to a transformed NCS descried by Fig. 13.6. Using the Cauchy-Schwarz inequality and $n_k \leq N$, it is obtained from (13.13) that

$$w^2(t_k) = \mathtt{T}_s^2\left(\sum_{i=k-n_k}^{k-1} z(t_i)\right)^2 \leq \mathtt{T}_s^2 N \sum_{i=k-n_k}^{k-1} z^2(t_i) \leq \mathtt{T}_s^2 N \sum_{i=k-N}^{k-1} z^2(t_i)$$

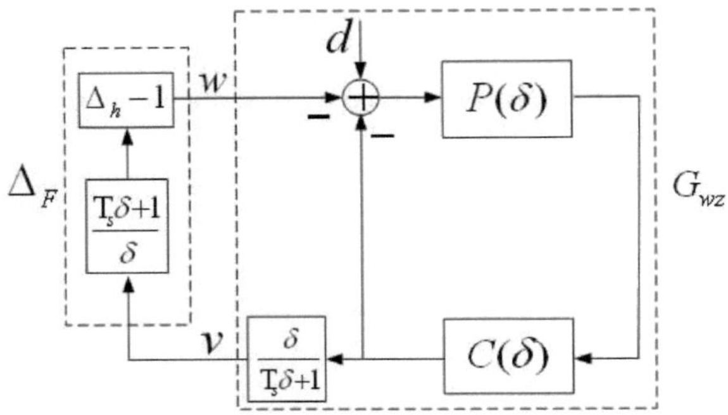

Fig. 13.6 Loop transformation of the delayed operator system.

The output energy of the transformed NCS descried by Fig. 13.6 is given as

$$\begin{aligned}\|w\|^2 &\leq \mathrm{T}_s^2 \sum_{k=0}^{\infty} N \sum_{i=k-N}^{k-1} z^2(t_i) = \mathrm{T}_s^2 N \sum_{k=0}^{\infty} \left(\sum_{i=k-N}^{k-1} z^2(t_i) \right) \\ &= \mathrm{T}_s^2 N \sum_{k=0}^{\infty} \left(\sum_{i=-N}^{-1} z^2(t_{i+k}) \right) = \mathrm{T}_s^2 N \sum_{i=-N}^{-1} \left(\sum_{k=0}^{\infty} z^2(t_{i+k}) \right) \\ &= \mathrm{T}_s^2 N^2 \left(\sum_{k=0}^{\infty} z^2(t_{i+k}) \right) = \mathrm{T}_s^2 N^2 \|z\|^2 \\ &= h_{\max} \|z\|^2\end{aligned}$$

The gain of the linear part of the transformed system from w to z in Fig. 13.6 is simply as

$$\|G_{wz}\| = \sup_{\theta \in [\vartheta, 2\pi - \vartheta]} \left| \frac{(1 - e^{j\theta}) P(\delta) C(\delta)}{\mathrm{T}_s \left(1 + P(\delta) C(\delta)\right)} \right|, \quad \delta = (e^{j\theta} - 1)/\mathrm{T}_s$$

Applying the small gain theorem to the transformed system descried by Fig. 13.6, Theorem 13.2 is proved.

By letting $h_k \equiv h_{\max}$ for all t_k, one can readily show the $\mathcal{L}_2[0,h)$-induced norm $\|\Delta_F\| = h_{\max}$. That is the $\mathcal{L}_2$-gain of Δ_F is bounded by $h_{\max}$ In the following, two lemmas will be given to compute a less conservative $\mathcal{L}_2[0,h)$-induced norm for the NCS in this chapter.

The goal of the first lemma is given to compute the norm of the compression $\underline{D}$ to $\mathcal{K}$ for a DOS descried by

$$\hat{g}(\delta) = \left[\begin{array}{c|c} A_\delta & B_\delta \\ \hline C_\delta & 0 \end{array} \right] \tag{13.14}$$

It is assumed that (A_δ, B_δ) is controllable and (C_δ, A_δ) is observable in (13.14). The operator $\underline{D}$: $\mathcal{K} \to \mathcal{K}$ is defined as follows:

$$y = \underline{D}u \Leftrightarrow y(t_k) = \mathtt{T}_s \sum_{j=0}^{k-1} C_\delta (I + \mathtt{T}_s A_\delta)^{k-j-1} B_\delta u(t_j)$$

It is described by the following state equations

$$\begin{aligned} \delta x_1(t_k) &= A_\delta x_1(t_k) + B_\delta u(t_k),\ x_1(0) = 0 \\ y(t_k) &= C_\delta x_1(t_k),\ 0 \le t_k \le h_{\max} = N\mathtt{T}_s \end{aligned}$$

Since $\underline{D}$ is a compact operator for the reason of $D_\delta = 0$, we have that $\|\underline{D}\|$ equals the square root of the largest eigenvalue of $\underline{DD}^*$. Therefore, $\|\underline{D}\|$ can be computed by characterizing the nonzero eigenvalues of $\underline{DD}^*$. It is given that $\underline{D}^*$: $u \mapsto y$ is described by the following state equations:

$$\begin{aligned} \delta x_2(t_k) &= -A_\delta^T x_1(t_k) - C_\delta^T u(t_k),\ x_2(h_{\max}) = 0 \\ y(t_k) &= B_\delta x_2(t_k),\ 0 \le t_k \le h_{\max} = N\mathtt{T}_s \end{aligned}$$

For $\gamma > 0$ we define the following matrix-valued function:

$$Q(t_k) = \begin{bmatrix} Q_{11}(t_k) & Q_{12}(t_k) \\ Q_{21}(t_k) & Q_{22}(t_k) \end{bmatrix} = \left\{ \begin{bmatrix} I & 0 \\ 0 & I \end{bmatrix} + \mathtt{T}_s \begin{bmatrix} -A_\delta^T & -C_\delta^T C_\delta \\ \gamma^{-2} B_\delta B_\delta^T & A_\delta \end{bmatrix} \right\}^{n_k} \tag{13.15}$$

where the partitioning of $Q(t_k)$ is conformable with that of the right hand matrix.

Lemma 13.3. *Assuming $\gamma > 0$, we have that γ^2 is an eigenvalue of $\underline{DD}^*$ if and only if det$[Q_{11}(h_{\max})] = 0$. Hence, $\|\underline{D}\|$ can be obtained as:*
1) Compute $Q_{11}(h_{\max})$ as a function of $\gamma > 0$,
2) $\|\underline{D}\|$ equals the largest γ such that $Q_{11}(h_{\max})$ has an eigenvalue at 0.

Proof. Sufficient: Let γ^2 be an eigenvalue of $\underline{DD}^*$. Then there exists a nonzero $f \in \mathcal{K}$ such that $\underline{DD}^* f = \gamma^2 f$. Define $g = \gamma^{-2}\underline{D}^* f$ to get the pair of equations

$$\underline{D}g = f,\ \underline{D}^* f = \gamma^2 g \tag{13.16}$$

In terms of state equations, there exist

$$\begin{aligned} \delta x_1(t_k) &= A_\delta x_1(t_k) + B_\delta g(t_k),\ x_1(0) = 0 \\ f(t_k) &= C_\delta x_1(t_k),\ 0 \le t_k \le h_{\max} = N\mathtt{T}_s \\ \delta x_2(t_k) &= -A_\delta^T x_1(t_k) - C_\delta^T f(t_k),\ x_2(h_{\max}) = 0 \\ g(t_k) &= B_\delta x_2(t_k),\ 0 \le t_k \le h_{\max} = N\mathtt{T}_s \end{aligned}$$

Eliminate f and g, we have that

$$\begin{bmatrix} \delta x_2(t_k) \\ \delta x_1(t_k) \end{bmatrix} = \begin{bmatrix} -A_\delta^T & -C_\delta^T C_\delta \\ \gamma^{-2} B_\delta B_\delta^T & A_\delta \end{bmatrix} \begin{bmatrix} x_2(t_k) \\ x_1(t_k) \end{bmatrix}, \begin{bmatrix} x_2(h) \\ x_1(0) \end{bmatrix} = 0$$

For $0 \leq t_k \leq h_{\max}$, it can be obtained that

$$\begin{bmatrix} x_2(t_k) \\ x_1(t_k) \end{bmatrix} = Q(t_k) \begin{bmatrix} x_2(0) \\ x_1(0) \end{bmatrix} = \begin{bmatrix} Q_{11}(t_k) x_2(t_k) \\ Q_{21}(t_k) x_2(t_k) \end{bmatrix}$$

Because of satisfying the boundary condition $x_2(h_{\max}) = 0$, we have

$$Q_{11}(h_{\max}) x_2(0) = 0.$$

From $f \neq 0$, it follows that $x_2(0) \neq 0$ and $\det[Q_{11}(h_{\max})] = 0$.

Necessary: Assume $\det[Q_{11}(h_{\max})] = 0$. We can choose a nonzero constant number x_{20} such that $Q_{11}(h_{\max})\mathrm{x}_{20} = 0$. By reversing the above steps, both f and g satisfying (13.16) are given as follows:

$$\begin{bmatrix} \delta x_2(t_k) \\ \delta x_1(t_k) \end{bmatrix} = \begin{bmatrix} -A_\delta^T & -C_\delta^T C_\delta \\ \gamma^{-2} B_\delta B_\delta^T & A_\delta \end{bmatrix} \begin{bmatrix} x_2(t_k) \\ x_1(t_k) \end{bmatrix}, \begin{bmatrix} x_2(0) \\ x_1(0) \end{bmatrix} = \begin{bmatrix} \mathrm{x}_{20} \\ 0 \end{bmatrix}$$
$$\begin{bmatrix} f(t_k) \\ g(t_k) \end{bmatrix} = \begin{bmatrix} 0 & C_\delta \\ \gamma^{-2} B_\delta^T & 0 \end{bmatrix} \begin{bmatrix} x_2(t_k) \\ x_1(t_k) \end{bmatrix}$$

The proof is completed if we show $f \neq 0$ or $\begin{bmatrix} f^T & g^T \end{bmatrix}^T \neq 0$. Since $\mathrm{x}_{20} \neq 0$, it follows from observability of the pair

$$\left(\begin{bmatrix} 0 & C_\delta \\ \gamma^{-2} B_\delta^T & 0 \end{bmatrix}, \begin{bmatrix} -A_\delta^T & -C_\delta^T C_\delta \\ \gamma^{-2} B_\delta B_\delta^T & A_\delta \end{bmatrix} \right)$$

which is a consequence of the minimality assumption of $(A_\delta, B_\delta, C_\delta, 0)$.

Lemma 13.4. *The operator Δ_F defined by (13.13) is bounded on $\mathcal{L}_2$ with the $\mathcal{L}_2$-induced norm*

$$\|\Delta_F\| = T_s \sqrt{2N(N-1)}/2 \tag{13.17}$$

Proof. The idea of lifting the discrete-time signal $\{t_k\}$ for $w(t_k)$ and $z(t_k)$ in system (13.11)-(13.12) is given as follows: There is an underlying clock with based period $h_i = n_i \mathtt{T}_s$ with n_i is some positive integer and the discrete-time signal $\{t_k\}$ is referred to the subperiod $\mathtt{T}_s$ of the based period. The lifted signal t_i is defined as:

$$t_i = h_1 + h_2 + \cdots + h_i = (n_1 + n_2 + \cdots + n_i)\mathtt{T}_s$$

Hence the sampling pattern $\{t_i\}$ is not uniform. That is, the lifting operator should also split the uniform discrete time axis nonuniformly.

The lifting of $\{w(t_k)\}$ and $\{z(t_k)\}$ is the two sequences $\{w(t_i)\}$ and $\{z(t_i)\}$, respectively. The delay operator Δ_F is clearly treated not only as an uniform discrete time operator $z(t_k) \to w(t_k)$ but also as a nonuniform discrete time lifted operator $z(t_i) \to w(t_i)$. This is merely a change of the viewpoint, so that many properties including the induced norm of Δ_F are preserved in the domain. The key property of Δ_F in the lifted domain is named static. That is, the output $z(t_k)$ in each time interval $[t_k - h_k, t_k)$ depends on the input $w(t_k)$ in this interval only. Therefore, the induced norm of the lifted Δ_F is the maximal induced norm of the $\mathcal{L}_2[0, h_k)$ restrictions of Δ_F integrator over all $k \in \mathcal{N}$. It means that the maximum is attained at the longest interval $h_{\max}$.

It is obtained that the sought is the $\mathcal{L}_2[0, h_{\max})$-induced norm of the plain integrator. This norm can be calculated by the result of Lemma 13.3. That is, $\|\Delta_F\| = \gamma$ if and only if $Q_{11}(h_{\max}) = 0$ and $Q_{11}(t_k) \neq 0$ for all $t_k \in [0, h_{\max})$. It is easy to get $Q_{11}(t_k) \neq 0$ from (13.15). By considering the delay operator $\Delta_F : z \mapsto w$ and the small sampling period T_s, the matrix function $Q(h_{\max})$ can be given as follows:

$$Q(h_{\max}) = \begin{bmatrix} 1 & -\mathrm{T}_s \\ \gamma^{-2}\mathrm{T}_s & 1 \end{bmatrix}^N \approx \begin{bmatrix} 1 - \frac{N(N-1)}{2}\mathrm{T}_s^2\gamma^{-2} & -N\mathrm{T}_s \\ N\mathrm{T}_s\gamma^{-2} & 1 - \frac{N(N-1)}{2}\mathrm{T}_s^2\gamma^{-2} \end{bmatrix}$$

Based on the above matrix function, $Q_{11}(h_{\max}) = 0$ yields (13.17).

By using the $\mathcal{L}_2[0, h)$-induced norm obtained in Lemma 13.4, a less conservative stability condition is given for the NCS (13.11)-(13.12) in the following theorem.

Theorem 13.5. *The NCS (13.11)-(13.12) with the high frequency constraint $\vartheta \leq |\theta| \leq \pi$ is asymptotically stable if there exist some symmetric positive definite matrices $X = X^* > 0$ and $Y = Y^* > 0$, as well as a positive scalar $\epsilon > 0$ satisfying*

$$\begin{bmatrix} G(\delta) \\ I \end{bmatrix}^* \begin{bmatrix} \rho_F^2 X & -Y \\ -Y & -X \end{bmatrix} \begin{bmatrix} G(\delta) \\ I \end{bmatrix} \leq -\epsilon I, \ \forall \vartheta \leq |\theta| \leq \pi \qquad (13.18)$$

where $\delta = (e^{j\theta} - 1)/T_s$ and $\rho_F = \|\Delta_F\| = T_s\sqrt{2N(N-1)}/2$.

Proof. Because $z(t_i)$ makes sense only from $i = k - n_k$ to $i = k - 1$. For getting $\langle z, \Delta_F z\rangle \leq 0$, it is sufficient to show that

$$\sum_{i=k-n_k}^{k-1} z^*(t_i) \left(-\mathrm{T}_s \sum_{i=k-n_k}^{k-1} z(t_i)\right) \leq 0$$

for all $i \in \{k - n_k, k - n_k + 1, \cdots, k - 1\}$. By the definition of (13.13), we have that

$$\mathrm{T}_s \sum_{i=k-n_k}^{k-1} z^*(t_i) \left(-\mathrm{T}_s \sum_{i=k-n_k}^{k-1} z(t_i)\right) = -\mathrm{T}_s^2 \sum_{i=k-n_k}^{k-1} \left(z^*(t_i) \sum_{i=k-n_k}^{k-1} z(t_i)\right)$$

$$
\begin{aligned}
&= -\frac{1}{2}\left(-\mathtt{T}_s \sum_{l=k-n_k}^{k-1} z^*(t_l)\right)\left(-\mathtt{T}_s \sum_{i=k-n_k}^{k-1} z(t_i)\right) \\
&= -\frac{1}{2}\cdot \Delta_F^* z(t_k)\cdot \Delta_F z(t_k) \\
&\leq 0
\end{aligned}
$$

From the above inequality, we have that $(-\Delta_F)$ is passive. That is, if there exists $\epsilon > 0$ satisfying $G^*(\delta) + G(\delta) \geq \epsilon I$ for all $\vartheta \leq |\theta| \leq \pi$, we have $G(\delta)$ is stable. From the inner product inequality $\langle z, \Delta_F z\rangle \leq 0$, it also can be obtained that $\langle z, Y\Delta_F z\rangle \leq 0$ for $Y = Y^* > 0$. Consider the closed-loop system in Fig. 6 with a square transfer matrix $G(\delta)$ which is another expression of G_{zw}. This is passive for $\langle z, Y\Delta_F z\rangle \leq 0$, hence the following inequality holds:

$$G^*(\delta)Y^* + YG(\delta) \geq 0,\ \forall\vartheta \leq |\theta| \leq \pi \tag{13.19}$$

Furthermore, the frequency inequality (13.18) can be rewritten as:

$$\rho_F^2 G^*(\delta)XG(\delta) - G^*(\delta)Y - YG(\delta) - X \leq -\epsilon I,\ \forall\vartheta \leq |\theta| \leq \pi \tag{13.20}$$

Considering (13.19) and (13.20), for arbitrary small $\epsilon > 0$ one has that

$$\rho_F^2 G^*(\delta)XG(\delta) - X < 0,\ \forall\vartheta \leq |\theta| \leq \pi \tag{13.21}$$

By using $\rho_F = \|\Delta_F\|$, (13.21) is rewritten as

$$[\|\Delta_F\|G(\delta)]^* X(\|\Delta_F\|G(\delta)) < X,\ \forall\vartheta \leq |\theta| \leq \pi \tag{13.22}$$

which is equivalent to

$$\|G(\delta)\| \cdot \|\Delta_F\| < 1,\ \forall\vartheta \leq |\theta| \leq \pi$$

Considering the NCS in Fig. 13.6 and using the small gain theorem, we have that the NCS (13.11)-(13.12) with the high frequency constraint $\vartheta \leq |\theta| \leq \pi$ is asymptotically stable. Theorem 13.5 is proved.

Because the $\mathcal{L}_2$-gain of Δ_F has the relation of that $\mathtt{T}_s\sqrt{2N(N-1)}/2$ is smaller than $h_{\max} = N\mathtt{T}_s$, the stability condition obtained in Theorem 13.5 is less conservative than Theorem 13.2. However, the condition of Theorem 13.5 also requires to be verified at all frequencies $\vartheta \leq |\theta| \leq \pi$, which is an infinite dimension condition. Hence, we have to turn the infinite dimension condition to a finite dimension condition which can be checked numerically.

The following lemma which captures certain properties of dynamical system expressed in terms of rational transfer function will be used in this chapter to obtain a high frequency case for DOSs.

Lemma 13.6. *[114] Let complex matrices A, B, $\Theta = \Theta^*$ and $(\Phi, \Psi) \in \Omega$ be given where*

$$\Omega := \left\{ (\Phi, \Psi) \middle| \Phi = M^* \begin{bmatrix} 0 & \alpha \\ \alpha & 0 \end{bmatrix} M,\ \Psi = M^* \begin{bmatrix} -1 & \beta \\ \beta & 1 \end{bmatrix} M \right\} \tag{13.23}$$

with $\alpha \det(M) \neq 0$, $\alpha, \beta \in R$ and $M \in \mathbf{C}^{2\times 2}$. Define

$$\Lambda := \{\lambda \in \mathbf{C} \,|\sigma(\lambda, \Phi) = 0, \sigma(\lambda, \Psi) \geq 0\} \tag{13.24}$$

Suppose (A, B) is controllable and A has no eigenvalues λ such that $\sigma(\lambda, \Phi) =$ 0. Then, the following statements are equivalent.

(i) The following frequency domain condition

$$\begin{bmatrix} (\lambda I - A)^{-1} B \\ I \end{bmatrix}^* \Theta \begin{bmatrix} (\lambda I - A)^{-1} B \\ I \end{bmatrix} < 0, \tag{13.25}$$

holds for all $\lambda \in \Lambda$.
(ii)There exist Hermitian matrices P and $Q > 0$ such that

$$\begin{bmatrix} A & B \\ I & 0 \end{bmatrix}^T L(P, Q) \begin{bmatrix} A & B \\ I & 0 \end{bmatrix} + \Theta < 0 \tag{13.26}$$

where $L(P, Q) = \Phi \otimes P + \Psi \otimes Q$.

The set Λ in (13.24) in delta domain is written as

$$\Lambda_\delta := \left\{ \frac{e^{j\theta} - 1}{\mathrm{T}_s} \middle| \theta \in R, \sigma(\lambda, \Phi) = 0, \sigma(\lambda, \Psi) \geq 0 \right\} \tag{13.27}$$

which is the intersection of the following two sets

$$\Lambda_\phi := \left\{ \frac{e^{j\theta} - 1}{\mathrm{T}_s} \middle| \theta \in R, \sigma\left(\frac{e^{j\theta} - 1}{\mathrm{T}_s}, \Phi\right) = 0 \right\} \tag{13.28}$$

$$\Lambda_\psi := \left\{ \frac{e^{j\theta} - 1}{\mathrm{T}_s} \middle| \theta \in R, \sigma\left(\frac{e^{j\theta} - 1}{\mathrm{T}_s}, \Psi\right) \geq 0 \right\} \tag{13.29}$$

By the same way, the set Ω in delta domain is also the intersection of the following two sets

$$\Omega_\phi := \left\{ \Phi \middle| \Phi = M^* \begin{bmatrix} 0 & \alpha \\ \alpha & 0 \end{bmatrix} M,\ \alpha \det(M) \neq 0 \right\} \tag{13.30}$$

$$\Omega_\psi := \left\{ \Psi \middle| \Psi = M^* \begin{bmatrix} -1 & \beta \\ \beta & 1 \end{bmatrix} M \right\} \tag{13.31}$$

with $\alpha, \beta \in R$ and $M \in \mathbf{C}^{2\times 2}$. Then, Φ and Ψ are parametrized by

$$M = NK, \ N \in \mathcal{N}_\gamma$$

in which $\mathcal{N}_\gamma$ is defined as

$$\mathcal{N}_\gamma := \left\{ J^* F J e^{j\theta} \left| F \in R^{2\times 2}, \theta \in R, \det(F) = \gamma \right. \right\}$$

where $\gamma := 1/\alpha$ and $J = diag\{1, j\}$, and K is an arbitrary matrix satisfying

$$\Phi = K^* \begin{bmatrix} 0 & 1 \\ 1 & 0 \end{bmatrix} K \tag{13.32}$$

Choosing appropriate Φ and Ψ such that $(\Phi, \Psi) \in \Omega$ is the key problem of using Lemma 13.6. However, it is difficult to choose Φ and Ψ for the reason of freedom in the choice of α and β in (13.23). The following lemma gives an easily verifiable necessary and sufficient condition for $(\Phi, \Psi) \in \Omega$.

Lemma 13.7. *[114] Letting $\Phi, \Psi \in \mathbf{H}$ be given and considering the set Ω in (13.23), it is obtained that $\Phi, \Psi \in \Omega$ holds if and only if both the following two inequalities hold*

$$\det(\Phi) < 0, \ \det(\Phi_o) < 0$$

with

$$\Phi_o := \mathrm{Re}((JK)^{-*} \Psi (JK)^{-1})$$

where K is an arbitrary matrix satisfying (13.32).

Lemma 13.6 provides a general framework for conversion of restricted frequency domain into state space condition involving linear matrix inequalities. In the following lemma, we will specialize general result of Lemma 13.6 to obtain a generalized KYP lemma for the DOS (13.11)-(13.12) with the high frequency constraint $\vartheta \le |\theta| \le \pi$.

Lemma 13.8. *Consider the DOS (13.11)-(13.12) with the high frequency constraint $\vartheta \le |\theta| \le \pi$. Suppose that Ω represents curves on the complex plant and the pair (A_F, B) is controllable. Give a symmetric matrix $\Theta \in \mathbf{H}_{n+m}$ and define Ω by (13.30) and (13.31) in delta domain. The following statements are equivalent:*

(i) The following finite frequency inequality with $\delta = (e^{j\theta} - 1)/T$

$$\begin{bmatrix} (\delta I - A_F)^{-1} B \\ I \end{bmatrix}^* \Theta \begin{bmatrix} (\delta I - A_F)^{-1} B \\ I \end{bmatrix} < 0, \ \vartheta \le |\theta| \le \pi \tag{13.33}$$

holds for all $\delta \in \Lambda_\delta$ in (13.27).

(ii)There exist matrices $P, Q \in \mathbf{H}_n$ satisfying $Q > 0$ and

$$\begin{bmatrix} A_F & B \\ I & 0 \end{bmatrix}^T \begin{bmatrix} T_s^2 P & T_s P - T_s Q \\ T_s P - T_s Q & (2\cos\vartheta - 2)Q \end{bmatrix} \begin{bmatrix} A_F & B \\ I & 0 \end{bmatrix} + \Theta < 0 \quad (13.34)$$

Proof. By an appropriate choice of Φ in (13.30), the delta domain frequency variable $\delta = (e^{j\theta} - 1)/\mathrm{T}$ in the set Λ_ϕ of (13.28) can be characterized by

$$\sigma(\delta, \Phi) = 0, \ \Phi := \begin{bmatrix} \mathrm{T}_s^2 & \mathrm{T}_s \\ \mathrm{T}_s & 0 \end{bmatrix}$$

Note that

$$\det(\Phi) = \mathrm{T}_s^2 \cdot 0 - \mathrm{T}_s \cdot \mathrm{T}_s = -\mathrm{T}_s^2 < 0$$

The matrix K in (13.32) is chosen as

$$K = \begin{bmatrix} \mathrm{T}_s & 0 \\ \frac{\mathrm{T}_s}{2} & 1 \end{bmatrix} \quad (13.35)$$

by which we have that

$$JK = \begin{bmatrix} \mathrm{T}_s & 0 \\ \frac{\mathrm{T}_s}{2} j & j \end{bmatrix}, \ (JK)^{-1} = \begin{bmatrix} \frac{1}{\mathrm{T}_s} & 0 \\ -\frac{1}{2} & -j \end{bmatrix}, \ (JK)^{-*} = \begin{bmatrix} \frac{1}{\mathrm{T}_s} & -\frac{1}{2} \\ 0 & j \end{bmatrix}$$

By an appropriate choice of Ψ in (13.31), the set Λ_ψ is specialized to define a certain range of the frequency variable $\delta = (e^{j\theta} - 1)/\mathrm{T}$ in the set Λ_ψ of (13.29). By the high frequency constraint $\vartheta \leq |\theta| \leq \pi$ in system (13.11)-(13.12), we define the following high frequency set

$$\Lambda_h = \left\{ \frac{e^{j\theta} - 1}{\mathrm{T}_s} \middle| \theta \in R, \vartheta \leq |\theta| \leq \pi \right\} \quad (13.36)$$

Note that $\vartheta \leq |\theta| \leq \pi$ if and only if $\delta = (e^{j\theta} - 1)/\mathrm{T}$ satisfies

$$\delta + \bar{\delta} = \frac{2\cos\theta - 2}{\mathrm{T}_s} \leq \frac{2\cos\vartheta - 2}{\mathrm{T}_s}$$

which means that $\cos\vartheta \geq \cos\theta$ holds. If define $\gamma_h := 2\cos\vartheta - 2$ and choose

$$\Psi = \begin{bmatrix} 0 & -\mathrm{T}_s \\ -\mathrm{T}_s & \gamma_h \end{bmatrix}$$

there exists

$$\sigma(\delta, \Psi) = -e^{j\theta} - e^{-j\theta} + 2 + \gamma_h = 2(\cos\vartheta - \cos\theta) \geq 0$$

In this case, we have that the set Λ_h in (13.36) is equal to Λ_ψ in (13.29). Furthermore, we have that

$$(JK)^{-*}\Psi(JK)^{-1} = \begin{bmatrix} \frac{1}{2}(1+\cos\vartheta) & j\cos\vartheta \\ -j\cos\vartheta & \gamma_h \end{bmatrix}$$

Hence, it can be obtained

$$\Psi_o = \mathrm{Re}[(JK)^{-*}\Psi(JK)^{-1}] = \begin{bmatrix} \frac{1}{2}(1+\cos\vartheta) & 0 \\ 0 & \gamma_h \end{bmatrix}$$

There exists

$$\det(\Psi_o) = \frac{1}{2}(1+\cos\vartheta)\gamma_h = \cos^2\vartheta - 1 < 0$$

if and only if $0 < \vartheta < \pi$. Based on Lemma 13.7, we obtain that $(\Phi, \Psi) \in \Omega$ satisfies Lemma 13.6 and the set Λ_h in (13.36) is equal to Λ in (13.27). We have chosen Φ and Ψ by using (13.28) and (13.29), respectively. The corresponding state space condition is given by (13.26) with $L(P,Q)$ in delta domain as

$$L(P,Q) = \Phi \otimes P + \Psi \otimes Q = \begin{bmatrix} \mathrm{T}_s^2 P & \mathrm{T}_s P - \mathrm{T}_s Q \\ \mathrm{T}_s P - \mathrm{T}_s Q & (2\cos\vartheta - 2)Q \end{bmatrix}$$

Because of the term $L(P,Q) = \Phi \otimes P + \Psi \otimes Q$, we have that Φ and Ψ satisfy (13.27). Then, this lemma is easily obtained from Lemma 13.6 and the above results. This proof is finished.

By Lemma 13.8, we turn the infinite dimension condition in Theorem 13.5 to a finite dimension LMI condition. The main result of this chapter is given in the following theorem.

Theorem 13.9. *The NCS (13.11)-(13.12) with the high frequency constraint $\vartheta \le |\theta| \le \pi$ is asymptotically stable, if there exist some symmetric positive definite matrices $Q = Q^* > 0$, $X = X^* > 0$ and $Y = Y^* > 0$, as well as a symmetric matrix $P = P^*$ and the constant $\rho_F = T_s\sqrt{2N(N-1)}/2$, satisfying the following LMI:*

$$\begin{bmatrix} \Omega(1,1) & T_s^2 A_F^T PB + T_s(P-Q)B + \rho_F^2 C^T XD - C^T Y \\ * & T_s^2 B^T PB + \rho_F^2 D^T XD - D^T Y - YD - X \end{bmatrix} < 0 \quad (13.37)$$

where

$$\Omega(1,1) = T_s^2 A_F^T P A_F + T_s A_F^T (P-Q) + T_s (P-Q) A_F + 2(\cos\vartheta - 1)Q + \rho_F^2 C^T X C$$

Proof. It is known from the generalized KYP lemma that if we give a designated matrix Θ in Lemma 13.8, we can get the corresponding gain property.

For the linear DOS (13.11)-(13.12), matrix Θ in (13.34) can be chosen as follows

$$\Theta = \begin{bmatrix} C & D \\ 0 & I \end{bmatrix}^* \Pi \begin{bmatrix} C & D \\ 0 & I \end{bmatrix} \tag{13.38}$$

Based on $G(\delta) = C(\delta I - A_F)B + D$, the frequency inequality (13.33) in Lemma 13.8 is rewritten as

$$\begin{bmatrix} G(\delta) \\ I \end{bmatrix}^* \Pi \begin{bmatrix} G(\delta) \\ I \end{bmatrix} < 0, \ \vartheta \leq |\theta| \leq \pi \tag{13.39}$$

Considering Theorem 13.5 for all $\epsilon > 0$ and choosing

$$\Pi = \begin{bmatrix} \rho_F^2 X & -Y \\ -Y & -X \end{bmatrix}$$

we have that inequality (13.39) is changed to (13.18). By using Lemma 13.8, inequality (13.18) is equivalent to the LMI (13.34). Considering (13.38), the LMI (13.34) can be rewritten as

$$\begin{bmatrix} A_F & B \\ I & 0 \end{bmatrix}^T \begin{bmatrix} \mathrm{T}_s^2 P & \mathrm{T}_s P - \mathrm{T}_s Q \\ \mathrm{T}_s P - \mathrm{T}_s Q & (2\cos\vartheta - 2)Q \end{bmatrix} \begin{bmatrix} A_F & B \\ I & 0 \end{bmatrix} + \begin{bmatrix} C & D \\ 0 & I \end{bmatrix}^T \begin{bmatrix} \rho_F^2 X & -Y \\ -Y & -X \end{bmatrix} \begin{bmatrix} C & D \\ 0 & I \end{bmatrix} < 0$$

which is equivalent to the LMI (13.37). This completes the proof.

13.4 Experimental Results

In the following, we provide some experiment results to demonstrate the effectiveness of the proposed methods in this chapter.

Experiment: The real inverted-pendulum model shown in Fig. 13.7 consists of the cart that is moving on a rail and the pole that is mounted on the cart via the pole shaft so that the pendulum can swing freely in the vertical plane. The physical structure of the inverted pendulum is shown in Fig. 13.8 All the constant parameters of the inverted pendulum model in Fig. 13.8 are given in Table 13.1. In Fig. 13.8, ϕ, $\boldsymbol{x}$ and $\boldsymbol{u}$ are variables which are given in Table 13.2. By the parameters and variables, the following nonlinear inverted-pendulum model can be obtained:

$$\begin{aligned} (M+m)\ddot{x} + b\dot{x} + ml\ddot{\phi}\cos\phi - ml\dot{\phi}^2\sin\phi &= \boldsymbol{u} \\ -(I+ml^2)\ddot{\phi} + mgl\sin\phi &= ml\ddot{x}\cos\phi \end{aligned}$$

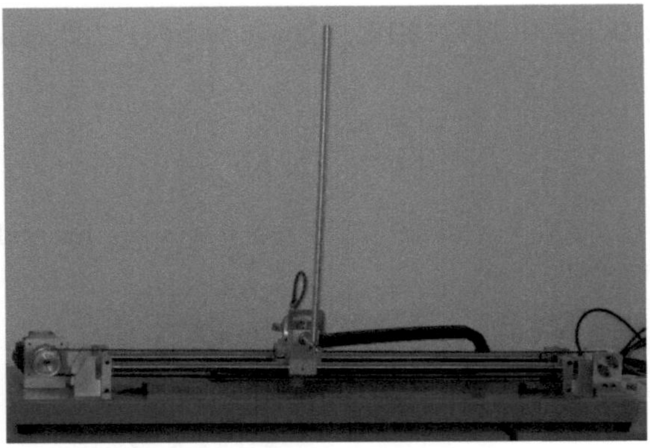

Fig. 13.7 The real inverted-pendulum model.

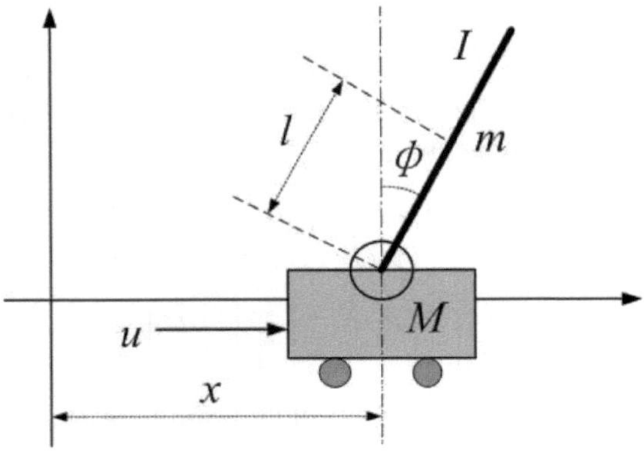

Fig. 13.8 The physical structure of inverted pendulum model.

Table 13.1 The parameters in the inverted pendulum model

Symbol	Value	Meaning
M	1.096 Kg	Mass of the cart
m	0.109 Kg	Mass of the pendulum rod
l	0.25 m	Length from the pendulum's axis to its centroid
b	0.1 $N/m/sec$	The friction coefficient of the cart
I	$0.0034kg*m*m$	Inertia of the rod
g	$9.81m/s^2$	Gravitational constant

Table 13.2 The variables in the inverted pendulum model

Symbol	Meaning
ϕ	Angle between the pendulum rod and the vertical
x	Horizontal displacement of the cart
u	Control input

Linearize the nonlinear dynamics equations by using the $\cos\phi = -1, \sin\phi = \phi, (\frac{d\phi}{dt})^2 = 0$, we get:

$$
\begin{aligned}
-(I+ml^2)\ddot{\phi} + mgl\phi &= ml\ddot{x} \\
(M+m)\ddot{x} + b\dot{x} + ml\ddot{\phi} &= \boldsymbol{u}
\end{aligned}
$$

Let the state variables be $\boldsymbol{x} = x_1(t)$, $\dot{\boldsymbol{x}} = x_2(t)$, $\phi = x_3(t)$ and $\dot{\phi} = x_4(t)$, and input variable be $u(t) = \boldsymbol{u}$. The state variables correspond to the horizontal position, horizontal velocity of the cart, angle and angle velocity of the pole, respectively. Because the translation of the cart is enabled by a dc motor, acceleration variable has to be used as a control input variable, which means $\boldsymbol{u} = -\ddot{\boldsymbol{x}}$, where "−" is based on the turning direction of dc motor. We obtain the continuous system (13.1)-(13.2) with the following parameters

$$
A_s = \begin{bmatrix} 0 & 1 & 0 & 0 \\ 0 & 0 & 0 & 0 \\ 0 & 0 & 0 & 1 \\ 0 & 0 & 29.4 & 0 \end{bmatrix}, \quad B_s = \begin{bmatrix} 0 \\ -1 \\ 0 \\ 3 \end{bmatrix}, \quad C_s = \begin{bmatrix} 1 & 0 & 0 & 0 \\ 0 & 0 & 1 & 0 \end{bmatrix}.
$$

In the following, the LQR is used for the calculation of the optimal gain matrix F such that the following state-feedback control law $u(t) = Fx(t)$ minimizes the cost function

$$
J_{LQG} = \int (x^T(t)\mathrm{Q}x(t) + u^T(t)\mathrm{R}u(t))dt.
$$

For choosing the appropriate matrices as

$$
\mathrm{Q} = \begin{bmatrix} 1000 & 0 & 0 & 0 \\ 0 & 0 & 0 & 0 \\ 0 & 0 & 200 & 0 \\ 0 & 0 & 0 & 0 \end{bmatrix}, \quad \mathrm{R} = [1]
$$

we can adjust the controller F as follows

$$
F = \begin{bmatrix} -31.623 & -20.151 & -72.718 & -13.155 \end{bmatrix} \tag{13.40}
$$

To fit for the high frequency case, the sampling period is chosen as $\mathrm{T}_s = 0.02$s. The high frequency constraint of the inverted-pendulum system in this experiment is not less than 1Hz, which means that $\vartheta = 0.02$. By using delta operator approach, system (13.11)-(13.12) is obtained with the following parameters

$$
A_F = \begin{bmatrix} 0.3162 & 1.2015 & 0.7272 & 0.1316 \\ 31.6230 & 20.1510 & 72.7180 & 13.1550 \\ -0.9496 & -0.6051 & -1.8894 & 0.6069 \\ -95.0551 & -60.5716 & -189.1242 & -39.2481 \end{bmatrix}, \quad B = \begin{bmatrix} -0.0100 \\ -1.0000 \\ 0.0300 \\ 3.0059 \end{bmatrix},
$$

$$
C = \begin{bmatrix} 0 & -31.6230 & -408.9155 & -76.7320 \end{bmatrix}, \quad D = \begin{bmatrix} -21.2588 \end{bmatrix}.
$$

In this experiment, the output and input signals were physically transmitted between two Intranet IP addresses 192.168.108.104 and 192.168.108.105 which were both located on the campus network of the University of Beijing Institute of Technology. It was measured that the maximum network delay between the two embedded boards was 0.04s, so the sum of feedback and forward channel time delays is less than 0.08s. Because the sample time is $\mathtt{T}_s = 0.02$s, we have the upper bound $N = 2$ in forward channel. It is obtained that $\rho_F = 0.02\sqrt{2 \times 2 \times 1}/2 = 0.02$. Using the Matlab LMI Control Toolbox to solve LMI (13.37) in Theorem 13.9, we obtain the solution as follows:

$$P = 10^6 \times \begin{bmatrix} 3.1065 & 1.3492 & 3.1414 & 0.4971 \\ 1.3492 & 0.9042 & 1.8181 & 0.3442 \\ 3.1414 & 1.8181 & 4.8686 & 0.7968 \\ 0.4971 & 0.3442 & 0.7968 & 0.1904 \end{bmatrix}, \quad X = \begin{bmatrix} 1323.6 \end{bmatrix},$$

$$Q = 10^4 \times \begin{bmatrix} 1.6138 & 0.0000 & 0.0038 & -0.0000 \\ 0.0000 & 1.6858 & 0.0000 & -0.3720 \\ 0.0038 & 0.0000 & 1.6094 & 0.0000 \\ -0.0000 & -0.3720 & 0.0000 & 3.1357 \end{bmatrix}, \quad Y = \begin{bmatrix} 1.3965 \end{bmatrix}.$$

Control algorithm and data acquisition and visualization are realized with MathWorks Matlab, Simulink, and Real-Time xPC target products on a personal computer with the 12-bit AD/DA converter. The two blocks Dac9 and Adc9 are the driver of the A/D and D/A channels in the embedded system and are designed in Matlab S-Function. The control block diagram of the real-time experiment is shown in Fig. 13.9. It is shown that a compensator

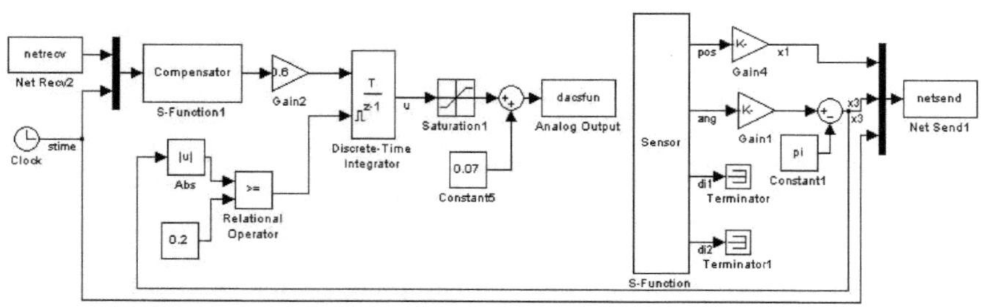

Fig. 13.9 Inverted pendulum experiment diagram.

module is used for the network compensator program. Based on the compensation method in this chapter, the compensator module can select the optimal sequence of control output to the controller. Because of the limited pendulum-rail length (0.6m), the pendulum-cart movement is also limited (0.3m). The real-time experiment results are shown in Fig. 13.10. The cart position in the first figure of Fig. 13.10 is denoted by CartPos. The pendulum

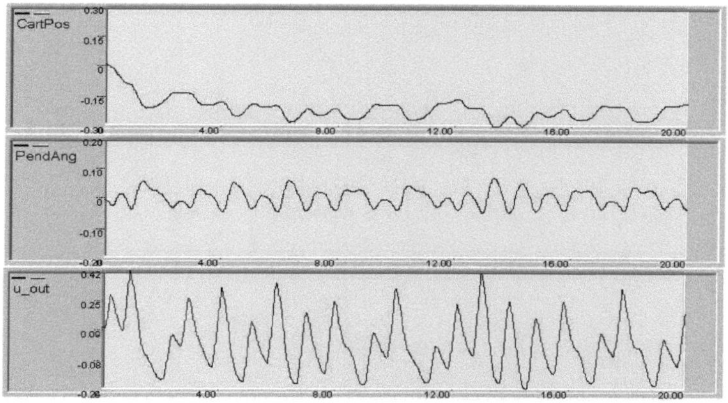

Fig. 13.10 The results of this experiment.

angle in the second figure of Fig. 13.10 is denoted by PendAng. The control signal in the third figure of Fig. 13.10 is denoted by o_out.

Remark 13.10. The inverted pendulum control problem is challenging because it may go unstable when sampling frequencies are not sufficiently high. Therefore, the practical inverted pendulum model is presented as an interesting application of the high frequency NCSs in this chapter. The experimental setup in this section has shown the developed techniques of the designed stability criterion in Theorem 13.9 is practicable.

13.5 Summary

This chapter has generalized the stability analysis for a class of high frequency NCSs. The high frequency NCS has been described as a kind of DOSs with another nonuniform sampling pattern and a high frequency constraint. Some stability conditions have been given for the systems. By developing the generalized KYP lemma, improved stability conditions have also been presented in terms of LMIs. Some experiment results of an inverted pendulum on a laboratory-scale setup have been provided to illustrate the effectiveness and potential for the developed techniques.

References

1. Aldeen, M., Sharma, R.: Robust detection of faults in frequency control loops. IEEE Transactions on Power Systems 22(1), 413–422 (2007)
2. Amato, F., Cosentino, C., Mattei, M., Paviglianiti, G.: A direct/functional redundancy scheme for fault detection and isolation on an aircraft. Aerospace Science and Technology 10(4), 338–345 (2006)
3. Anderson, B.D.O., Vongpanitlerd, S.: Network analysis and synthesis: a modern systems theory approach. Prentice-Hall, Upper Saddle River (1973)
4. Anderson, S.R., Kadirkamanathan, V.: Modelling and identification of non-linear deterministic systems in the delta-domain. Automatica 43(11), 1859–1868 (2007)
5. Assawinchaichote, W., Nguang, S.K.: H_∞ filtering for nonlinear singularly perturbed systems with pole placement constraints: an LMI approach. IEEE Transactions on Signal Processing 34(6), 1659–1667 (2004)
6. Assawinchaichote, W., Nguang, S.K., Shi, P.: Robust H_∞ fuzzy filter design for uncertain nonlinear singularly perturbed systems with Markovian jumps: an LMI approach. Information Sciences 177(7), 1699–1714 (2007)
7. Atkinson, D.J., Acarnley, P.P., Finch, J.W.: Observers for induction motor state and parameter stimation. IEEE Transactions on Industry Applj-cations (6), 1119–1127 (1991)
8. Barut, M., Bogosyan, S., Gokasan, M.: Speed-sensorless estimation for induction motors using extended Kalman filters. IEEE Transactions on Industrial Electronics 54(1), 272–280 (2007)
9. Basin, M., Perez, J., Martinez-Zuniga, R.: Optimal filtering for nonlinear polynomial systems over linear observations with delay. International Journal of Innovative Computing, Information and Control 2(4), 863–874 (2006)
10. Basin, M., Rodriguez-Gonzalez, J.: A closed-form optimal control for linear systems with equal state and input delays. Automatica 41(5), 915–920 (2005)
11. Basin, M., Rodriguez-Gonzalez, J., Fridman, L.: Optimal and robust control for linear state-delay systems. Journal of the Franklin Institute 344(6), 830–845 (2007)
12. Basin, M., Sanchez, E., Martinez-Zuniga, R.: Optimal linear filtering for systems with multiple state and observation delays. International Journal of Innovative Computing, Information and Control 3(5), 1309–1320 (2007)

13. Basin, M.V., Ferreira de Loza, A.D., Fridman, L.M.: Sliding mode identification and control for linear uncertain stochastic systems. International Journal of Systems Science 38(11), 861–870 (2007)
14. Bergsten, P., Palm, R., Driankov, D.: Observers for Takagi-Sugeno fuzzy systems. IEEE Transactions on Systems, Man and Cybernetics, Part B 32(1), 114–121 (2002)
15. Blachura, M.J.: On approximate pulse transfer functions. IEEE Transactions on Automatic Control 44(11), 2062–2067 (1999)
16. Bobtsov, A.A., Nagovitsina, A.G.: Adaptive control of linear nonstationary objects output. Automation and Remote Control 67(12), 2010–2020 (2006)
17. Bobtsov, A.A., Nagovitsina, A.G.: Adaptive output control of linear time-varying systems. In: IFAC Workshop on Adaptation and Learning in Control and Signal Processing, Imperial Anichkov Palace, Russia, vol. 9 (2007)
18. Boukas, E.K.: State feedback stabilization of nonlinear discrete-time systems with time-varying time delay. Nonlinear Analysis 66(6), 1341–1350 (2007)
19. Boukas, E.K., Al-Muthairi, N.F.: Delay-dependent stabilization of singular linear systems with delays. International Journal of Innovative Computing, Information and Control 2(2), 283–291 (2006)
20. Boukas, E.K., Liu, Z.K.: Robust H_∞ filtering for polytopic uncertain time-delay systems with Markov jumps. Computers and Electrical Engineering 28(3), 171–193 (2003)
21. Boutayeb, M., Rafaralahy, H., Darouach, M.: Convergence analysis of the extended Kalman filter used as an observer for nonlinear deterministic discrete-time systems. IEEE Transactions on Automatic Control 42(4), 581–586 (1997)
22. Bugeja, M.: Non-linear swing-up and stabilizing control of an inverted pendulum system. In: Proceedings of the IEEE Region 8 EUROCON 2003: Computer as a Tool, vol. 2, pp. 437–441 (2003)
23. Cao, Y., Frank, P.: Stability analysis and synthesis of nonlinear time-delay systems via linear Takagi-Sugeno fuzzy models. Fuzzy Sets and Systems 124(2), 213–229 (2001)
24. Cao, Y., Ying, M., Chen, G.: State-based control of fuzzy discrete-event systems. IEEE Transactions on Systems, Man, and Cybernetics, Part B: Cybernetics 37(2), 410–424 (2007)
25. Cao, Y.Y., Sun, Y.X., Cheng, C.: Delay dependent robust stabilisation of uncertain systems multiple state delays. IEEE Transactions on Automatic Control 43(11), 1608–1612 (1998)
26. Cao, Y.Y., Sun, Y.X., Lam, J.: Delay dependent robust H_∞ control for uncertain systems with time-varying delays. IEE Proceedings of Control Theory and Applications 145(3), 338–344 (1998)
27. Chan, C.Y.: Discrete adaptive sliding mode control of a state-space system with a bounded disturbance. Automatica 34(12), 1631–1635 (1998)
28. Chang, S., Peng, T.: Adaptive guaranteed cost control of systems with uncertain parameters. IEEE Transactions on Automatic Control 17(4), 474–483 (1972)
29. Chen, B., Lam, J., Xu, S.: Memory state feedback guaranteed cost control for neutral delay systems. International Journal of Innovative Computing, Information and Control 2(2), 293–303 (2006)
30. Chen, J., Patton, R.J.: Robust model-based fault diagosis for dynamic systems. Kluwer Academic Publishers (1999)

31. Chen, J., Yang, C., Lien, C., Horng, J.: New delay-dependent non-fragile H_∞ observer-based control for continuous time-delay systems. Information Sciences 178(24), 4699–4706 (2008)
32. Chen, S., Istepanian, R.H., Wu, J., Chu, J.: Comparative study on optimizing closed-loop stability bounds of finite-precision controller structures with shift and delta operators. Systems & Control Letters 40(3), 153–163 (2000)
33. Chen, S., Wu, J., Istepanian, R.H., Chu, J., Whidborne, J.F.: Optimising stability bounds of finite-precision controller structures for sampled-data systems in the δ-operator domain. IEE Proceedings-Control Theory and Applications 146(6), 517–526 (1999)
34. Chen, T., Francis, B.A.: Optimal sampled-data control system. Springer, Tokyo (1995)
35. Chen, X., Hisayama, T.: Adaptive sliding-mode position control for piezo-actuated stage. IEEE Transactions on Industrial Electronics 55(11), 3927–3934 (2008)
36. Chen, X., Yu, J., Gu, M., Tang, X.: A new GPS/SINS integrated navigation fusion algorithm based on RBF neural network and Kalman filtering. ICIC Express Letters 3(3(B)), 841–846 (2009)
37. Cheng, C., Chien, S.: Adaptive mode controller design based on T-S fuzzy system models. Automatica 42(6), 1005–1010 (2006)
38. Choi, J.Y., Chwa, D., Kim, M.S.: Adaptive control for feedback-linearized missiles with uncertainties. IEEE Transactions on Aerospace and Electronic Systems 36(2), 467–481 (2000)
39. Chow, M., Tipsuwan, Y.: Gain adaptation of networked dc motor controllers based on qos variations. IEEE Transactions on Industrial Informatics 50(5), 936–943 (2003)
40. Cloosterman, M.B.G.: Stability of networked control systems with uncertain time-varying delays. IEEE Transactions on Automatic Control 54(7), 1575–1580 (2009)
41. Colandairaj, J., Irwin, G.W., Scanlon, W.G.: Wireless networked control systems with qos-based sampling. IET Control Theory & Applications 1(1), 430–438 (2007)
42. Collins Jr., E.G., Song, T.: A delta operator approach to discrete-time H_∞ control. International Journal of Control 72(4), 315–320 (1999)
43. Davila, J., Fridman, L., Levant, A.: Second-order sliding-mode observer for mechanical systems. IEEE Transactions on Automatic Control 50(11), 1785–1789 (2005)
44. Davvaz, B., Corsini, P., Leoreanu-Fotea, V.: Atanassov's intuitionistic (s,t)-fuzzy n-ary sub-hypergroups and their properties. Information Sciences 179(5), 654–666 (2009)
45. Ding, B., Sun, H., Yang, P.: Further studies on LMI-based relaxed non-quadratic stabilization conditions for nonlinear systems in the Takagi-Sugeno's form. Automatica 42(3), 503–508 (2006)
46. Dong, J., Wang, Y., Yang, G.: Control synthesis of continuous-time T-S fuzzy systems with local nonlinear models. IEEE Transactions on Systems, Man, and Cybernetics, Part B: Cybernetics 39(5), 1245–1258 (2009)
47. Dong, J., Yang, G.: H_∞ control design for fuzzy discrete-time singularly perturbed systems via slow state variables feedback: an LMI-based approach. Information Sciences 179(17), 3041–3058 (2009)

48. Donkers, M., Heemels, W., van de Wouw, N., Hetel, L.: Stability analysis of networked control systems using a switched linear systems approach. IEEE Transactions on Automatic Control (99), 1 (2011)
49. Ebert, W.: Optimal filtered predictive control a delta operator approach. Systems & Control Letters 42(1), 69–80 (2001)
50. Edwards, C., Spurgeon, S.K.: Sliding mode control: theory and applications. Spurgeon (1998)
51. Edwards, C., Spurgeon, S.K., Patton, R.J.: Sliding mode observers for fault detection and isolation. Automatica 36(4), 541–553 (2000)
52. Emmanuel, G., Collins, J., Song, T.: A delta operator approach to discrete-time H_∞ control. International Journal of Control 72(4), 315–320 (1999)
53. Fan, H.: On delta-operator Schur-Cohn zero-location tests for fast sampling. IEEE Transactions on Signal Processing 46(7), 1851–1860 (1998)
54. Fan, H.: Singular root distribution problem for delta-operator based real polynomials. Automatica 35(5), 791–807 (1999)
55. Fang, X., Wang, J.: Stochastic observer-based guaranteed cost control for networked control systems with packet dropouts. IET Control Theory and Applications 2(11), 980–989 (2008)
56. Fong, K.F., Loh, A.P., Tan, W.W.: A frequency domain approach for fault detection. International Journal of Control 81(2), 264–276 (2007)
57. Fridman, E.: New Lyapunov-Krasovskii functionals for stability of linear retarded and neutral type systems. Systems & Control Letters 43(4), 309–319 (2001)
58. Fridman, E., Shaked, U.: An improved stabilisation method for linear time-delay systems. IEEE Transactions on Automatic Control 47(11), 1931–1937 (2002)
59. Fridman, E., Shaked, U.: Delay-dependent stability and control: constant and time-varying delays. International Journal of Control 76(1), 48–60 (2003)
60. Fujioka, H.: Stability analysis for a class of networked/embeded control systems: a discrete-time approach. In: American Control Conference, Seattle, pp. 4997–5002 (2008)
61. Fujioka, H.: Stability analysis of systems with aperiodic sample-and hold devices. Automatica 45(3), 771–775 (2009)
62. Gahinet, P., Apkarian, P.: A linear matrix inequality approach to H_∞ control. International Journal of Robust Nonlinear Control 4(4), 421–448 (1994)
63. Gahinet, P., Nemirovski, A., Laub, A., Chilali, M.: LMI control toolbox for use with Matlab. The Math Works Inc. (1995)
64. Gao, H., Chen, T.: New results on stability of discrete-time systems with time-varying state delay. IEEE Transactions on Automatic Control 52(2), 328–334 (2007)
65. Gao, H., Chen, T.: Stabilization of nonlinear systems under variable sampling: a fuzzy control approach. IEEE Transactions on Fuzzy System 15(5), 972–983 (2007)
66. Gao, H., Chen, T.: Network-based H_∞ output tracking control. IEEE Transactions on Automatic Control 53(3), 655–667 (2008)
67. Gao, H., Chen, T.: A new approach to quantized feedback control systems. Automatica 44(2), 534–542 (2008)
68. Gao, H., Chen, T., Lam, J.: A new delay system approach to network-based control. Automatica 44(1), 39–52 (2008)

69. Gao, H., Lam, J., Chen, T.: New criteria for synchronization stability of general complex dynamical networks with coupling delays. Physics Letters A 360(2), 263–273 (2006)
70. Gao, H., Lam, J., Wang, C.: Delay-dependent output feedback stabilization of discrete-time systems with time-varying state delay. IEE Proceedings on Control Theory and Applications 151(6), 691–698 (2004)
71. Gao, H., Lam, J., Wang, C.: Induced l_2 and generalized H_2 filtering for systems with repeated scalar nonlinearities. IEEE Transactions on Signal Processing 53(11), 4215–4226 (2005)
72. Gao, H., Lam, J., Xie, L., Wang, C.: New approach to mixed H_2/H_∞ filtering for polytopic discrete-time systems. IEEE Transactions on Signal Processing 53(8), 3183–3192 (2005)
73. Gao, H., Meng, X., Chen, T.: Stabilization of networked control systems with a new delay characterization. IEEE Transactions on Automatic Control 53(9), 2142–2148 (2008)
74. Gao, H., Wang, C.: Delay-dependent robust H_∞ and L_2-L_∞ filtering for a class of uncertain nonlinear time-delay systems. IEEE Transactions on Automatic Control 48(9), 1661–1666 (2003)
75. Gao, H., Wang, C.: A delay-dependent approach to robust H_∞ filtering for uncertain discrete-time state-delayed systems. IEEE Transactions on Signal Processing 52(6), 1631–1640 (2004)
76. Gao, H., Wang, Z., Wang, C.: Improved H_∞ control of discrete-time fuzzy systems: a cone complementarity linearization approach. Information Sciences 175(1-2), 57–77 (2005)
77. Gao, H., Wu, J., Shi, P.: Robust sampled-data H_∞ control with stochastic sampling. Automatica 45(7), 1729–1736 (2009)
78. García-Rivera, M., Barreiro, A.: Analysis of networked control systems with drops and variable delays. Automatica 43(12), 2054–2059 (2007)
79. Gershon, E., Limebeer, D.J.N., Shaked, U., Yaesh, I.: Robust H_∞ filtering of stationary continuous time linear systems with stochastic uncertainties. IEEE Transactions on Automatic Control 46(11), 1788–1793 (2001)
80. Gobbo, D.D., Napolitano, M., Famouri, P., Innocenti, M.: Experimental application of extended Kalman filtering for sensor validation. IEEE Transactions on Control Systems Technology 9(2), 376–380 (2001)
81. Gonzalez-Rodriguez, G., Colubi, A., Trutschnig, W.: Simulation of fuzzy random variables. Information Sciences 179(5), 642–653 (2009)
82. Goodwin, G.C., Lozano Leal, R., Mayne, D.Q., Middleton, R.H.: Rapprochement between continuous and discrete model reference adaptive control. Automatica 22(2), 199–207 (1986)
83. Green, M., Limebeer, D.J.N.: Linear robust control. Prentice-Hall, Inc., Englewood Cliffs (1995)
84. Gu, K., Kharitonov, V.L., Chen, J.: Stability of time-delay systems. Birkhäuser (2003)
85. Guerra, T.M., Vermeiren, L.: LMI-based relaxed nonquadratic stabilization conditions for nonlinear systems in the Takagi-Sugeno's form. Automatica 40(5), 823–829 (2004)
86. Hale, J., Lunel, S.M.V.: Introduction to functional differential equations. Springer, New York (1993)
87. Han, Q.L.: Absolute stability of time-delay systems with sector-bounded nonlinearity. Automatica 41(12), 2171–2176 (2005)

88. Hassan, M., Boukas, K.: Multilevel technique for large scale LQR with time-delays and systems constraints. International Journal of Innovative Computing, Information and Control 3(2), 419–434 (2007)
89. He, Y., Wang, Q., Lin, C., Wu, M.: Delay-range-dependent stability for systems with time-varying delay. Automatica 43(2), 371–376 (2007)
90. He, Y., Wu, M., She, J.H., Liu, G.P.: Delay dependent robust stability criteria for uncertain neutral systems with mixed delays. Systems & Control Letters 51(1), 57–65 (2004)
91. He, Y., Wu, M., She, J.-H., Liu, G.-P.: Parameter-dependent Lyapunov functional for stability of time-delay systems with polytopic-type uncertainties. IEEE Transactions on Automatic Control 49(5), 828–832 (2004)
92. Heui-Wook, K., Seung-Ki, S.: A new motor speed estimator using Kalman filter in low-speed range. IEEE Transactions on Industrial Electronics 43(4), 498–504 (1996)
93. Hou, M., Patton, R.J.: An LMI approach to H_-/H_∞ fault detection observers. In: Proceedings of the UKACC International Conference on Control, Cambridge, UK, pp. 305–310 (1996)
94. Hu, L., Bai, T., Shi, P., Wu, Z.: Sampled-data control of networked linear control systems. Automatica 43(5), 903–911 (2007)
95. Hu, L., Shi, P., Frank, P.M.: Robust sampled-data control for Markovian jump linear systems. Automatica 42(11), 2025–2030 (2006)
96. Hu, Q., Ma, G., Xie, L.: Robust and adaptive variable structure output feedback control of uncertain systems with input nonlinearity. Automatica 44(2), 552–559 (2008)
97. Hu, S., Yan, W.: Stability of networked control systems under a multiple-packet transmission policy. IEEE Transactions on Automatic Control 53(7), 1706–1711 (2008)
98. Hua, C., Ding, S.X.: Model following controller design for large-scale systems with time-delay interconnections and multiple dead-zone inputs. IEEE Transactions on Automatic Control 56(4), 962–968 (2011)
99. Hua, C., Guan, X., Shi, P.: Adaptive fuzzy control for uncertain interconnected time-delay systems. Fuzzy Sets and Systems 153(3), 447–458 (2005)
100. Hua, C., Guan, X., Shi, P.: Robust adaptive control for uncertain time-delay systems. International Journal of Adaptive Control and Signal Processing 19(7), 531–545 (2005)
101. Hua, C., Guan, X., Shi, P.: Robust backstepping control for a class of time delay systems. IEEE Transactions on Automatic Control 50(6), 894–899 (2005)
102. Hua, C., Guan, X., Shi, P.: Decentralized robust model reference adaptive control for interconnected time-delay systems. Journal of Computational and Applied Mathematics 193(2), 383–396 (2006)
103. Hua, C., Guan, X., Shi, P.: Robust output feedback tracking control for time delayed nonlinear systems using neural networks. IEEE Transactions on Neural Networks 18(2), 495–505 (2007)
104. Hua, C., Liu, P.X.: Teleoperation over the Internet with/without velocity signal. IEEE Transactions on Instrumentation & Measurement 60(1), 4–13 (2011)
105. Hua, C., Liu, P.X., Guan, X.: Backstepping control for nonlinear systems with time delays and applications to chemical reactor systems. IEEE Transactions on Industrial Electronics 56(9), 3723–3732 (2009)

106. Hua, C., Wang, Q., Guan, X.: Adaptive fuzzy output feedback controller design for nonlinear time delay systems with unknown control direction. IEEE Transactions on Systems, Man and Cybernetics-B: Cybernetics 39(2), 363–374 (2009)
107. Hua, C., Wang, Q., Guan, X.: Robust adaptive controller design for nonlinear time-delay systems via T-S fuzzy approach. IEEE Transactions on Fuzzy Systems 7(4), 901–910 (2009)
108. Huang, D., Nguang, S.K.: State feedback control of uncertain networked control systems with random time delays. IEEE Transactions on Automatic Control 53(3), 829–834 (2008)
109. Huang, M., Dey, S.: Stability of Kalman filtering with Markovian packet losses. Automatica 43(4), 598–607 (2007)
110. Huang, Y., Kuo, T., Chang, S.: Adaptive sliding-mode control for nonlinear systems with uncertain parameters. IEEE Transactions on Systems, Man and Cybernetics 38(2), 534–539 (2008)
111. Huh, S.H., Bien, Z.: Robust sliding mode control of a robot manipulator based on variable structure-model reference adaptive control approach. IET Control Theory and Applications 1(5), 1355–1363 (2007)
112. Ioannou, P., Tao, G.: Frequency domain conditions for strictly positive real functions. IEEE Transactions on Automatic Control 32(1), 53–54 (1987)
113. Istepanian, R.H., Wu, J., Chu, J., Whidborne, J.F.: Maximizing lower bound stability measures of finite precision PID controller realizations by nonlinear programming. In: Proceedings of the American Control Conference, Philadelphia, pp. 2596–2600 (1998)
114. Iwasaki, T., Hara, S.: Generalization of Kalman-Yakubovic-Popov lemma for restricted frequency inequalities. In: Proceedings of the American Control Conference, American, vol. 5 (2003)
115. Iwasaki, T., Hara, S.: Generalized KYP lemma: unified frequency domain inequalities with design applications. IEEE Transactions on Automatic Control 50(1), 41–59 (2005)
116. Iwasaki, T., Hara, S., Fradkov, A.L.: Restricted frequency inequalities equivalent to restricted dissipativity. In: The 43rd IEEE Conference on Decision and Control, Atlantis, Paradise Island, Bahamas, vol. 14-17, pp. 426–431 (2004)
117. Iwasaki, T., Hara, S., Fradkov, A.L.: Time domain interpretations of frequency domain inequalities on (semi)finite ranges. Systems & Control Letters 54(7), 681–691 (2005)
118. Iwasaki, T., Hara, S., Yamauchi, H.: Dynamical system design from a control perspective: Finite frequency positive realness approach. IEEE Transactions on Automatic Control 48(8), 1337–1354 (2003)
119. Iwasaki, T., Meinsma, G., Fu, M.: Generalized S-procedure and finite frequency KYP lemma. Malhernarical Problems in Engineering 6(2-3), 305–320 (2000)
120. Iwasaki, T., Skelton, R.E.: All controllers for the general H_∞ control problem: LMI existence conditions and state space formulas. Automatica 30(8), 1307–1317 (1994)
121. Jaimoukha, I.M., Li, Z., Papakos, V.: A matrix factorization solution to the H_-/H_∞ fault detection problem. Automatica 42(11), 1907–1912 (2006)
122. Jana, A., Samanta, A., Ganguly, S.: Observer-based control algorithms for a distillation column. Chemical Engineering Science 61(12), 4071–4085 (2006)

123. Janecki, D.: Model reference adaptive control using the delta operator. IEEE Transactions on Automatic Control 33(8), 771–775 (1988)
124. Jiang, X., Ha, Q., Yu, X.: Stability criteria for linear discerte-time systems with interval-like time-varying delay. In: Proceedings of the American Control Conferenc, Portland, OR, USA, pp. 2817–2822 (2005)
125. Jiang, X., Han, Q.: New stability criteria for linear systems with interval time-varying delay. Automatica 44(10), 2680–2685 (2008)
126. Jiang, X., Han, Q.-L.: On H_∞ control for linear systems with interval time-varying delay. Automatica 41(12), 2099–2106 (2005)
127. Jing, X.J., Tan, D.L., Wang, Y.C.: An LMI approach to stability of systems with severe time-delay. IEEE Transactions on Automatic Control 49(7), 1192–1195 (2004)
128. Jo, N.H., Seo, J.H.: Input output linearization approach to state observer design for nonlinear system. IEEE Transactions on Automatic Control 45(12), 2388–2393 (2000)
129. Joshi, S.M.: Control of large flexible space structures. Springer, New York (1989)
130. Jun, S., Yang, M., Yao, X., Zhong, R.: A new two-stage Kalman filter method for chaos system. ICIC Express Letters 4(2), 539–546 (2010)
131. Jun-Keun, J., Seung-Ki, S.: Kalman filter and LQ based speed controller for torsional vibration suppression in a 2-mass motor drive system. IEEE Transactions on Industrial Electronics 42(6), 564–571 (1995)
132. Kalman, R.E.: A new approach to linear filtering and prediction problems. Journal of Basic Engineering 82(1), 35–46 (1960)
133. Kalman, R.E.: Lyapunov functions for the problem of lur'e in automatic control. Proc. Nat. Acad. Sci. 49, 201–205 (1963)
134. Khoo, I.H., Reddy, H.C., Rajan, P.K.: Symmetry study for delta-operator-based 2-D digital filters. IEEE Transactions on Circuits and Systems I: Regular Papers 53(9), 2036–2047 (2006)
135. Kia, S.H., Henao, H., Capolino, G.A.: A high-resolution frequency estimation method for three-phase induction machine fault detection. IEEE Transactions on Industrial Electronics 54(4), 2305–2314 (2007)
136. Lam, J., Gao, H., Wang, C.: Stability analysis for continuous systems with two additive time-varying delay components. System & Control Letter 56(1), 16–24 (2007)
137. Lee, H.J., Park, J.B., Joo, Y.H.: Further refinement on LMI-based digital redesign: delta-operator approach. IEEE Transactions on Circuits and Systems II: Express Briefs 53(6), 473–474 (2006)
138. Lee, K.C., Lee, S., Lee, M.H.: Qos-based remote control of networked control systems via Profibus token passing protocol. IEEE Transactions on Industrial Informatics 1(3), 183–191 (2005)
139. Lee, Y.S., Moon, Y.S., Kwon, W.H., Park, P.G.: Delay-dependent robust H_∞ control for uncertain systems with a state-delay. Automatica 40(1), 65–72 (2004)
140. Li, D., Fei, S.: Robust stabilization for delta operator based systems. Fuzzy Systems and Mathematics 19(2), 140–145 (2005)
141. Li, G.: A polynomial-operator-based DFIIt structure for IIR filters. IEEE Transactions on Circuits and Systems II: Express Briefs 51(3), 147–151 (2004)
142. Li, G., Gevers, M.: Comparative study of finite wordlength effects in shift and delta operator parameterizations. IEEE Transactions on Automatic Control 38(5), 803–807 (1993)

143. Li, H., Chow, M.-Y., Sun, Z.: EDA-based speed control of a networked DC motor system. IEEE Transactions on Industrial Electronics 56(5), 1727–1735 (2009)
144. Li, H., Chow, M.-Y., Sun, Z.: Optimal stabilizing gain selection for networked control systems with time delays and packet losses. IEEE Transactions on Control Systems Technology 17(5), 1154–1162 (2009)
145. Li, H., Chow, M.-Y., Sun, Z.: State feedback stabilisation of networked control systems. IET Control Theory & Applications 3(7), 929–940 (2009)
146. Li, H., Chow, M.-Y., Sun, Z., Sun, F.: Gain scheduling based state feedback integral control for networked control systems. IEEE Transactions on Industrial Electronics 58(6), 2465–2472 (2011)
147. Li, H., Sun, Z., Liu, H., Sun, F.: State feedback integral control of networked control systems with external disturbance. IET Control Theory & Applications 5(2), 283–290 (2011)
148. Li, L., Liu, X.: New results on delay-dependent robust stability criteria of uncertain fuzzy systems with state and input delays. Information Sciences 179(8), 1134–1148 (2009)
149. Li, X., de Sonza, C.E.: Criteria for robust stability and stabilisation of uncertain linear systems with state-delay. Automatica 33(9), 1657–1662 (1997)
150. Li, X., de Sonza, C.E.: Delay-dependent robust stability and stabilization of uncertain linear delay systems: a linear matrix inequality approach. IEEE Transactions on Automatic Control 42(8), 1144–1148 (1997)
151. Li, X., Zhou, K.: A time domain approach to robust fault detection of linear time-varying systems. Automatica 45, 94–102 (2009)
152. Lin, C., Wang, Q., Lee, T.H., He, Y., Chen, B.: Observer-based H_∞ fuzzy control design for T-S fuzzy systems with state delays. Automatica 44(3), 868–874 (2008)
153. Lin, C., Wang, Z., Yang, F.: Observer-based networked control for continuous-time systems with random sensor delays. Automatica 45(2), 578–584 (2009)
154. Lin, F., Ying, H.: Modeling and control of fuzzy discrete event systems. IEEE Transactions on Systems, Man, and Cybernetics, Part B: Cybernetics 32(4), 408–415 (2002)
155. Liu, G.-P., Xia, Y., Chen, J., Rees, D., Hu, W.: Networked predictive control of systems with random network delays in both forward and feedback channels. IEEE Transactions on Industrial Electronics 54(3), 1282–1297 (2007)
156. Liu, G.-P., Xia, Y., Rees, D., Hu, W.: Design and stability criteria of networked predictive control systems with random network delay in the feedback channel. IEEE Transactions on Systems, Man, and Cybernetics 37(2), 173–184 (2007)
157. Liu, M., Ho, D.W., Niu, Y.: Stabilization of Markovian jump linear system over networks with random communication delay. Automatica 45(2), 416–421 (2009)
158. Liu, X., Zhang, Q.: New approaches to H_∞ controller designs based on fuzzy observers for T-S fuzzy systems via LMI. Automatica 39(9), 1571–1582 (2003)
159. Luo, R.C., Chung, L.Y.: Stabilization for linear uncertain system with time latency. IEEE Transactions on Industrial Electronics 49(4), 905–910 (2002)
160. Mahmoud, M.S., Shi, P.: Robust Kalman filtering for continuous time-lag systems with Markovian jump parameters. IEEE Transactions on Circuits and Systems I: Fundamental Theory and Applications 50(1), 98–105 (2003)
161. Mahmoud, M.S., Soh, Y.C., Xie, L.: Observer-based positive real control of uncertain linear systems. Automatica 35(4), 749–754 (1999)

162. Mahmoud, M.S., Xie, L.: Positive real analysis and synthesis of uncertain discrete time systems. IEEE Transactions on Circuits and Systems I 47(3), 403–406 (2000)
163. Mahmoud, M.S., Zribi, M.: Design of delay observer-based controllers for uncertain time-lag systems. Journal of Math Problems in Engineering 5(2), 121–137 (1999)
164. Mahmoud, M.S., Zribi, M.: Stabilizing controllers using observers for uncertain systems with delays. International Journal of Systems Science 32(6), 767–773 (2001)
165. Mahmoud, M.S., Zribi, M.: Guaranteed cost observer-based control of uncertain time-lag systems. Journal of Computers and Electrical Engineering 29(1), 193–212 (2003)
166. Mao, Z., Jiang, B., Shi, P.: H_∞ fault detection filter design for networked control systems modelled by discrete Markovian jump systems. IET Control Theory and Applications 1(5), 1336–1343 (2007)
167. Mao, Z., Jiang, B., Shi, P.: Protocol and fault detection design for nonlinear networked control systems. IEEE Transactions on Circuits and Systems–II: Express Briefs 56(3), 255–259 (2009)
168. Margolis, E., Eldar, Y.C.: Nonuniform sampling of periodic bandlimited signals. IEEE Transactions on Signal Processing 56(7), 2728–2745 (2008)
169. Martin, D., Toro, R., Haber, R., Dorronsoro, J.: Optimal tuning of a networked linear controller using a multi-objective genetic algorithm and its application to one complex electromechanical process. International Journal of Innovative Computing, Information and Control 5(10B), 3405–3414 (2009)
170. Middleton, R.-H., Goodwin, G.C.: Improved finite word length characteristics in digital control using delta operators. IEEE Transactions on Automatic Control 31(11), 1015–1021 (1986)
171. Middleton, R.H., Goodwin, G.C.: Digital control and estimation: a unified approach. Prentice-Hall, Englewood Cliffs (1990)
172. Moon, Y.S., Park, P., Kwon, W.H., Lee, Y.S.: Delay-dependent robust stabilisation of uncertain state-delayed systems. International Journal of Control 74(14), 1447–1455 (2001)
173. M'Sirdi, N.K., Rabhi, A., Fridman, L., Davila, J., Delanne, Y.: Second order sliding-mode observer for estimation of vehicle dynamic parameters. International Journal of Vehicle Design 48(3-4), 190–207 (2008)
174. Narendra, K., Tripathi, S.S.: Identification and optimization of aircraft dynamics. AIAA Journal of Aircraft 10(4), 193–199 (1973)
175. Narendra, K.S., Taylor, J.H.: Frequency domain criteria for absolute stability. Academic, New York (1973)
176. Neuman, C.P.: Properties of the delta operator model of dynamic physical systems. IEEE Transactions on Systems, Man and Cybernetics 23(1), 296–301 (1993)
177. Neuman, C.P.: Transformations between delta and forward shift operator transfer function models. IEEE Transactions on Systems, Man and Cybernetics 23(1), 295–296 (1993)
178. Nguang, S.K., Shi, P.: On designing filters for uncertain sampled-data nonlinear systems. Systems and Control Letters 41(5), 305–316 (2000)
179. Nguang, S.K., Shi, P.: Fuzzy H_∞ output feedback control of nonlinear systems under sampled measurements. Automatica 39(12), 2169–2174 (2003)

180. Nguang, S.K., Shi, P.: H_∞ fuzzy output feedback control design for nonlinear systems: an LMI approach. IEEE Transactions on Fuzzy Systems 11(3), 331–340 (2003)
181. Nguang, S.K., Shi, P., Ding, S.: Fault detection for uncertain fuzzy systems: an LMI approach. IEEE Transactions on Fuzzy Systems 15(6), 1251–1262 (2007)
182. Niu, Y., Lam, J., Wang, X., Ho, D.W.C.: Observer-based sliding mode control for nonlinear state-delayed systems. International Journal of Systems Science 35(2), 139–150 (2004)
183. Ohsumi, A., Kimura, T., Kono, M.: Kalman filter-based identification of unknown exogenous input of stochastic linear systems via pseudomeasurement approach. International Journal of Innovative Computing, Information and Control 5(1), 1–16 (2009)
184. Ohta, T., Murakami, T.: A stabilization control of bilateral system with time delay by vibration index-application to inverted pendulum control. IEEE Transactions on Industrial Electronics 56(5), 1595–1603 (2009)
185. Palhares, R.M., de Souza, C.E., Peres, P.L.D.: Robust H_∞ filtering for uncertain discrete-time state-delayed systems. IEEE Transactions on Signal Processing 49(8), 1696–1703 (2001)
186. Park, P.: A delay-dependent stability criterion for systems with uncertain time-invariant delays. IEEE Transactions on Automatic Control 44(4), 876–877 (1999)
187. Peng, C., Yue, D., Gu, Z., Xia, F.: Sampling period scheduling of networked control systems with multiple-control loops. Mathematics and Computers in Simulation 79(5), 1502–1511 (2009)
188. Peng, Z.K., Lang, Z.Q., Billings, S.A.: The nonlinear output frequency response function and its application to fault detection. In: 6th IFAC Symposium on Fault Detection, Supervision and Safety of technical Process, Beijing, P. R China, pp. 36–41 (2006)
189. Petersen, I.R.: A stabilization algorithm for a class of uncertain linear systems. Systems and Control Letters 8(4), 351–357 (1987)
190. Ploplys, N.J., Kawka, P.A., Alleyne, A.G.: Closed-loop control over wireless networks. IEEE Control Systems Magazine 24(3), 58–71 (2004)
191. Popov, V.M.: Absolute stability of nonlinear systems of automatic control. Automation and Remote Control 22(8), 857–875 (1962)
192. Popov, V.M.: Hyper stability of Control Systems. Springer, Berlin (1973)
193. Premaratne, K., Jury, E.I.: Tabular method for determining root distribution of delta-operator formulated real polynomials. IEEE Transactions on Automatic Control 39(2), 352–355 (1994)
194. Premaratne, K., Salvi, R., Habib, N.R., LeGall, J.P.: Delta-operator formulated discrete-time approximations of continuous-time systems. IEEE Transactions on Automatic Control 39(3), 581–585 (1994)
195. Proca, A.B., Keyhani, A., Miller, J.M.: Sensorless sliding mode control of induction motors using operating condition dependent models. IEEE Transaction on Energy Conversion 18(2), 205–212 (2003)
196. Qiu, J., Xia, Y., Yang, H., Zhang, J.: Robust stabilisation for a class of discrete-time systems with time-varying delays via delta operators. IET Control Theory and Applications 2(1), 87–93 (2008)
197. Qiu, Z., Gertler, J.: Robust FDI systems and H_∞-optimization. In: Proceedings of the 32nd Conference on Decision Control, San Antonio, Texas, pp. 1710–1715 (1993)

198. Rantzer, A.: On the Kalman-Yakubovich-Popov lemma. Systems & Control Letters 28(1), 7–10 (1996)
199. Rigatos, G.G.: Fault detection and isolation based on fuzzy automata. Information Sciences 179(12), 1893–1902 (2009)
200. Sahebsara, M., Chen, T., Shah, S.: Optimal H_∞ filtering in networked control systems with multiple packet dropouts. Systems and Control Letters 57(9), 696–702 (2008)
201. Sala, A.: Computer control under time-varying sampling period: An LMI gridding approach. Automatica 41(12), 2077–2082 (2005)
202. Salgado, M., Middleton, R., Goodwin, G.C.: Connection between continuous and discrete Riccati equations with applications to Kalman filtering. IEE Proceedings Control Theory and Applications 135(1), 28–34 (1988)
203. Sarpturk, S.Z., Istefanopulos, Y., Kaynak, O.: On the stability of discrete-time sliding mode control systems. Transactions on Automatic Control 32(10), 930–932 (1987)
204. Schenato, L.: Optimal estimation in networked control systems subject to random delay and packet drop. IEEE Transactions on Automatic Control 53(5), 1311–1317 (2008)
205. Seiler, P., Sengupta, R.: An H_∞ approach to networked control. IEEE Transactions on Automatic Control 50(3), 356–364 (2005)
206. Shao, H.: Improved delay-dependent stability criteria for systems with a delay varying in a range. Automatica 44(12), 3215–3218 (2008)
207. Shao, H.: New delay-dependent stability criteria for systems with interval delay. Automatica 45(3), 744–749 (2009)
208. Shi, P., Boukas, E., Agarwal, R.K.: Control of Markovian jump discrete-time systems with norm bounded uncertainty and unknown delay. IEEE Transactions on Automatic Control 44(11), 2139–2144 (1999)
209. Shi, P., Mahmoud, M., Nguang, S.K., Ismail, A.: Robust filtering for jumping systems with mode-dependent delays. Signal Processing 86(1), 140–152 (2006)
210. Shi, P., Nguang, S.K.: H_∞ output feedback control of fuzzy system models under sampled measurements. Computers and Mathematics with Application 46(5-6), 705–717 (2003)
211. Shi, P., Xia, Y., Liu, G.-P., Rees, D.: On designing of sliding-mode control for stochastic jump systems. IEEE Transactions on Automatic Control 51(1), 97–103 (2006)
212. Simon, D.: Optimal state estimation: Kalman, H_∞, and nonlinear approaches. John Wiley & Son, Inc., Hoboken (2006)
213. Sira-Ramirez, H., Colina-Morles, E.: A sliding mode strategy for adaptive learning in adalines. IEEE Transactions on Circuits and Systems I: Fundamental Theory and Applications 42(12), 1001–1012 (1995)
214. Soh, C.B.: Robust discrete time systems using delta operator. International Journal of Control 54(2), 453–464 (1991)
215. Soh, C.R.: Robust stability of discrete-time systems using delta operators. IEEE Transactions on Automatic Control 36(3), 377–380 (1991)
216. Song, X., Xu, S., Shen, H.: Robust H_∞ control for uncertain fuzzy systems with distributed delays via output feedback controllers. Information Sciences 178(22), 4341–4356 (2008)
217. Srinivasan, B., Huguenin, P., Bonvin, D.: Global stabilization of an inverted pendulum-control strategy and experimental verification. Automatica 45(1), 265–269 (2009)

218. Suchomski, P.: A J-lossless coprime factorisation approach to H_∞ control in delta domain. Automatica 38(10), 1807–1814 (2002)
219. Suchomski, P.: Numerically robust delta-domain solutions to discrete-time lyapunov equations. Systems & Control Letters 47(4), 319–326 (2002)
220. Suh, Y.S.: Stability and stabilization of nonuniform sampling system. Automatica 44(12), 3222–3226 (2008)
221. Sun, Y., Wang, L., Xie, G.: Delay-dependent robust stability and stabilization for discrete-time switched systems with mode-dependent tim-varying delays. Applied Mathematics and Computation 180(2), 428–435 (2006)
222. Tadjine, M., M'Saad, M., Dugard, L.: Discrete-time compensators with loop transfer recovery. IEEE Transactions on Automatic Control 39(6), 1259–1262 (1994)
223. Takagi, T., Sugeno, M.: Fuzzy identification of systems and its application to modeling and control. IEEE Transactions on Systems, Man, and Cybernetics 15(1), 116–132 (1985)
224. Takei, T., Imamura, R., Yuta, S.: Baggage transportation and navigation by a wheeled inverted pendulum mobile robot. IEEE Transactions on Industrial Electronics 56(10), 3985–3994 (2009)
225. Takeuchi, Y., Hirata, A.: A method of optimization of linear observations for the Kalman filter based on a generalized water filling theorem. International Journal of Innovative Computing, Information and Control 5(1), 75–86 (2009)
226. Tan, C.P., Edwards, C.: An LMI approach for designing sliding mode observers. International Journal of Control 74(16), 1559–1568 (2001)
227. Tanaka, K., Sano, M.: A robust stabilization problem of fuzzy control systems and its application to backing up control of a truck-trailer. IEEE Transactions on Fuzzy Systems 2(1), 119–134 (1994)
228. Tao, C.W., Chang, J.H., Taur, J., Su, S.: Adaptive fuzzy switched swing-up and sliding control for the double-pendulum-and-cart system. IEEE Transactions on Systems, Man, and Cybernetics, Part B: Cybernetics 40(1), 241–252 (2010)
229. Tiano, A., Sutton, R., Lozowicki, A., Naeem, W.: Observer Kalman filter identification of an autonomous underwater vehicle. Control Engineering Practice 15(6), 727–739 (2007)
230. Tondu, B., Lopez, P.: Modeling and control of mckibben artificial musclerobot actuators. IEEE Control System Magazine 20(2), 15–38 (2000)
231. Tong, S., Li, H., Chen, G.: Adaptive fuzzy decentralized control fora class of large-scale nonlinear systems. IEEE Transactions on Systems, Man and Cybernetics, Part B 34(1), 770–775 (2004)
232. Utkin, V.I.: Variable structure systems with sliding modes. IEEE Transactions on Automatic Control 22(2), 212–222 (1977)
233. Verriest, E.I., Fan, M.K.H., Kullstam, J.: Frequency domain robust stability criteria for linear delay systems. Automatica 42(1), 183–188 (2006)
234. Wai, R., Yang, Z.: Adaptive fuzzy neural network control design via a T-S fuzzy model for a robot manipulator including actuator dynamics. IEEE Transactions on Systems, Man, and Cybernetics, Part B: Cybernetics 38(5), 1326–1346 (2008)
235. Wang, H., Yang, G.-H.: A finite frequency domain approach to fault detection for linear discrete-time systems. International Journal of Control 81(7), 1162–1171 (2008)

236. Wang, H., Yang, G.-H.: A finite frequency domain approach to fault detection for linear discrete-time systems. International Journal of Control 81(7), 1162–1171 (2008)
237. Wang, H., Yang, G.-H.: A finite frequency domain approach to fault detection observer design for linear continuous-time systems. Asian Journal of Control 10(5), 1–10 (2008)
238. Wang, H., Yang, G.-H.: A finite frequency domain approach to fault detection observer design for linear continuous-time systems. Asian Journal of Control 10(5), 1–10 (2008)
239. Wang, J., Yang, G.-H., Liu, J.: An LMI approach to H_- index and mixed H_-/H_∞ fault detection observer design. Automatica 43(9), 1656–1665 (2007)
240. Wang, W., Chen, Y., Sun, C.: Relaxed stabilization criteria for discrete-time T-S fuzzy control systems based on a switching fuzzy model and piecewise lyapunov function. IEEE Transactions on Systems, Man, and Cybernetics, Part B: Cybernetics 37(3), 551–559 (2007)
241. Wang, Y., Yang, G.: H_∞ controller design for networked control systems via active-varying sampling period method. Acta Automatica Sinica 34(7), 814–818 (2008)
242. Wang, Z., Yang, F., Ho, D.W.C., Liu, X.: Robust finite-horizon filtering for stochastic systems with missing measurements. IEEE Signal Processing Letters 12(6), 437–440 (2005)
243. Wang, Z., Yang, F., Ho, D.W.C., Liu, X.: Robust H_∞ filtering for stochastic time-delay systems with missing measurements. IEEE Transactions on Signal Processing 54(7), 2579–2587 (2006)
244. Wu, J., Chen, T.: Design of networked control systems with packet dropouts. IEEE Transactions on Automatic Control 52(7), 1314–1319 (2007)
245. Wu, L., Ho, D.W.: Sliding mode control of singular stochastic hybrid systems. Automatica 64(4), 779–783 (2010)
246. Wu, L., Shi, P., Gao, H.: State estimation and sliding mode control of Markovian jump singular systems. IEEE Transactions on Automatic Control 55(5), 1213–1219 (2010)
247. Wu, L., Shi, P., Gao, H., Wang, C.: H_∞ filtering for 2-D Markovian jump systems. Automatica 44(7), 1849–1858 (2009)
248. Wu, L., Shi, P., Wang, C., Gao, H.: Delay dependent robust H_∞ and l_2-l_8 filtering for LPV systems with both discrete and distributed delays. IEE Proceedings of Control Theory and Applications 153(4), 483–492 (2006)
249. Wu, L., Su, X., Shi, P., Qiu, J.: Model approximation for discrete-time state-delay systems in the T-S fuzzy framework. IEEE Transactions on Fuzzy Systems 19(2), 366–378 (2011)
250. Wu, L., Wang, C., Zeng, Q.: Observer-based sliding mode control for a class of uncertain nonlinear neutral delay systems. Journal of the Franklin Institute 345(3), 233–253 (2008)
251. Wu, M., He, Y., She, J.-H.: New delay dependent stability criteria and stabilising method for neutral systems. IEEE Transactions on Automatic Control 49(12), 2266–2271 (2004)
252. Xia, Y., Fu, M., Liu, G.-P.: Analysis and synthesis of networked control systems. Springer, Heidelberg (2010)
253. Xia, Y., Fu, M., Shi, P.: Analysis and synthesis of dynamical systems with time-delays. Springer, Heidelberg (2009)

254. Xia, Y., Fu, M., Yang, H., Liu, G.-P.: Robust sliding mode control for uncertain time-delay systems based on delta operator. IEEE Transactions on Industrial Electronics 56(9), 3646–3655 (2009)
255. Xia, Y., Ge, S., Liu, G.-P., Shi, P., Rees, D.: Robust adaptive sliding mode control for uncertain time-delay systems. In: Proceedings of the 25th American Control Conference, Minneapolis, USA, pp. 3855–3860 (2006)
256. Xia, Y., Han, J.: Robust Kalman filtering for systems under norm bounded uncertainties in all system matrices and error covariance constraints. Journal of Systems Science and Complexity 18(4), 439–445 (2005)
257. Xia, Y., Jia, Y.: Robust stability analysis for time-delay systems with polytopic uncertainties via parameter-dependent Lyapunov functionals. International Journal of Control 75(16-17), 1427–1434 (2002)
258. Xia, Y., Jia, Y.: Robust control of state delayed systems with polytopic type uncertainties via parameter-dependent lyapunov functionals. Systems & Control Letters 50(3), 183–193 (2003)
259. Xia, Y., Jia, Y.: Robust sliding-mode control for uncertain time-delay systems: an LMI approach. IEEE Transactions on Automatic Control 48(6), 1086–1091 (2003)
260. Xia, Y., Liu, G.-P., Chen, J., Rees, D., Liang, J.: Sliding mode control of uncertain linear discrete time systems with input delay. IET Control Theory and Applications 1(4), 1169–1175 (2007)
261. Xia, Y., Liu, G.-P., Fu, M., Rees, D.: Predictive control of networked systems with random delay and data dropout. IET Control Theory and Applications 3(11), 1476–1486 (2009)
262. Xia, Y., Shang, J., Chen, J., Liu, G.-P.: Networked data fusion with packet losses and variable delays. IEEE Transactions on Systems, Man, and Cybernetics, Part B: Cybernetics 39(5), 1107–1120 (2009)
263. Xia, Y., Shi, P., Rees, D., Liu, G.-P., Han, J.: Active disturbance rejection control for uncertain multivariable systems with time-delay. IET Control Theory and Applications 1(1), 75–81 (2007)
264. Xia, Y., Zhang, J., Boukas, E.K.: Control for discrete singular hybrid systems. Automatica 44(10), 2635–2641 (2008)
265. Xia, Y., Zhu, Z., Li, C., Yang, H., Zhu, Q.: Robust adaptive sliding mode control for uncertain discrete-time systems with time delay. Journal of the Franklin Institute 347(1), 339–357 (2010)
266. Xiang, Z., Chen, Q., Zhang, R.: Robust stability analysis and control for fuzzy systems with uncertainties using the delta operator. Control and Decision 18(6), 720–723 (2003)
267. Xie, J., Fan, B., Lee, Y., Yang, J.: Guaranteed cost controller design of networked control systems with state delay. Acta Automatica Sinica 33(2), 170–174 (2007)
268. Xie, J., Tao, G., Zhou, H.: Sliding mode tracking control of penumatic muscle joint actuated by high-speed on-off sonlenoid valve. Chinese Journal of Mechanical Engneering 18(5), 540–544 (2007)
269. Xie, L.: Output feedback H_∞ control of systems with parameter uncertainty. International Journal of Control 63(4), 741–750 (1996)
270. Xie, L., Soh, Y.C.: Positive real control problem for uncertain linear time-invariant systems. Systems & Control Letters 24(4), 265–271 (1995)
271. Xiong, J., Lam, J.: Stabilization of discrete-time Markovian jump linear systems via time-delayed controllers. Automatica 42(5), 747–753 (2006)

272. Xiong, J., Lam, J.: Stabilisation of networked control systems with a logic ZOH. IEEE Transactions on Automatic Control 54(2), 358–363 (2009)
273. Xiong, Y., Saif, M.: Sliding mode observer for nonlinear uncertain systems. IEEE Transactions on Automatic Control 46(12), 2012–2017 (2001)
274. Xu, S., Chen, T.: Robust H_∞ control for uncertain discrete-time systems with time-varying delays via exponential output feedback controllers. Systems & Control Letters 51(3-4), 171–183 (2004)
275. Xu, S., Chen, T., Lam, J.: Robust H_∞ filtering for uncertain Markovian jump systems with mode-dependent time delays. IEEE Transactions on Automatic Control 48(5), 900–907 (2003)
276. Xu, S., Dooren, P.V.: Robust H_∞ filtering for a class of nonlinear systems with state delay and parameter uncertainty. International Journal of Control 75(10), 766–774 (2002)
277. Xu, S., Lam, J.: Exponential H_∞ filter design for uncertain Takagi-Sugeno fuzzy systems with time delay. Engineering Applications of Artificial Intelligence 17(6), 645–695 (2004)
278. Xu, S., Lam, J.: On equivalence and efficiency of certain stability criteria for time-delay systems. IEEE Transactions on Automatic Control 52(1), 95–101 (2007)
279. Xu, S., Lam, J., Zou, Y.: Simplified descriptor system approach to delay-dependent stability and performance analyses for time-delay systems. IEE Proceedings of Control Theory and Applications 152(2), 147–151 (2005)
280. Xu, S., Lu, J., Zhou, S., Yang, C.: Design of observers for a class of discrete-time uncertain nonlinear systems with time delay. Journal of the Franklin Institute 341(3), 295–308 (2004)
281. Xu, Y., Hespanha, J.P.: Estimation under uncontrolled and controlled communications in networked control systems. In: IEEE Conference on Decision and Control 2005 and 2005 European Control Conference, China Shanghai, pp. 842–847 (2005)
282. Yakubovich, V.A.: Solution of certain matrix inequalities in the stability theory of nonlinear control systems. Dokl. Akad. Nauk. SSSR 143, 1304–1307 (1962)
283. Yan, X., Edwards, C.: Nonlinear robust fault reconstruction and estimation using a sliding mode observer. Automatica 43(9), 1605–1614 (2007)
284. Yang, H., Xia, Y., Fu, M., Shi, P.: Robust adaptive sliding mode control for uncertain delta operator systems. International Journal of Adaptive Control and Signal Processing 24(8), 623–632 (2010)
285. Yang, H., Xia, Y., Qiu, J., Zhang, J.: Filtering for a class of discrete-time systems with time-delays via delta operator approach. International Journal of Systems Science 4(423-433), 41 (2010)
286. Yang, R.H., Lim, Y.C.: A dynamic frequency grid allocation scheme for the efficient design of equiripple FIR filters. IEEE Transactions on Signal Processing 44(9), 2335–2339 (1996)
287. Yang, T.C.: Networked control system: a brief survey. IEE Proceedings-Control Theory and Applications 153(4), 403–412 (2006)
288. Ye, H., Ding, S.X., Wang, G.: Integrated design of fault detection systems in time-frequency domain. IEEE Transactions on Automatic Control 47(2), 384–390 (2002)

289. Yue, D.: Robust stabilization of uncertain systems with unknown input delay. Automatica 40(2), 331–336 (2004)
290. Yue, D., Han, Q.L.: Delay-dependent exponential stability of stochastic systems with time-varying delay, nonlinearity, and Markovian switching. IEEE Transactions on Automatic Control 50(2), 217–222 (2003)
291. Yue, D., Han, Q.L.: Delayed feedback control of uncertain systems with time-varying input delay. Automatica 41(2), 233–240 (2005)
292. Yue, D., Han, Q.L., Lam, J.: Network-based robust H_∞ control of systems with uncertainty. Automatica 41(6), 999–1007 (2005)
293. Zhang, H., Feng, G.: Stability analysis and controller design of discrete-time fuzzy large-scale systems based on piecewise lyapunov functions. IEEE Transactions on Systems, Man, and Cybernetics, Part B: Cybernetics 38(5), 1390–1401 (2008)
294. Zhang, H., Yang, D., Chai, T.: Guaranteed cost networked control for T-S fuzzy systems with time delays. IEEE Transactions on Systems, Man, and Cybernetics 37(2), 160–172 (2007)
295. Zhang, J., Shi, P., Xia, Y.: Robust adaptive sliding mode control for fuzzy systems with mismatched uncertainties. IEEE Transactions on Fuzzy Systems 18(4), 700–711 (2010)
296. Zhang, J., Xia, Y.: New LMI approach to fuzzy H_∞ filter designs. IEEE Transactions on Circuits and Systems II: Express Briefs 56(9), 739–743 (2009)
297. Zhang, J., Xia, Y., Tao, R.: New results on H_∞ filtering for fuzzy time-delay systems. IEEE Transactions on Fuzzy Systems 17(1), 128–137 (2009)
298. Zhang, L., Boukas, E.K.: Stability and stabilization of Markovian jump linear systems with partly unknown transition probabilities. Automatica 45(2), 463–468 (2009)
299. Zhang, L., Boukas, E.K., Haidar, A.: Delay-range-dependent control synthesis for time-delay systems with actuator saturation. Automatica 44(10), 2691–2695 (2008)
300. Zhang, L., Boukas, E.K., Lam, J.: Analysis and synthesis of Markov jump linear systems with time-varying delays and partially known transition probabilities. IEEE Transactions on Automatic Control 53(10), 2458–2464 (2008)
301. Zhang, L., Shi, Y., Chen, T., Huang, B.: A new method for stabilization of networked control systems with random delays. IEEE Transactions on Automatic Control 50(8), 1177–1181 (2005)
302. Zhang, P., Ding, S.X., Wang, G.Z., Zhou, D.H.: A frequency domain approach to fault detection in sampled-data systems. Automatica 39(7), 1303–1307 (2003)
303. Zhang, W., Wang, T., Tong, S.: Delay-dependent stabilization conditions and control of T-S fuzzy systems with time-delay. ICIC Express Letters 3(4(A)), 871–876 (2009)
304. Zhang, W., Yu, L.: Modelling and control of networked control systems with both network-induced delay and packet-dropout. Automatica 44(12), 3206–3210 (2008)
305. Zhang, Z.G., Chan, S.C., Tsui, K.M.: A recursive frequency estimator using linear prediction and a Kalman-filter-based Iterative algorithm. IEEE Transactions on Circuits and Systems II: Express Briefs 55(6), 576–580 (2008)

306. Zhao, Y., Liu, G.-P., Rees, D.: Networked predictive control systems based on the Hammerstein model. IEEE Transactions on Circuits and Systems II: Express Briefs 55(5), 469–473 (2008)
307. Zhong, Q.: On distributed delay in linear control laws-part II: rational implementations inspired from the δ-operator. IEEE Transaction on Automatic Control 50(5), 729–734 (2005)
308. Zhu, X., Hua, C., Wang, S.: State feedback controller design of networked control systems with time delay in the plant. International Journal of Innovative Computing, Information and Control 4(2), 283–290 (2008)

Index

MIX
Papier aus verantwortungsvollen Quellen
Paper from responsible sources
FSC® C105338

If you have any concerns about our products,
you can contact us on
ProductSafety@springernature.com

In case Publisher is established outside the EU,
the EU authorized representative is:
Springer Nature Customer Service Center GmbH
Europaplatz 3, 69115 Heidelberg, Germany

Printed by Libri Plureos GmbH
in Hamburg, Germany